A-level Study Guide

Physics

Revis[illegible] [illegible]nd [illegible]pda[illegible] by Russell Wallington

Wendy Brown

Basil Donnelly

Jeremy Spackman

Tony Winzor

Series Consultants: Geoff Black and Stuart Wall
Project Manager: Gillian Ragsdale

Pearson Education Limited
Edinburgh Gate, Harlow
Essex CM20 2JE, England
and Associated Companies throughout the world

British Library Cataloguing in Publication Data
A catalogue entry for this title is available from the British Library.

ISBN 0-582-78418-2

First edition published 2000
Reprinted 2001, 2002, 2003
Updated 2004

Set by 35 in Univers, Cheltenham
Printed by Ashford Colour Press, Gosport, Hants

Mechanics

Mechanics – the study of forces and their effects – is the backbone of classical physics.

Exam themes

- *Measurements and units* All ideas in physics are tested by experiment (by measuring things). Measurements need appropriate, clearly stated units. Familiarity with SI units is required. Equations and definitions always give clues to the correct units to use.
- *Vector quantities* Direction often matters in mechanics. You need to know how forces acting in different directions add up. You need to be confident in dealing with other vector quantities too.
- *Forces in equilibrium* In static structures, the forces and moments on each and every point must be balanced. Moving bodies tend to keep moving with constant velocity.
- *Unbalanced forces* can cause acceleration, changes in momentum, circular motion, simple harmonic motion etc.
- *Use of free-body force diagrams* You will be expected to be able to draw force diagrams to isolate the forces acting on a body.
- *Interpreting graphs and gradients* You should be able to interpret graphs of motion – in particular, to understand the significance of gradient and area.

Topic checklist

○ AS ● A2	AQA/A	AQA/B	CCEA	EDEXCEL	OCR/A	OCR/B	WJEC
Scalars and vectors	○	○	○	○	○	○	○
Forces and moments in equilibrium	○	○	○	○	○	○	○
Ways of describing motion	○	○	○	○	○	○	○
Equations of motion	○	○	○	○	○	○	○
Projectiles	○	○	○	○	○	○	○
Newton's laws of motion	○	○	○	○	○	○	○
Some important forces	○	○	○	○	○	○	○
Density and pressure	○	●		○	○		
Drag and lift		○			○		
Work, energy and power	○	○●	○	○	○	○	○●
Momentum and impulse	○	●	○	○		●	●
Stress, strain and Hooke's law	○	●	○	●	○	○	●
Vibrations and resonance	●	●	○	●	●	●	●
Circular motion	●	●	○	●	●	●	○
Simple harmonic motion	●	●	○	●	●	●	●

Scalars and vectors

Scalar quantities like mass and energy have *size* (scale) and that's about it. *How many?* (a number) and *of what?* (a unit) and your measurement is fully defined. Vector quantities like force and velocity also have *direction*. You have to define not only *how many?*, and *of what?*, but also *in which direction?*

Checkpoint 1

Sort the following list into scalar and vector quantities: mass, weight, temperature, energy, acceleration.

Vector addition by scale drawing

Vectors can be represented by arrows. The length of the arrow gives the size of the vector. The sum or **resultant** of any number of vectors can be found by joining vector arrows together in any order. The vectors must all follow on from start to finish, with the tail of each arrow starting from the tip of the last.

Action point

Convince yourself that the order of addition does not matter by adding three vectors in any order and measuring the resultant.

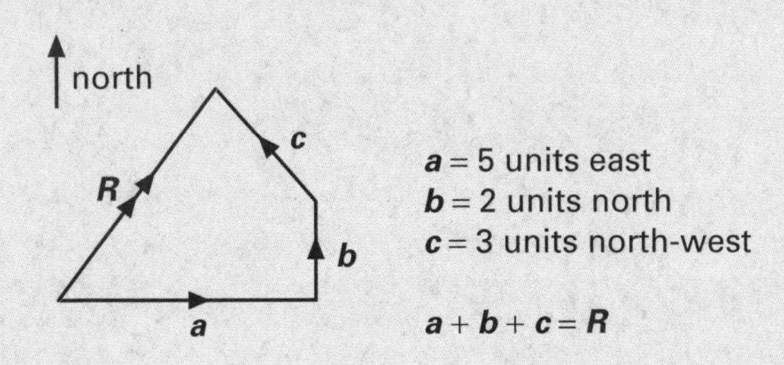

The resultant *R*

- The **resultant *R*** is the single vector that could replace all the others.
- ***R*** points from the starting point to the finishing point.
- ***R*** is normally given a double arrow to distinguish it from its components.
- The size (magnitude) and direction of ***R*** are measured from the diagram using a ruler and a protractor.
- Don't forget to use the appropriate units. If you are adding forces, your answer must be in newtons (not centimetres!).

Examiner's secrets

You can use Pythagoras' theorem to calculate the resultant of two perpendicular vectors. For anything more complicated, stick to scale drawing. The bigger the drawing, the better.

Vector notation

Vector quantities are given symbols written in bold type (***a***), or underlined ($\underline{a}$), or with an arrow over the top ($\overrightarrow{OA}$). This notation should give you a clue that simple addition may not be appropriate.

Vector subtraction

The negative version of a vector is simply the vector reversed, so ***a*** – ***a*** = 0. To subtract any vector, reverse its direction and add it in the usual way.

Checkpoint 2

Why does the size of the drawing matter?

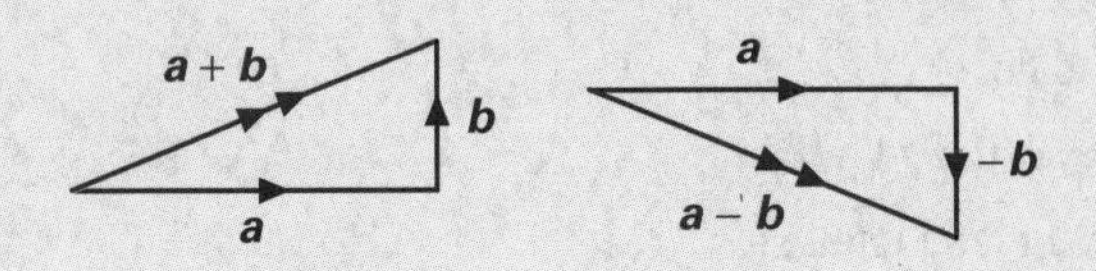

Action point

Adding scalar quantities is simple, *providing* they all have the same units. For example, find the sum of 2 kg, 4.5 kg, 300 g and 750 g.

The parallelogram of forces 

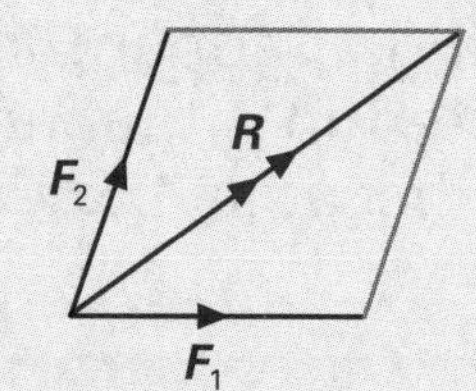

The resultant of two forces is given by the diagonal of the parallelogram. (Be sure to choose the right diagonal!) The resultant must start from the point of action of the forces. This method is entirely equivalent to the previous one (applied to just two vectors). Its main benefits are:

- ➜ the angle between the forces (which is normally given) is more easily marked and measured (it's less easy to make mistakes)
- ➜ it shows the two forces acting on the same body at the same time (which makes sense)

Resolution of vectors 

A vector can be *resolved* into two perpendicular components, such that the sum of the components equals the vector itself. This is a very useful trick. It allows you to separate horizontal motion, which is not affected by gravity, from vertical motion, which is. It allows you to separate out the forces acting along a slope from the forces acting at right angles to it.

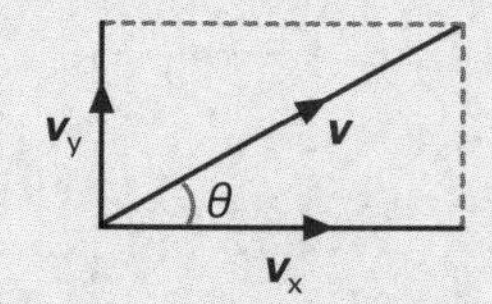

$$\mathbf{v}_x = \mathbf{v}\cos\theta$$
$$\mathbf{v}_y = \mathbf{v}\sin\theta$$
$$\mathbf{v}_x + \mathbf{v}_y = \mathbf{v}$$

Exam questions answers: page 34

1 The diagram below shows the two forces produced by a spoiler on a racing car when it moves.

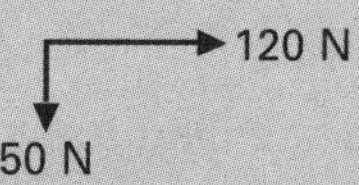

(a) Calculate the magnitude of the resultant force on the car.

(b) Calculate the direction the resultant force makes with the 120 N force. (5 min)

2 (a) A boat's speed through still water is 2 m s^{-1}. It heads due east across a river. The river runs north to south and is 20 m wide. If the river flows south at 1 m s^{-1}, how far downstream does the boat reach the other shore?

(b) In which direction should the boat aim in order to get straight across in the shortest possible distance? (20 min)

3 A rope is used to pull a narrow boat along a canal. The rope is pulled by a horse. The tension in the rope is 600 N and the rope makes an angle of 30° with the canal bank. What force must be provided (by the rudder and keel) to keep the boat travelling parallel to the bank? (15 min)

Examiner's secrets

The whole point about vector quantities is that they have both *magnitude* and *direction*. Make sure you give *both* in your answers.

Examiner's secrets

Don't forget to *show* how you calculated both the magnitude and the direction of the resultant vector. There will be at least one mark for a clear method.

Watch out!

Don't forget to give the answer! A drawing is not enough. You must show that you can use your scale diagram to find the resultant. Use appropriate units and state the angle relative to one (or more) of the forces given in the question.

Check the net

Learn how to describe motion in words by going to www.glenbrook.k12.il.us/gbssci/phys/Class/1DKin/U1L1b.html

Checkpoint 3

Learn your trigonometry. *SOHCAHTOA!* In the figure opposite, $\mathbf{v}_x$ is clearly *adjacent* to angle θ, $\mathbf{v}_y$ is (not quite so clearly) equal to the side *opposite* θ. Check the formulae given for $\mathbf{v}_x$ and $\mathbf{v}_y$ (*derive them*).

Watch out!

Sometimes you may be able to give the angle of the resultant in terms of a bearing. For example, an angle of 27° is the same as 153°, but watch out – it is easy to go wrong.

Forces and moments in equilibrium

Forces can make things speed up, slow down, change direction, change shape, or spin faster or slower. Forces can make things happen! In this section, however, we are interested in situations where forces do none of the above – because they are in equilibrium.

Links

See *Newton's first law of motion*, pages 14–15 for more information on moving objects in equilibrium.

Equilibrium

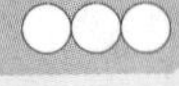

An object is in **equilibrium** if the forces acting on it are balanced.

- The forces acting on each and every point must be balanced (otherwise the body will be deformed).
- The turning moments about each point must be balanced (otherwise the body will tend to twist or turn).

Static, solid objects are always in equilibrium. We can often use this fact to calculate the sizes and directions of unknown forces.

Links

See *vector addition*, page 4.

The jargon

A *free-body force diagram* is simply a diagram of the object of interest with all the forces acting on it.

Examiner's secrets

Follow these steps to avoid disaster:
(a) Sketch the set-up described. Find the angles at which all forces act.
(b) Draw a *free-body force diagram* with all the forces acting away from the body.
(c) Draw the force polygon to solve the problem.

Force polygons

If you add a series of force vectors tail to nose, the resultant is found by joining the start to the finish (and measuring). If the forces are balanced, you will end up with a closed polygon. The arrows will lead back to the start because the resultant equals zero. The most common situations dealt with in A-level Physics exams involve three forces (in which case the polygon is a triangle).

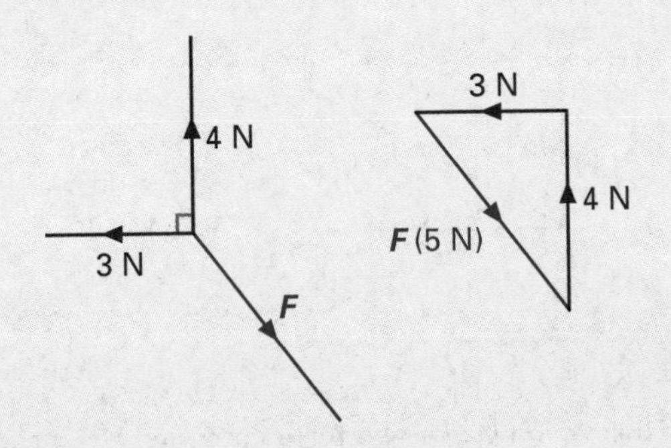

The jargon

Torque is another word for moment.

Moments

The turning effect of a force is called its **moment**. The size of a moment depends on the size and direction of the force and the distance from its point of action to the axis of rotation.

moment = Fd

Where F is the force and d is the perpendicular distance from the line of action of the force to the pivot. Moments are measured in newton metres (N m).

Checkpoint 1

The product of force and perpendicular distance is exactly the same as the product of perpendicular force and actual distance to the point of action. Check for yourself.

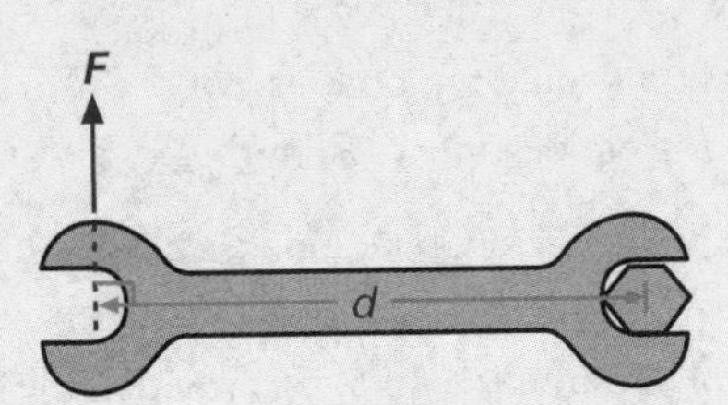

Principle of moments 

If a body is in equilibrium, the sum of clockwise moments is equal (and opposite to) the sum of anticlockwise moments *about each and every point.*

Couples

A single force can cause a body to turn, but it will also cause linear acceleration. A **couple** is a pair of equal and opposite forces along different lines of action. Couples cause angular acceleration, but not linear acceleration.

→ The moment of a couple is equal to one of the forces multiplied by the perpendicular separation of the lines of action of the forces.

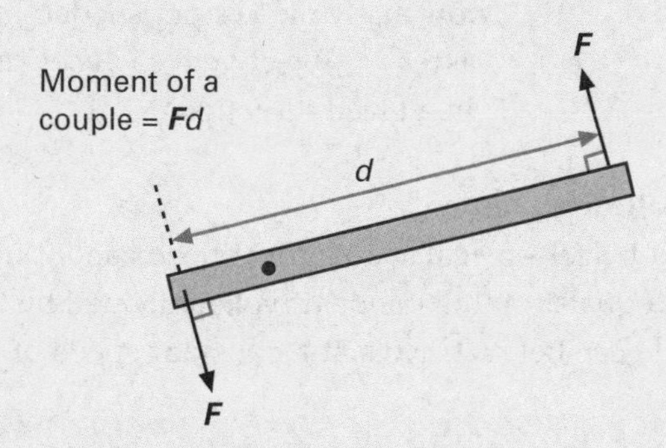

Centre of gravity

The **centre of gravity** of a rigid body is the point at which its weight can be considered to act. This concept allows us to simplify and answer questions where we need to work out the moments caused by a body's weight. For simple symmetrical objects like ladders, beams and doors (the kinds of objects that appear most often in A-level Physics papers), the centre of gravity is also the geometrical centre of the body.

> *"Give me but one firm place on which to stand, and I will move the earth."*
>
> Archimedes, who discovered the principle of moments

Examiner's secrets

Moments are vector quantities. You must show clockwise and anticlockwise moments with opposite signs.

The jargon

Direction A moment can cause clockwise *or* anticlockwise acceleration. The net turning effect of all moments acting about a point is found by adding up all clockwise moments and subtracting all anticlockwise moments.

Don't forget

The *centre of mass* of a body is the average position of its mass (rather than its *weight*). A body's centre of mass is in exactly the same position as its centre of gravity – provided the gravitational field strength doesn't vary within the body itself.

Action point

It is easier to lift any object at the centre of gravity. Can you use *moments* to explain why? Try lifting a ladder in different positions.

Don't forget

For an object in static equilibrium, if the forces are either side of the pivot, they act in the same direction. If the forces are on the same side of the pivot, they act in opposite directions.

Exam questions

answers: page 34

1 A strong, light rope spans a 15 m gap. A tightrope walker weighing 500 N sets off across it. The rope sags a little due to the walker's weight. The backward and forward tension forces in the rope (T_1 and T_2) change as she makes her journey. Half way across, T_1 equals T_2 in magnitude.

Given that the centre of the rope dips 0.6 m below the horizontal, calculate the tension in the rope. (15 min)

2

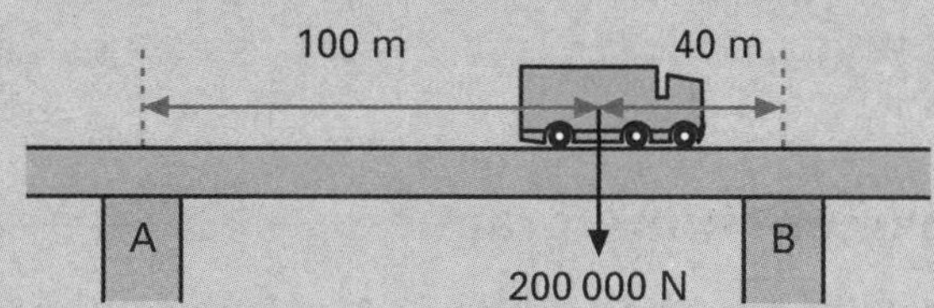

A bridge is made by resting a concrete beam on two pillars (A and B). A lorry crosses the bridge. Using the weights and dimensions given in the diagram above, calculate the extra load on each pillar due to the lorry. (30 min)

Ways of describing motion

The easiest way to describe motion is in the language of mathematics, using symbols, graphs and equations.

The jargon

The prefix *SI* stands for *Système International* – an internationally agreed set of standard units.

Definitions, symbols and units

Quantity (symbol)	*Description*	*SI unit*
Displacement (s)	The vector version of distance. The distance an object has moved in a certain direction.	m
Velocity (v)	The vector version of speed. An object's speed or rate of progress in a given direction.	m s^{-1}
Acceleration (a)	The rate of increase in velocity; how many metres per second faster an object gets each second (in a given direction).	m s^{-2}

Checkpoint 1

A racing car may have an average speed of 150 mph over the course of a lap, but its average velocity over one lap is always zero. Explain why.

Examiner's secrets

Are your drawing skills up to scratch? You gain valuable marks with clear, legible, well-labelled diagrams and sketches.

Speed and velocity

Speed is rate of travel – a scalar quantity; speed cannot be negative. Average speed equals total distance travelled divided by the time taken. The car in the figure below travels at a constant speed of 20 m s^{-1}.

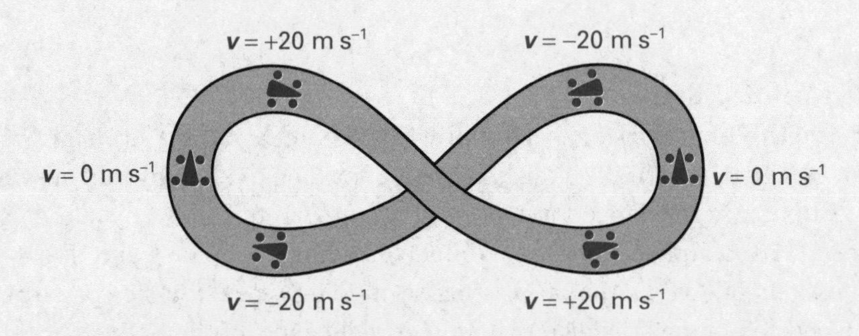

Action point

Use the information given in the figure opposite to sketch graphs of the car's speed and velocity against time.

Distance–time graphs

The gradient of a distance–time graph is equal to the speed of the body at that particular instant. Note that distance travelled never decreases.

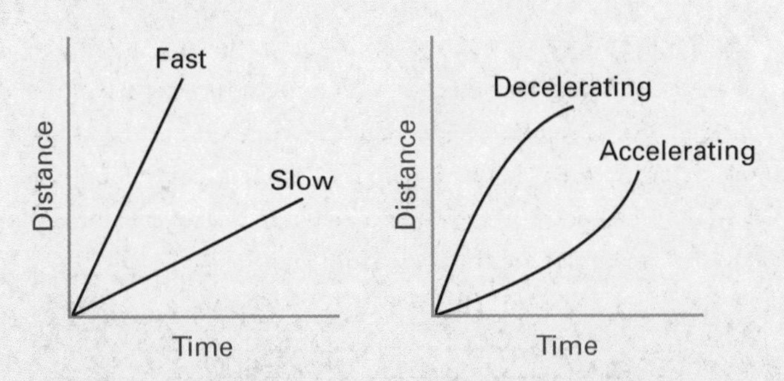

Examiner's secrets

If you are asked to *sketch* a graph: *don't* relax, get out your charcoal, and demonstrate your understanding of light and shade; *do* draw a graph showing all the information you are given. Show the shape of the graph, show where the line cuts the axes (if it does), label axes properly.

Displacement–time graphs

Displacement–time graphs can show a body's position and its direction of motion. The gradient of a displacement–time graph is equal to the *instantaneous velocity* of the body.

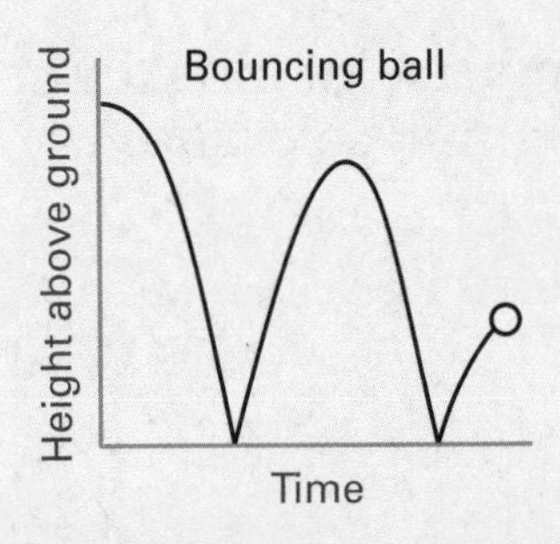

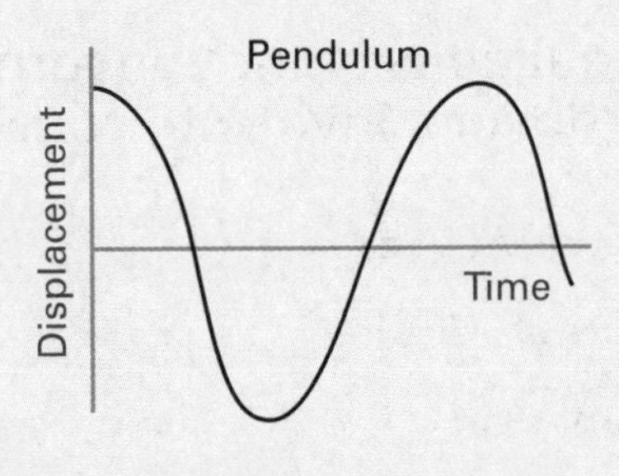

Speed–time graphs

Speed–time graphs can be used to show the distance travelled, the speed and the rate of acceleration of a body at any instant. A *tachograph* is a type of speed–time graph used in lorries to check on their drivers.

➜ The area under a speed–time graph is equal to the distance travelled.

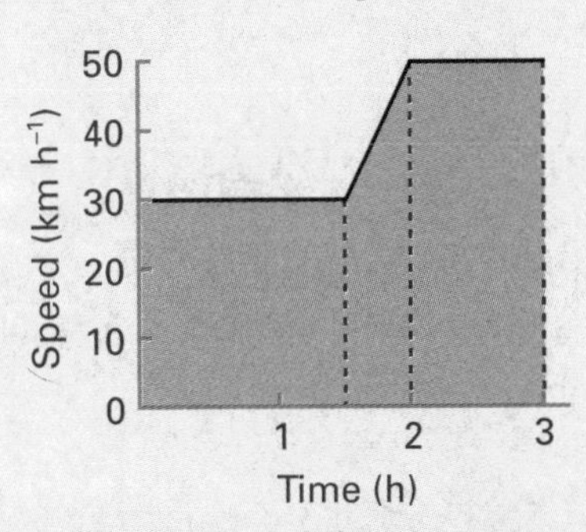

distance = $30 \times 1.5 + 40 \times 0.5 + 50 \times 1$
= 115 km

Velocity–time graphs

These are probably the most versatile graphs of motion. You can use them to find the displacement, the velocity and the acceleration of a body at any given instant.

Displacement

The displacement (from a given starting point) is equal to the area under a velocity–time graph.

Acceleration

The gradient of a velocity–time graph is equal to the instantaneous acceleration of the body.

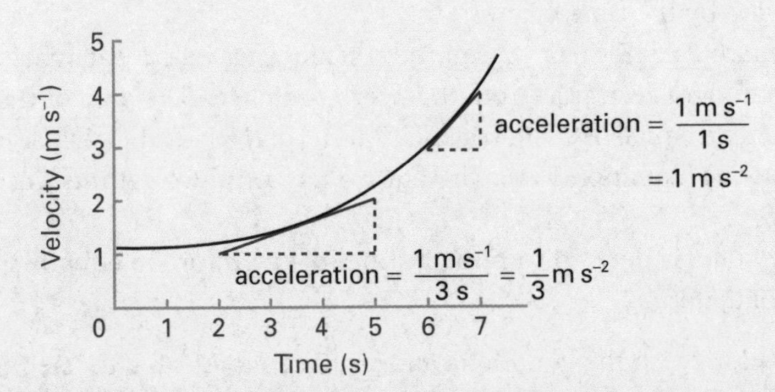

Exam question answer: page 34

A train accelerates from rest at a constant rate of 0.2 m s^{-2} for the first minute of its journey; it then carries on at constant speed for the next two minutes. Graphically or otherwise, find the train's top speed and the distance it travels in the first three minutes. (15 min)

Action point

Copy the graphs from the figure opposite and mark on your sketches the positions of maximum positive velocity, maximum negative velocity and zero velocity.

Examiner's secrets

You do not need to know how to do calculus for A-level Physics, but you should be familiar with the notation. If distance travelled is x and time taken is t, the speed at any instant is written dx/dt said 'dx by dt' in calculus notation.

The jargon

The *differential* of any graph is equal to its gradient. The *integral* of any graph is equal to the area under the graph. The differential of distance with respect to time is speed. The integral of speed with respect to time is distance.

Checkpoint 2

Sketch a velocity–time graph for a stone thrown upward. How is its displacement represented by this graph?

Links

See also *simple harmonic motion*, pages 32–3, where the links between displacement, velocity and acceleration are explored in more detail.

Examiner's secrets

To find the gradient of a graph, you need to draw the tangent at a point on the graph. You may find it easier to draw a line at right angles to the graph at this point and then use a protractor to draw the tangent. Always draw a long tangent and construct a large triangle from it. Small triangles are less accurate and lose marks.

Equations of motion

The equations of motion form a useful tool kit for solving problems involving constant acceleration.

The equations

$$v = u + at \quad [1]$$

$$s = ut + \tfrac{1}{2}at^2 \quad [2]$$

$$v^2 = u^2 + 2as \quad [3]$$

Where

s = displacement
u = initial velocity
v = final velocity
a = acceleration
t = time taken

Derivation

A body accelerates steadily from an initial velocity u to a final velocity v in time t seconds. Its rate of acceleration is a and the distance it travels is s. We want to derive equations that link these quantities.

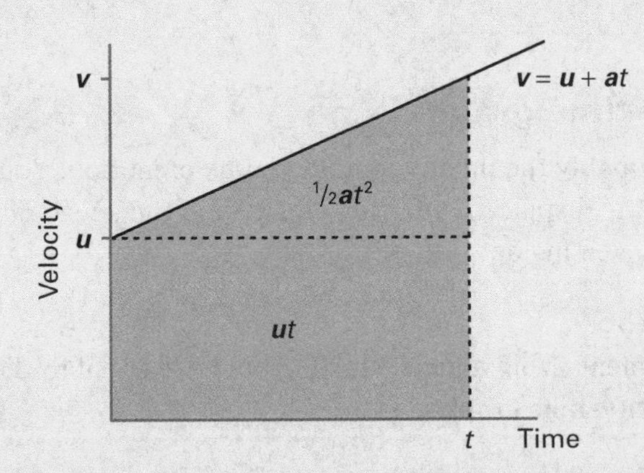

- *Equation 1* comes from the definition of acceleration. Acceleration is the rate of increase in velocity, i.e. the increase in velocity $(v - u)$ divided by the time taken (t).
- *Equation 2* comes from the graph. Distance travelled, s, equals the area under the graph (from time zero to time t). The area of the rectangle equals ut. The triangle's area (half base × height) equals $\tfrac{1}{2}t(v - u)$. Equation 1 tells us that $v - u = at$, so we get the $\tfrac{1}{2}at^2$ term.
- *Equation 3* is derived by substituting $(v - u)/a$ for t in equation 2 and shuffling terms.

You need to learn the equations and you need to be able to use them. You do *not* need to learn the derivations.

Watch out!

These equations of motion assume constant acceleration. If acceleration varies, they won't work!

Examiner's secrets

You must get used to using **u** for initial velocity, **v** for final velocity and **s** for displacement. Some questions will expect you to manipulate the equations of motion without using numbers.

Checkpoint 1

The three equations given are all you need to learn provided you are reasonably competent at algebra. Each equation contains four of the five quantities. Two others may be learned if you want to avoid the prospect of having to deal with simultaneous equations. They are:

$$s = vt - \tfrac{1}{2}at^2 \quad [4]$$

$$s = (u + v)t/2 \quad [5]$$

Try to derive them from the others. If you succeed, you needn't bother to learn them!

Checkpoint 2

Referring to the graph opposite, if $u = 10\ \mathrm{m\,s^{-1}}$, $v = 18\ \mathrm{m\,s^{-1}}$ and $t = 4$ s, work out the rate of acceleration and the distance travelled.

Checkpoint 3

Sketch a velocity–time graph for a ball thrown up vertically in the air. Assume $g = 10\ \mathrm{m\,s^{-2}}$. If the ball is thrown upwards at $10\ \mathrm{m\,s^{-1}}$, how high will it get?

Watch out!

If an object decelerates (slows down), it has *negative* acceleration. You must use *negative* values for a in your calculations.

Changing the subject

Basic algebra is all you need when tackling problems involving the equations of motion. You must know how to change the subject of an equation. It pays to be slow and methodical. Work one step at a time and write everything down. There are no prizes for getting the wrong answer in record time!

The rules

You can do anything to one side of the equation, provided you do exactly the same thing to the other side.

- To move a term from one side to the other, subtract it from both sides

  e.g. $\boldsymbol{v} = \boldsymbol{u} + \boldsymbol{a}t \Rightarrow \boldsymbol{v} - \boldsymbol{u} = \boldsymbol{a}t$

- To remove an unwanted factor, divide both sides of the equation by it

  e.g. $\boldsymbol{v} - \boldsymbol{u} = \boldsymbol{a}t \Rightarrow \boldsymbol{a} = (\boldsymbol{v} - \boldsymbol{u})/t$

- To remove an unwanted denominator, multiply both sides by it

  e.g. $\boldsymbol{s} = \boldsymbol{u}t + \tfrac{1}{2}\boldsymbol{a}t^2 \Rightarrow 2\boldsymbol{s} = 2\boldsymbol{u}t + \boldsymbol{a}t^2$

If you find yourself making mistakes, slow down! Try enclosing each side of the equation in brackets before you do anything. This makes it clear that any multiplication or division must involve every term.

Problem solving 

- Write down everything you have been given (there are five quantities: $\boldsymbol{s}$, $\boldsymbol{u}$, $\boldsymbol{v}$, $\boldsymbol{a}$ and t) and identify the unknowns.
- Look for equations that contain just one unknown quantity – preferably the one asked for in the question – and solve them.
- *An equation cannot be solved if there is more than one unknown factor.*
- If all else fails use simultaneous equations, i.e. find one unknown factor in terms of another and then substitute into a second equation to get rid of one of the two unknowns!

Exam questions answers: page 35

1 Calculate the steady acceleration needed to take a cyclist from rest to $10\ \text{m s}^{-1}$ in a distance of 20 m. (5 min)

2 A sprinter accelerates at $2.5\ \text{m s}^{-2}$ for the first 4 s of a 100 m race, then keeps going at a steady pace.
(a) What is her top speed?
(b) What is her average speed over the 100 m race? (10 min)

3 A stone is thrown up into the air. It reaches a height of 8 m. Assuming $\boldsymbol{g} = 10\ \text{m s}^{-2}$, calculate its initial upward velocity and the time taken before it returns to its starting position. (10 min)

4 A car approaches a junction at $20\ \text{m s}^{-1}$. Traffic lights at the junction suddenly turn red when the car is 15 m away. The driver's reaction time is 0.5 s and the brakes decelerate the car steadily at $10\ \text{m s}^{-2}$. Where does the car stop? (15 min)

Action point

Skip this section if you already feel confident about manipulating equations. You can always come back to it if you hit problems.

The jargon

Algebra has its own symbolic shorthand:
$\Rightarrow$ means *implies*
$\therefore$ means *therefore*.

Checkpoint

Displacement equals average velocity, $\boldsymbol{s} = \left(\frac{\boldsymbol{u} + \boldsymbol{v}}{2}\right)\boldsymbol{t}$, × time.
Acceleration is change in velocity divided by time, $\boldsymbol{a} = \frac{\boldsymbol{v} - \boldsymbol{u}}{\boldsymbol{t}}$.
Combine these two equations to give $\boldsymbol{s} = \boldsymbol{u}\boldsymbol{t} + \tfrac{1}{2}\boldsymbol{a}\boldsymbol{t}^2$.

Watch out!

$\boldsymbol{s}$, $\boldsymbol{u}$, $\boldsymbol{v}$ and $\boldsymbol{a}$ are all vector quantities. Choose a positive direction before you start on a problem. This is particularly important when dealing with projectiles (rising and) falling under gravity.

Examiner's secrets

There is very often more than one way to solve problems using the equations of motion. The most important thing to remember is to follow through your calculation and show a clear method.

The jargon

The data given in the question is usually only what is required for the answers. Do look carefully for clues in the question, such as *accelerates from rest*, which means $\boldsymbol{u} = 0\ \text{m s}^{-1}$.

Projectiles

Projectiles are subject to gravity. As they move along, they also accelerate downwards. This added layer of complexity makes projectiles questions a favourite among examiners.

Action point

If you are at all sceptical, check for yourself by dropping objects simultaneously from the same height. Provided the objects you drop are heavy enough that air resistance is insignificant, they will hit the ground at the same time.

Gravitational acceleration

Near the Earth's surface, the rate of acceleration due to gravity is constant for all objects, regardless of their mass (as Galileo discovered). The rate of gravitational acceleration is given the symbol $\boldsymbol{g}$. Near the Earth's surface: $\boldsymbol{g} = 9.81 \text{ m s}^{-2}$.

Provided air resistance is negligible, any free-falling object will accelerate (towards the centre of the Earth) at 9.81 m s^{-2}.

- ➜ Acceleration is a vector quantity.
- ➜ Gravitational acceleration is directed towards the centre of the Earth. *It has no effect at all on horizontal motion.*

Two falling balls are shown below.

Vertical fall $\boldsymbol{v}_x = 0$

Fall with constant horizontal velocity $\boldsymbol{v}_x = 1 \text{ m s}^{-1}$

Ground

The jargon

Free fall is a term used to describe bodies falling and accelerating (due to gravity) at a steady rate. Air resistance is ignored and the equations of motion can be applied to free-fall problems.

Resolving horizontal and vertical motion

Projectile motion must always be broken down into horizontal and vertical components. Air resistance complicates matters, so you will normally be told to ignore it (very handy)! With air resistance out of the way, it's easy:

- ➜ horizontal velocity is constant for the duration of the projectile's flight
- ➜ vertical motion is subject to a constant downward acceleration of 9.81 m s^{-2}

Watch out!

You should be prepared for some problems to yield two possible solutions (a squared term can have positive and negative roots). In such cases, you should give both answers and mention the significance of each (for example, positive and negative velocity might signify upward and downward motion).

The first step in every projectiles question is to calculate initial vertical and horizontal velocities:

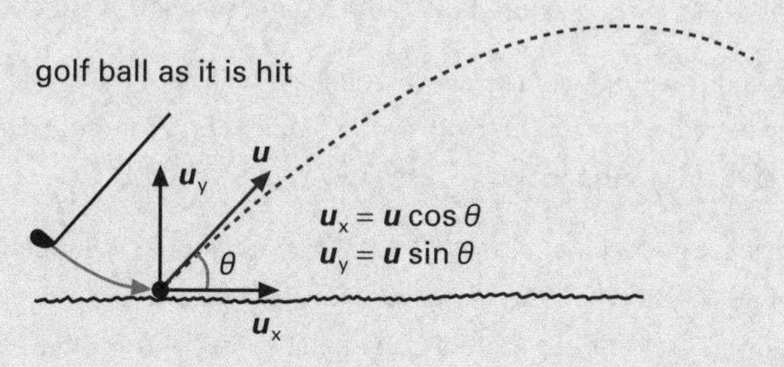

Don't forget

The curved path of a projectile is called a *parabola*.

The next steps depend on the question you have been asked.

Action point

Can you show that the time of flight of a projectile depends on the vertical velocity?
Can you also show that range of a projectile depends on the horizontal velocity and the time of flight?

Solving projectiles problems

You may be asked to:

→ *find the maximum height reached by a projectile*
The maximum height attained has nothing to do with horizontal motion, so just work out initial vertical velocity $\boldsymbol{u}_y$ and then apply the equations of motion, with $\boldsymbol{v}_y = 0\ \text{m s}^{-1}$ and $\boldsymbol{a} = -9.81\ \text{m s}^{-2}$ to find out the height reached

→ *find the range of a projectile*
The range is the product of horizontal velocity and time in the air. The tricky bit is working out the time from the vertical (gravitationally accelerated) motion.

Split the problem into two halves:
Going up The projectile decelerates from $\boldsymbol{u}_y$ to $0\ \text{m s}^{-1}$ at its peak ($\boldsymbol{a} = -9.81\ \text{m s}^{-2}$).
Going down The projectile accelerates from $0\ \text{m s}^{-1}$ (at $9.81\ \text{m s}^{-2}$).

If the starting height is the same as the stopping height, the path will be symmetrical and the total time in the air will simply be twice the time taken to reach the peak. If not, you will have to calculate the distance the projectile must fall and work out the time taken from that!

→ *find the trajectory of a projectile at some point*
The trajectory of a projectile is found using trigonometry. The tangent of the angle, relative to the horizontal, is equal to $\boldsymbol{v}_y/\boldsymbol{v}_x$. $\boldsymbol{v}_x$ is constant, but $\boldsymbol{v}_y$ varies over time, so you will have to use the equations of motion to find it!

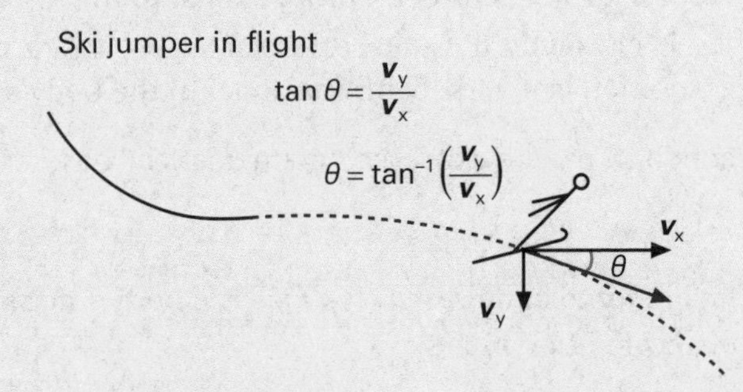

Exam questions answers: page 35

1 A car is accidentally driven off a (level) cliff at $8\ \text{m s}^{-1}$. It lands (in the sea) 3 s later.
(a) How far out to sea does it 'land'?
(b) How high is the cliff? (Assume $\boldsymbol{g} = 10\ \text{m s}^{-2}$)
(c) At what angle (relative to the horizontal) does it land? (15 min)

2 Find the range of a stone fired at an angle of 45° to the horizontal at a speed of $20\ \text{m s}^{-1}$. (Assume air resistance is negligible.) (10 min)

3 A helicopter flying horizontally at 500 m altitude and at a constant velocity of $50\ \text{m s}^{-1}$ drops a package. If air resistance is insignificant,
(a) how long does it take for the package to hit the ground?
(b) how far does it travel?
(c) at what *speed* does the package hit the ground? (Difficult) (20 min)

Test yourself

You must *learn* the equations of motion given in *equations of motion*, page 10. Make a conscious effort now.
$\boldsymbol{v}$ = ?
$\boldsymbol{s}$ = ?
$\boldsymbol{v}^2$ = ?

Examiner's secrets

In many questions you may be expected to state any assumptions you make.

The jargon

Subscripts are very handy. Rather than writing out *horizontal component of velocity* = . . . , we can write $\boldsymbol{v}_x$ = . . .

Check the net

To test your understanding of projectiles go to www.crocodile-clips.com/absorb/AP5/sample/010105.html where you will find questions based on animations.

Newton's laws of motion

"Nature and nature's laws lay hid in night; God said 'Let Newton be!' and all was light."

Alexander Pope

"If I have seen farther, it is by standing on the shoulders of giants."

Isaac Newton

Isaac Newton was a genius who changed the world. Every physicist in the world climbs up first on Newton's shoulders before scrambling onwards and upwards!

Newton's first law of motion

Newton's first law tells us what happens when the forces acting on a body are balanced (i.e. when the resultant force is zero).

→ A body will remain at rest, or will keep travelling with constant velocity, unless acted on by unbalanced external forces.

In exams, you may be told that a body is travelling with constant velocity; this is always a clue to use Newton's first law to find the size of any unknown forces.

Mass and inertia

Inertia is a disinclination to move or act; it is the tendency of a body to remain at rest, or to keep its uniform motion.

→ Mass *is* inertia.

Newton's second law of motion

Newton's second law describes what happens to a body acted upon by unbalanced forces.

→ unbalanced forces cause acceleration
→ the rate of acceleration is directly proportional to the net force
→ acceleration is in exactly the same direction as the net force
→ rate of acceleration is inversely proportional to the body's mass

Links

Newton's second law can also be expressed in terms of a body's change in momentum. See *momentum and impulse*, pages 24–5.

Newton's second law can be expressed mathematically as:

$$\boldsymbol{F} = m\boldsymbol{a}$$

Where $\boldsymbol{F}$ is the net force in newtons (N), m is the body's mass in kg and $\boldsymbol{a}$ is its rate of acceleration in m s^{-2}.

→ acceleration is always in the same direction as the net force
→ 1 N is defined as the force necessary to accelerate a 1 kg mass by $1\ \text{m s}^{-2}$ – with any other units, a conversion factor would be required

Checkpoint 1

What is the pair to each of these forces?
(a) the pull of the Earth's gravity on the Moon
(b) the weight of a man
(c) the impact of a dart on a dartboard.

Newton's third law of motion

Newton realized that you cannot have an isolated force. Pushes and pulls *always* involve two different objects.

→ When body A exerts a force on body B, body B exerts an equal and opposite force on body A.
→ Forces act in pairs.
→ The paired forces (called *action* and *reaction*) always act on different bodies. There is no way one of them can balance the other one!

Newton's third law becomes most obvious when we are dealing with two bodies of comparable mass; when one object is much more massive than the other, its acceleration may be too small to notice.

Watch out!

Spotting a force pair is not always easy. Use this checklist:
(a) They have the same magnitude (size).
(b) They act along the same line but in opposite directions.
(c) They act on different objects.
(d) They are the same type of force.

Rockets and jets

The rocket pushes itself forwards by pushing the exhaust jet backwards.

Links

See *momentum and impulse*, pages 24–5.

Free-body force diagrams

A free-body force diagram should be the starting point of every exam answer on forces. It should simplify a problem, by showing *only* the forces acting on the body you are interested in.

→ A free-body force diagram should show a body and the forces acting on it – and nothing else!

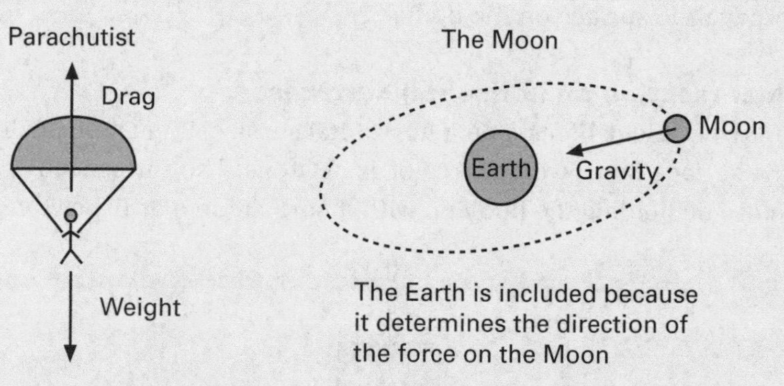

Checkpoint 2

Draw a free-body force diagram for:
(a) a space-rocket
(b) its exhaust jet.

"I do not know what I may appear to the world, but to myself I seem to have been only like a boy playing on the seashore . . . whilst the great ocean of truth lay all undiscovered before me."

Isaac Newton

You may notice slight variations in the ways free-body diagrams are treated in different books. Some people like to draw the object realistically, showing where each force acts (as well as showing each force's direction and size). This approach is right and proper, but it complicates things unnecessarily. Unless you are actually going to consider each part separately, or calculate turning moments, it is simpler and more sensible to treat the entire body as a single point, with force arrows pointing outwards wherever possible.

Exam questions answers: pages 35–6

1. An aircraft carrier catapults a 12 000 kg jet from rest to 50 m s^{-1} in a distance of 50 m. Find the average force on the jet. (15 min)

2. An arrow is fired from a bow. When the average force on the arrow is F, the arrow accelerates from rest to a top speed of 20 m s^{-1}. What speed would the arrow reach if the average force on it was doubled (to $2F$)? (15 min)

3. A truck of mass 2 000 kg pulls a trailer of mass 3 000 kg along a level road. The truck and trailer accelerate at 1.2 m s^{-2}. Friction and air resistance act on both the truck and the trailer. The total drag force is a constant 3 000 N, with a half of this force acting on the trailer. Draw free-body force diagrams for the truck and for the trailer and work out the drive force, D, and the tension, T, in the tow-bar. (25 min)

Some important forces

Some forces deserve special attention, simply because we come across them so often – in A-level Physics questions and in life!

Links

See *Newton's third law of motion*, page 14.

The normal-reaction force

When faced with a choice of words, physicists often choose the least obvious. 'Normal to', 'at right angles to', 'perpendicular to' and 'at 90° to' all mean the same thing. A *normal* force is not an everyday force; it is a force that acts at 90° to a surface.

If you press against a surface, the surface pushes against you with an equal and opposite *reaction* force. The **normal reaction** is the component of the surface's reaction which is normal to (at 90° to) the surface (the normal reaction is independent of friction and other forces that act *along* the surface).

Watch out!

The normal force sometimes causes unnecessary confusion. If the surface is not strong enough to hold the body up, it will break, but for as long as two surfaces remain in contact, the action of the body on the surface equals the reaction of the surface on the body.

→ The force of the body on the surface is equal and opposite to the force of the surface on the body.

Normal reaction on horizontal surfaces

If a body of weight $\boldsymbol{W}$ rests on a horizontal surface, it must be pushing the surface down with a force of $\boldsymbol{W}$ newtons. The surface must therefore be holding the body up with a normal force of $\boldsymbol{W}$ newtons.

Checkpoint 1

The normal reaction is not always equal in size to a body's weight. Racing cars use aerodynamics to increase the surface contact force. What happens to the normal force as a racing car's speed increases? What are the benefits?

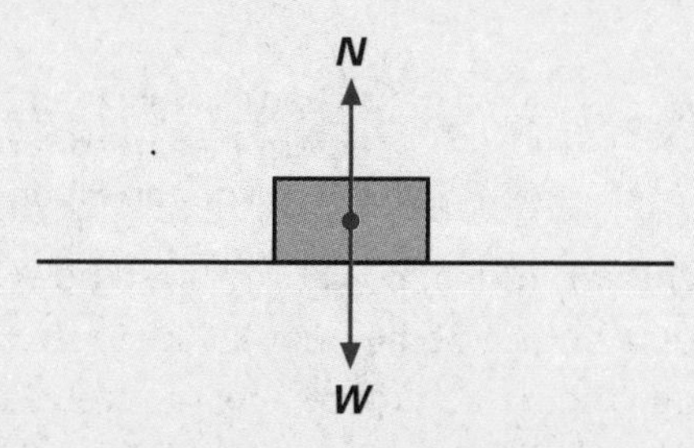

The block is not moving, so the normal reaction $\boldsymbol{N}$ must be equal and opposite to $\boldsymbol{W}$. Note that $\boldsymbol{N}$ and $\boldsymbol{W}$ are *not* action and reaction. They are different types of force: gravitational and electrostatic/contact.

Watch out!

The forces $\boldsymbol{N}$ and $\boldsymbol{W}$ are not force pairs as they act on the *same* object.

Normal reaction on sloping surfaces

The only complication that arises here is that the forces acting must be resolved into components acting along and at right angles to the surface. The normal reaction of the surface on the body is equal and opposite to the normal component of the body's force acting on the surface!

Watch out!

It is important that you remember that action and reaction forces always act on different bodies. Weight and the normal reaction both act on the same body, they are *not* action and reaction.

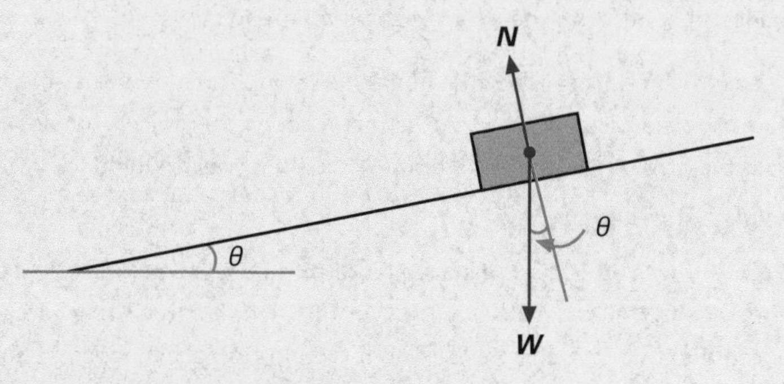

The block is not moving perpendicular to the plane, so the normal forces must balance: $\boldsymbol{N} = -\boldsymbol{W}\cos\theta$.

Checkpoint 2

Car tyres do not grip the road as well on hills as they do on horizontal roads. Why not? (*Hint* think about what happens to the normal force as the hill's gradient increases.)

Weight

Weight is the common name for the force of gravity on an object. Weight $\boldsymbol{W}$ (in newtons) is given by:

$$W = mg$$

Where m is the object's mass in kilograms and $\boldsymbol{g}$ is the Earth's gravitational field strength. At the surface of the Earth, a good average value for $\boldsymbol{g}$ is:

$$g = 9.81\ \mathrm{N\,kg^{-1}}$$

Notice that gravitational field strength determines the rate of acceleration due to gravity. Since every kilogram *weighs* 9.81 N, every kilogram *accelerates* at 9.81 $\mathrm{m\,s^{-2}}$ ($\boldsymbol{F} = m\boldsymbol{a}$).

➜ Weight always acts towards the centre of the Earth (on Earth!).

Friction

Friction is caused by surface roughness on a microscopic scale (and quite complex intermolecular forces). Frictional force $\boldsymbol{F}$ is given by the empirical formula:

$$F = \mu N$$

Where $\boldsymbol{N}$ is the normal force and μ is a constant called the **coefficient of friction**. μ's value depends on the surfaces, the presence or absence of any lubricant and whether or not the surfaces are in relative motion!

➜ Friction always opposes relative motion.
➜ Friction acts along the surface.
➜ Friction between *moving* surfaces generates thermal energy (heat).

Solving problems on forces

1 Always draw a free-body force diagram.
2 Resolve forces into components acting along and at right angles to the surface.
3 Solve the problem!

Forces perpendicular to a surface must balance (or the object will sink into or fly off the surface).

Exam questions answers: page 36

1 A car of mass 900 kg climbs a 15° hill. If the car's engine provides a forward thrust of 8 000 N and the drag and friction on the car add up to 800 N, calculate the car's rate of acceleration. (10 min)

2 A rope is used to pull a single skier up a friction-free incline of 25°. If the skier's mass is 50 kg, calculate the tension in the rope when the skier is moving steadily up the slope. (10 min)

3 A man of mass 700 kg stands in a lift on a set of weighing scales. The scales are calibrated in newtons. Assuming a value of 10 $\mathrm{N\,kg^{-1}}$ for $\boldsymbol{g}$,
(a) What weight will the scales register when the lift is static?
(b) What weight will the scales register when the lift accelerates downwards at 10 $\mathrm{m\,s^{-2}}$?
(c) What weight will the scales register when the lift accelerates upwards at 10 $\mathrm{m\,s^{-2}}$? (10 min)

Links

See *gravitational fields*, pages 134–5, to find out more about what $\boldsymbol{g}$ depends on and how it varies with location.

Watch out!

Mass and weight are not the same thing. Make sure you know the difference. It is easy to mix them up.

The jargon

An *empirical* formula, equation or constant is one which is derived from experimental data (without complete understanding) rather than from some elegant theory or by logic from first principles.

Checkpoint 3

Be careful with directions. In which direction does friction act when:
(a) a block is at rest on a slope?
(b) it is being dragged up a slope?
(c) it is sliding down a slope?

Examiner's secrets

Many questions involving friction require you to spot the *resultant* (or net) force acting.

Density and pressure

"Eureka!"

Archimedes

Checkpoint 1

Is sea water more dense or less dense than pure water? Is warm water more dense or less dense than cold water?

Checkpoint 2

What do you think will happen to the level of water in an icy drink as the ice melts? Explain your answer. (Try it if you are not sure.) The greenhouse effect may cause the polar ice caps to melt. The Arctic ice cap is a floating mass of ice. What effect will the melting of the Arctic ice cap have on sea levels?

Links

See *probing matter*, pages 54–5.

Watch out!

Upthrust depends on the density of the fluid, *not* the density of the object.

Checkpoint 3

Explain the position and (if appropriate) motion of the crab, the bubble and the submarine in the figure on the right.

The jargon

The term *fluid* covers liquids and gases and some powdered solids. Anything that *flows* is fluid.

Density and pressure must be very familiar quantities by now. Density measurements come up regularly in practical exams. Pressure and buoyancy calculations have too many applications to be ignored.

Density

$$\text{density} = \frac{\text{mass}}{\text{volume}} \qquad \rho = \frac{m}{V}$$

The SI unit of density is kg m^{-3}.

Density questions often involve unit conversions. These are not always easy, so take them one step at a time. For example, to convert from grams per cubic centimetre to kilograms per cubic metre, we need to work out how many kg in 1 g (10^{-3}) and how many m^3 in 1 cm^3 (10^{-6}):

$$1 \text{ g per cm}^3 = 10^{-3} \text{ kg per } 10^{-6} \text{ m}^3 = 10^3 \text{ kg m}^{-3}!$$

- You need to know practical methods for determining the densities of (regular and irregular) solids, liquids and gases.
- Density determines buoyancy.

Upthrust and Archimedes' principle

- The **upthrust** on a body immersed or floating in a fluid is equal to the weight of fluid displaced.
- Upthrust on a body depends on the volume of fluid it displaces, the fluid's density, and the gravitational field strength.
- Upthrust = ρVg where ρ is the fluid's density and V is the volume displaced (which also equals the submerged volume of the body).
- For water, $\rho = 1\,000$ kg m^{-3} (= 1 g cm^{-3}).

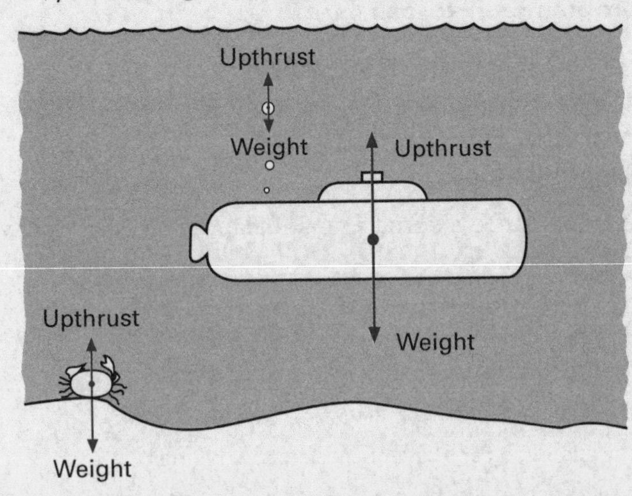

- Whether a body will float or sink depends upon its density.
- Sinkers are more dense than the fluid, floaters are less dense.
- Boats (and floaters in general) sit on the surface of the water, displacing exactly their own weight in water.
- Relative density (or specific gravity) = density ÷ density of water.

Buoyancy results from the vertical pressure gradient in the Earth's gravitational field. The air pressure on your feet is slightly greater than the air pressure on your head (at least it is when you stand up!). Water is about a thousand times more dense than air, so buoyancy forces are correspondingly greater in water.

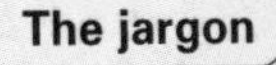

Pressure

Pressure is defined as the perpendicular force per unit area on a surface.

$$\text{pressure} = \frac{\text{force}}{\text{area}} \qquad p = \frac{F}{A}$$

The SI unit of pressure is the pascal (Pa). 1 Pa = 1 N m^{-2}.

Pressure in fluids

Pressure beneath a depth h of fluid is given by: $p = \rho gh$

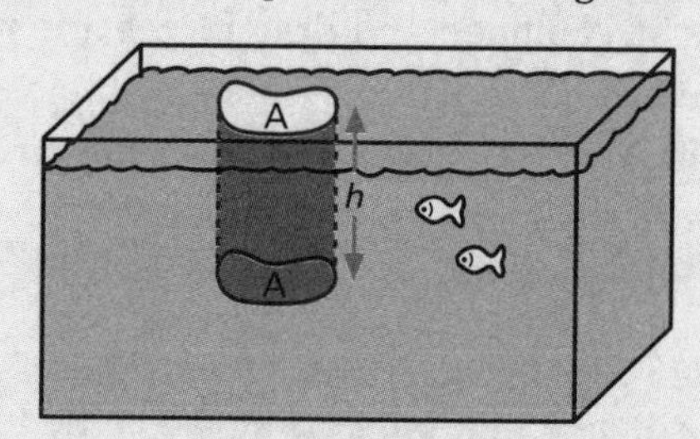

volume = Ah
mass = ρAh
weight = ρAgh
pressure = ρgh

Note The pressure exerted sideways on the container has no effect on the pressure beneath the column. The shape of the container has no effect on pressure in a static fluid.

Pressure in solids, liquids and gases

- Solids transmit forces.
- Liquids transmit pressure and are incompressible.
- Gases transmit pressure, but are compressible.

You can push or pull one end of a solid object and, provided no turning moments are involved, the same force is transmitted to the other end. Liquids and gases cannot transmit forces directly; they give. To make use of liquids and gases in hydraulic and pneumatic systems, we have to confine them and make use of the fact that they transmit pressure.

Atmospheric pressure

A **barometer** is used to measure atmospheric pressure. The greater the atmospheric pressure, the higher the column of mercury it will support. Mercury is used in barometers because it is such a dense liquid. Normal atmospheric pressure will support 76 cm of mercury.

A **manometer** can be used to measure the excess pressure over atmospheric pressure – useful for measuring water and gas pressures, and even for measuring blood pressure.

Units 1 atmosphere = 1 bar = 1 000 mbar = 76 cm Hg = 101 325 Pa.

Checkpoint 4

Why do bubbles in a beer glass expand as they rise towards the surface?

Checkpoint 5

How does a hydraulic jack work? (*Remember* The pressure in the fluid is the same at the effort piston as at the load piston, but the forces are not equal.)

Checkpoint 6

Scuba divers are told to breathe out all the way if they have to surface in a hurry. Why?

Checkpoint 7

Why are air bubbles dangerous in hydraulic brake systems?

Exam questions

answers: page 36

1 Ice has a density of 920 kg m^{-3}. Liquid water has a density of 1 000 kg m^{-3}. Calculate the volume occupied by 0.5 kg of (a) ice and (b) water. Use your answer to explain why water pipes often burst when they freeze. (6 min)

2 If the density of air in the atmosphere had a constant value of 1.2 kg m^{-3} throughout its depth, how deep would it be? (Density of mercury is 13 600 kg m^{-3} and 1 atmosphere = 76 cm Hg.) (10 min)

3 A water bed is 2 m long, 1.5 m wide and 0.25 m deep. Calculate both the pressure and the total force it exerts on the floor. (10 min)

Drag and lift

"Great whirls have
little whirls
That feed on their
velocity;
And little whirls have
lesser whirls
And so on to viscosity."

L. F. Richardson

The jargon

Laminar flow is generally used synonymously with *streamlined flow*.

Action point

Dimensionally speaking, 1 N is the same as 1 kg m s^{-2}. Check that the drag equation is dimensionally correct.

Examiner's secrets

Drag force varies as the square of the velocity ($D = kv^2$). To prove it you would need to show that k is *always* a constant.

Links

See *work, energy and power*, pages 22–3.

Checkpoint 1

What happens to the size of the drag force on a car when its speed doubles? What implications can you think of for fuel efficiency?

Checkpoint 2

Sketch a velocity–time graph for a sky diver who free-falls for two minutes before opening his or her parachute. (Variations on this theme are examiners' favourites.)

Some understanding of drag and lift is essential, but not all A-level specifications require the depth shown here. Check before devoting hours to the detailed and difficult bits!

Streamlined and turbulent flow

In streamlined flow, the fluid does not get mixed by the passing body. Every particle follows the streamline it is in. At any chosen point, every passing particle will have the same velocity. At low speeds, viscous forces dominate, fluid flow is streamlined and drag forces are small. As a body picks up speed, there will come a point where the fluid switches from streamlined to turbulent flow and drag forces become far greater.

Drag

There are two causes of **drag**, called *skin friction* and *form drag*. Skin friction is pretty similar to ordinary friction. Form drag is the force needed to move the fluid out of the way of a *bluff body* (a real body, which takes up space). Form drag depends on speed of relative motion v, frontal area A (maximum area of cross-section in the direction of travel), fluid density ρ_f and a dimensionless drag coefficient C_d, which depends on the body's shape (normally determined experimentally).

$$\text{drag force} = \tfrac{1}{2}AC_d\rho v^2$$

In most practical situations where you might need to calculate drag, form drag dominates and the drag equation holds true.

→ Drag always opposes the relative motion of a body in a fluid.

Terminal velocity

Terminal velocity is the steady downward speed a falling body reaches when the downward force of gravity is exactly balanced by the upward force of drag. Sky divers control their speed by changing their frontal area. At terminal velocity, weight = drag (= a constant value). A larger frontal area must reduce the terminal velocity!

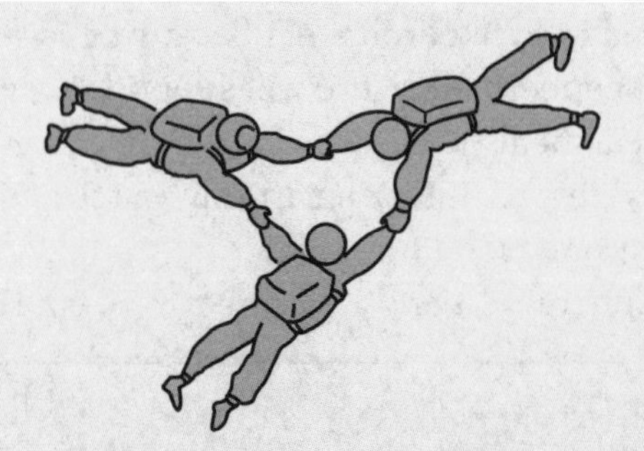

The terminal velocity v_{max} of a raindrop depends on its size (its radius). Big drops fall faster than small ones. Since a raindrop's weight is proportional to the cube of its radius and its frontal area is proportional to the square of its radius:

$$v_{max}^2 \propto r$$

Many animals have broadly similar densities which helps explain why small animals and insects can survive big falls!

Fluid flow in pipes

If a fluid flows through a pipe which has one point of entry and one point of exit, the mass flow rate *in* must equal the mass flow rate *out*. This is a direct result of the law of conservation of mass. It is true for every section of the pipe, so we get an *equation of continuity*:

$$\rho_1 A_1 v_1 = \rho_2 A_2 v_2$$

Mass flow rate ($kg\,s^{-1}$) is the product of density ρ, cross-sectional area of the pipe A and the speed of the fluid v. The subscripts 1 and 2 refer to any two points along the pipe. Liquids are incompressible (they have constant density whatever the pressure), and so the equation of continuity for a liquid in a pipe is simply $A_1 v_1 = A_2 v_2$.

→ Fluids travel fastest through the narrowest sections of a pipe. Individual particles *accelerate* as they enter a narrow section of pipe.

According to Newton's second law of motion, the forces acting on particles of fluid must therefore be unbalanced. This implies that:

→ the pressure is lowest at the narrowest points in a pipe (where the fluid is travelling fastest).

Checkpoint 3

Multiply the units of density, area and velocity and see if you get the right units for mass flow rate. What are the units of volume flow rate Av?

Examiner's secrets

This is analogous to the electric current through a wire: the electrons flow fastest through the thinnest bits of wire (e.g. through light bulb filaments).

Bernoulli equation

Assuming streamlined flow of an incompressible, non-viscous (!) fluid, Bernoulli came up with this famous equation:

$$p + \tfrac{1}{2}\rho v^2 + \rho g h = \text{constant} \quad or$$
$$p_1 + \tfrac{1}{2}\rho v_1^2 + \rho g h_1 = p_2 + \tfrac{1}{2}\rho v_2^2 + \rho g h_2$$

Where p is pressure, ρ is fluid density (assumed constant), v is speed of flow, g is gravitational field strength, and h is the height of the fluid above some arbitrary point (and subscripts 1 and 2 refer to two points in the same streamline).

→ The faster the air, the lower the pressure.
→ Wings are shaped to make air travel further and faster over their top surfaces.
→ The lift force on a wing or aerofoil is proportional to the square of the speed of flow.

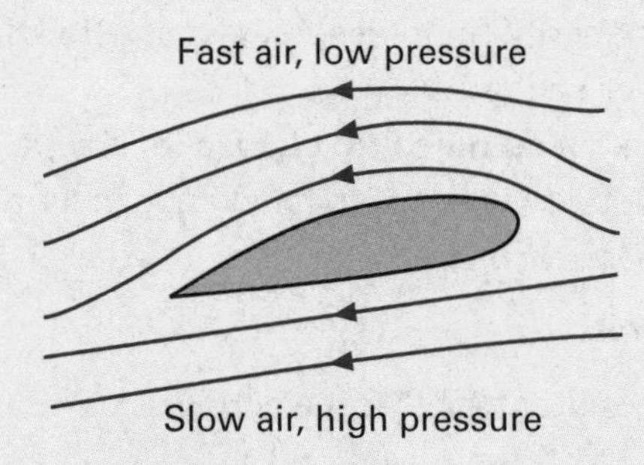

Checkpoint 4

Bernoulli's assumptions are very dodgy and as a result, the equations are rarely entirely accurate, but they do give us some useful predictions to work on. Which of Bernoulli's assumptions is always false: (a) for liquids and (b) for gases?

Checkpoint 5

Bernoulli's equation can be derived from the principle of the conservation of energy. The pressure drop is equal to the work done to raise the fluid's kinetic energy (per unit volume) and/or potential energy (also per unit volume). Which terms represent (a) kinetic and (b) potential energy *per unit volume*?

Checkpoint 6

How does an aerofoil work? How does its shape give rise (excuse the pun) to lift?

Exam question

answer: page 37

A football is kicked so that it spins clockwise. Explain why the ball's path curves and show with the aid of a diagram which way it will curve. (15 min)

Work, energy and power

Energy is quite a difficult concept. You only really notice it when it is being transferred from one place to another, or converted from one form to another, and yet it is all around us. Energy transfer drives every process!

Work

Work is done whenever a force is used to move something. Provided the force points in the direction of motion, work is defined by:

work done = force × distance moved $W = \boldsymbol{Fs}$

- Work is an energy transfer, measured in joules.
- 1 J = 1 newton metre (N m). (A joule is the amount of work done when a 1 newton force moves something 1 metre in the direction of the force.)

Work done by forces acting at angles to direction of motion

The only complication is that you must calculate the component of the force in the direction of motion and then use that to work out the work done. For a force $\boldsymbol{F}$ acting at angle θ to the direction of motion:

work = $\boldsymbol{F} \cos \theta \, \boldsymbol{s}$

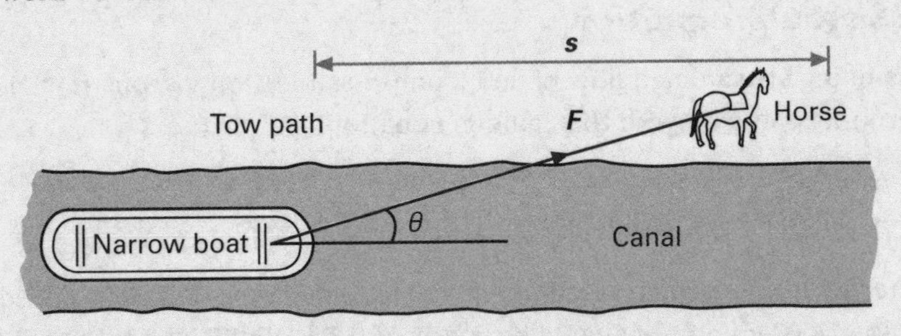

The boat travels a distance s parallel to the bank. Force in the direction of travel = $\boldsymbol{F} \cos \theta$ so the work done = $\boldsymbol{F} \cos \theta \, \boldsymbol{s}$.

Work and kinetic energy

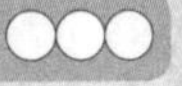

Kinetic energy E_k is the energy a body has by virtue of its motion. It is defined by the equation:

$$E_k = \tfrac{1}{2}m\boldsymbol{v}^2$$

Where E_k is kinetic energy (in joules), m is mass (in kilograms) and $\boldsymbol{v}$ is speed (in metres per second).

You can do work on an object to change its kinetic energy. In fact, the work done on a body is equal to its change in kinetic energy. (This is called the *work–energy theorem*.):

$$\boldsymbol{Fs} = \tfrac{1}{2}m\boldsymbol{v}^2 - \tfrac{1}{2}m\boldsymbol{u}^2$$

- *Note* $\boldsymbol{F}$ is the resultant (net) force acting.

This equation is particularly useful in calculations of minimum stopping force or minimum stopping distance. If a body is brought to a halt, the work done to it (e.g. by the braking force) is equal to the kinetic energy lost.

- Note the dependence on $\boldsymbol{v}^2$. If you double your speed, your minimum stopping distance is quadrupled (assuming the same retarding force). Speed limits save lives!

Links

See *binding energy and mass defect*, pages 50–1. According to Einstein's theory of special relativity, matter and energy are two forms of the same thing *mass–energy*!

Watch out!

If nothing is actually moving, no work is being done – no matter how great the force involved (since no energy is being *transferred*). Work is defined by the work equation. If $\boldsymbol{s} = 0$, $W = 0$.

The jargon

Work is a scalar quantity (so are both energy and power), but you can have positive and negative work. *Positive work* is where the force pulls in the same direction as the movement. *Negative work* is where the force is in the opposite direction (e.g. gravity does positive work to accelerate a cyclist down a hill and negative work to decelerate him up the other side).

Don't forget

Do you know your sines and cosines? It is important here to remember that *cos 90* = 0 and *cos 0* = 1.

Links

See *scalars and vectors*, pages 4–5.

Action point

If you enjoy algebra, try deriving the formula for kinetic energy (as the energy gained when a force does work to accelerate an object) by combining the equations $\boldsymbol{F} = m\boldsymbol{a}$, $W = \boldsymbol{Fs}$ and $\boldsymbol{v}^2 = \boldsymbol{u}^2 + 2\boldsymbol{as}$. If you don't, don't!

Watch out!

When a body is travelling at constant speed, the net force on it is zero. Any work done is *not* being done to the body in question.

Gravitational potential energy

The work that gravity can do to an object (if it should fall) is called **gravitational potential energy**. If you lift an object of mass m by a height $\boldsymbol{h}$, the gravitational potential energy E_p it gains is equal to $m\boldsymbol{gh}$ (i.e. the work done to lift its weight mg by a height $\boldsymbol{h}$ against the pull of gravity).

$$E_p = m\boldsymbol{gh}$$

- It is often useful to define h as the height above some reference point (usually the lowest point considered).
- It does not matter what path is taken; potential energy gained is equal to the work done against gravity (i.e. in the vertical direction).

Conservation of mechanical energy

The term *mechanical energy* refers to the sum of a body's kinetic and (gravitational) potential energy.

$$E = E_k + E_p$$

Provided that no energy is converted to other forms, mechanical energy is conserved. If we know the position and speed of a body at one point, we can use the formulae for E_p and E_k to work out its speed from its position (or vice versa) at any other point! We can apply conservation of mechanical energy to a roller-coaster ride:

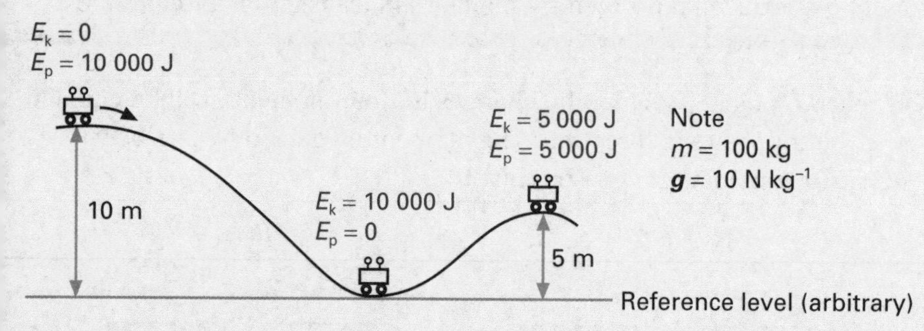

Law of conservation of energy

Energy can be neither created nor destroyed. It can only be converted from one form to others. The total amount of energy in any isolated system remains constant.

Power

Power is defined as rate of transfer of energy, or rate of doing work. It is measured in watts (W). 1 watt = 1 joule per second.

average power = work done (or energy transferred) ÷ time taken

For any vehicle travelling at speed v, power output is given by:

$P = Fv$ (F = *motive* or *drive* force and v = average speed)

Exam question answer: page 37

A boat of mass 8 000 kg approaches a jetty at a speed of 1.2 m s^{-1}. Its engine is put in reverse and the boat comes to a halt in a distance of 6 m. (a) What is the engine's (backward) thrust? (b) What is the engine's power? Assume that drag can be ignored(!) (20 min)

Checkpoint 1

Choosing an arbitrary reference point to measure h from yields $E_p = 0$ when $h = 0$, which is a bit of a nonsense, since gravity will happily go on doing work on a body until it reaches the centre of the Earth (if there's a deep enough hole), so why do we do it?

Checkpoint 2

Comets follow eccentric elliptical paths around the Sun. Where along its path does a comet have the greatest potential energy?

Action point

Check other sections to find formulae and graphical methods for calculating other energy transfers (electrical, heat, elastic, nuclear, etc.).

Don't forget

Power P is rate of work

$$P = \frac{\text{force} \times \text{distance}}{\text{time}}$$

Since speed = distance/time we get

$$P = Fv$$

Watch out!

In the *motive power* equation ($P = Fv$), F is the drive force, *not* the net force acting. If the velocity is constant, all the engine's power is being used to overcome resistive forces.

Examiner's secrets

Energy can *never* be lost or gained. It can be transferred from one form to another. Many students lose marks by stating that energy is *lost* without saying where it has gone!

Momentum and impulse

In life, we may often 'act on impulse', but in physics, impulse more often acts on us! Impulse is the product of force and time. Impulses always change the momentum of the body they act on.

Momentum

If a body of mass m has a velocity $\boldsymbol{v}$, then its **momentum $\boldsymbol{p}$** is:

$$\boldsymbol{p} = m\boldsymbol{v}$$

The units of momentum are kg m s^{-1} or newton seconds (N s).

Law of conservation of momentum

Momentum is conserved in *all* collisions, explosions and interactions! There are no exceptions to this law.

- The total momentum of a system before any interaction is exactly equal to the total momentum after it, provided no external forces act (external forces would allow momentum to be transferred to external bodies).
- In most real applications, you would have to calculate the components of momentum (of every body involved) in three dimensions (x-, y- and z-directions), but at A-level, you only have to deal with one-dimensional problems – momentum transfers in straight lines. Very handy.

When two objects collide, the changes in their momenta will be equal in size, but opposite in direction. The momentum gained by one body equals the momentum lost by the other.

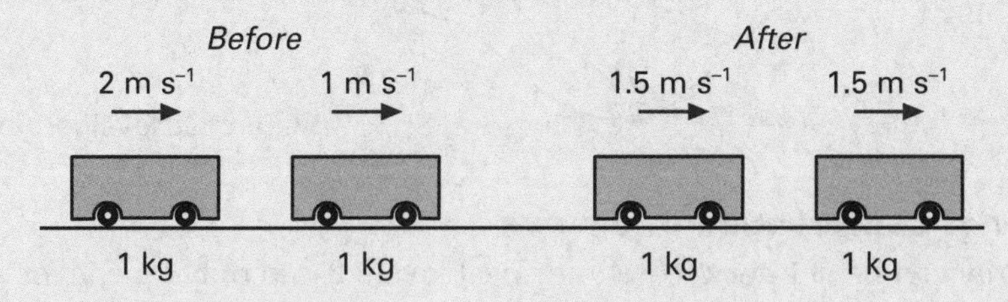

Tackling problems

1. Choose a positive direction.
2. Draw before and after sketches of the objects involved.
3. Calculate every momentum you can.
4. Apply the law of conservation of momentum for collisions involving two bodies:

$$(m_1\boldsymbol{v}_1 + m_2\boldsymbol{v}_2)_{\text{before}} = (m_1\boldsymbol{v}_1 + m_2\boldsymbol{v}_2)_{\text{after}}$$

Recoil and explosions

Guns and cannons *recoil* when fired because of the law of conservation of momentum. The positive momentum gained by the bullet or cannon ball is equal to the negative recoil momentum of the gun or cannon, and so the total momentum before and after the explosion is zero.

Examiner's secrets

A quantity's units can be derived from the equations they appear in. Mass is measured in kg, velocity is measured in m s^{-1}, so momentum can be expressed in kg m s^{-1}. If you substitute $\boldsymbol{F}/\boldsymbol{a}$ for m, you get an alternative unit for momentum – the *newton second*. Choose whichever unit suits your needs! (They are entirely equivalent.)

Watch out!

Momentum is a vector quantity. Don't forget to give it a direction. In problems involving motion in a straight line, you must state clearly which direction you are taking as the positive.

Checkpoint 1

The law of conservation of momentum is sometimes difficult to demonstrate convincingly with trolley experiments. Why? What could you do to improve things?

Examiner's secrets

Students often lose marks through carelessness with directions and signs. Be warned!

Elastic and inelastic collisions 

- → Kinetic energy is conserved in **elastic** collisions.
- → Kinetic energy is not conserved in **inelastic** collisions.
- → Momentum is conserved in *all* collisions.

In inelastic collisions some kinetic energy is converted to other forms of energy (usually mainly heat).

Coefficient of restitution *e*

The **coefficient of restitution** *e* describes how elastic a collision is.

$$e = \frac{\text{speed of separation}}{\text{speed of approach}}$$

- → For a *perfectly elastic* collision, $e = 1$.
- → For a *completely inelastic* collision, $e = 0$ (the two bodies stick together).
- → Most collisions lie between these two extremes and can be called *inelastic collisions*.

Links

See *work, energy and power*, pages 22–3. $E = \frac{1}{2}m\mathbf{v}^2$. Kinetic energy and momentum are calculated from the same quantities, but they are not the same thing. *Be clear* Momentum is always conserved, whereas kinetic energy is conserved only in special cases (in perfectly elastic collisions).

Impulse and Newton's second law of motion 

Most bodies have constant mass, so we normally (rightly) note that unbalanced forces cause acceleration. At a more fundamental level we can state that unbalanced forces cause a *change in momentum*. The change in momentum depends on the size and direction of the force and the period of time over which it is applied; i.e. it depends on its *impulse*.

- → Impulse is the product of force and time.
- → Impulse = change in momentum.
- → $\mathbf{F}t = m\mathbf{v} - m\mathbf{u}$ (Where $m\mathbf{u}$ is initial and $m\mathbf{v}$ is final momentum).
- → Force = rate of change of momentum = $(m\mathbf{v} - m\mathbf{u})/t$. This is another version of Newton's second law of motion.

Impulse is measured in either newton seconds (N s) or kg m s^{-1} (exactly the same units as momentum).

The jargon

Quantitative and *qualitative* If a question asks for a quantitative answer, you must give numbers and units; if it asks for a qualitative answer, you should give general trends ('When $F = 2$ N, $x = 8$ cm' is a *quantitative* answer. 'The extension of a spring is directly proportional to the force applied – until the elastic limit is reached' is a *qualitative* answer.)

Examiner's secrets

The area under a force–time graph is the change in momentum (impulse). To make a tennis ball leave the racket with a faster velocity you can hit it with a bigger force *or* for a longer time.

Exam questions answers: page 37

1 A bullet of mass 40 g is fired with a horizontal velocity of 500 m s^{-1} from a rifle of mass 2.5 kg.

(a) Find: (i) the bullet's forward momentum, (ii) the bullet's kinetic energy, (iii) the speed of recoil of the rifle, (iv) the rifle's kinetic energy after the explosion.

(b) Can explosions ever be perfectly elastic? Explain your answer. (15 min)

2 What driving force is necessary to accelerate a car of mass 1 400 kg from rest to a speed of 35 m s^{-1} in 20 s? (5 min)

3 Two skaters are skating together at a steady velocity of 8 m s^{-1}. Their masses are 80 kg and 50 kg. The lighter skater is pushed forwards and accelerates to 10 m s^{-1}. Calculate the new speed of her partner. (10 min)

4 Hard snowballs bounce back off you when they hit at 90°; soft snowballs don't. Explain why these hard snowballs exert the greater force. (10 min)

Stress, strain and Hooke's law

Robert Hooke was a brilliant scientist who had the misfortune to be a contemporary of Isaac Newton. He had theories on light and on gravity, but so did Newton. Hooke's lasting legacy to the world of science is his law of elastic deformation: that within bounds, strain is proportional to stress.

Springs

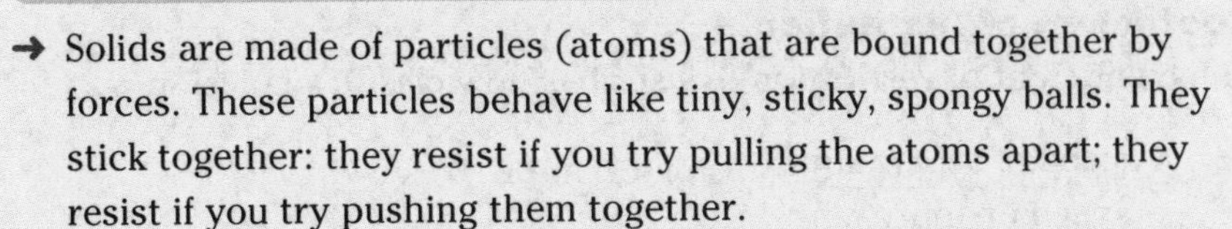

- Solids are made of particles (atoms) that are bound together by forces. These particles behave like tiny, sticky, spongy balls. They stick together: they resist if you try pulling the atoms apart; they resist if you try pushing them together.
- Springs behave like the forces between the atoms in a solid. Understanding how springs respond to forces is a useful first step towards a more general understanding of the behaviour of solid materials under stress.

Hooke's law applied to springs

Within the elastic limits of the spring:

$$F = -kx$$

Where F = force applied, x = extension (increase in length) and k = the *spring constant* – the force per unit extension.

- The minus sign shows that the force acts in the opposite direction to the extension (it is a *restoring force*); it can often be ignored!

The energy stored in a stretched spring

- Energy stored = work done in stretching the spring.
- Elastic potential energy E_p = average force applied × extension.

As you stretch a spring, the force required steadily increases from 0 to F, and so provided the spring obeys Hooke's law, the *average force* is half the maximum force. Average force = $F_{av} = \frac{1}{2}kx$, so:

$$E_p = \frac{1}{2}kx^2$$

Unstretched spring, zero extension

F

F

F

As the spring extends you have to work harder to extend it further

- The work done in stretching a spring is equal to the area under its force–extension graph (a graph with force plotted on the y-axis and extension plotted along the x-axis).
- Provided no energy is lost in heating the surroundings, the work done on the spring also equals the elastic potential energy it gains.

Links

See *kinetic theory*, page 91 and the *materials 1* option, pages 170–1, for more information.

The jargon

A substance's behaviour is *elastic* if it springs back into its original shape after a force has been applied. *Plastic* is the opposite to elastic. *Plastic deformation* occurs when intermolecular bonds are repeatedly broken and remade, changing a solid's shape (think Plasticine!).

Check the net

Check out the new breed of super-elastic materials that remember their shape, no matter what you do to them! Go to www.cs.ualberta.ca/~database/MEMS/sma_mems/sma.html

Action point

How would you determine the spring constant of the springs in the shock absorbers of a car? (*Hint* You need a metre ruler, lots of friends and a helpful car owner!)

The jargon

The term *elastic limit* is used fairly loosely to mean the limit beyond which further extension causes permanent stretching or deformation. Under compression, Hooke's law breaks down when a spring's coils touch – even without any permanent deformation.

Watch out!

To calculate the work done on a spring graphically, you must plot force on the y-axis and extension along the x-axis – so that work done equals the area under the graph. (Getting unnecessarily technical, work done = the integral of force with respect to extension!)

Links

See *work, energy and power*, pages 22–3 and consider the energy conversions in a weight oscillating on a spring.

Stress and strain 

$$\text{stress} = \frac{\text{force}}{\text{cross-sectional area}} \qquad \sigma = \frac{F}{A}$$

Units N m^{-2} or Pa (same units as pressure).

→ *Tensile stress* tends to stretch things. *Compressive stress* tends to squash and crush things.

$$\text{strain} = \frac{\text{extension}}{\text{unstressed length}} \qquad \varepsilon = \frac{x}{l}$$

Units None (strain is a dimensionless ratio).

Young's modulus

Young's modulus is a useful measure of a material's rigidity (or elasticity). It is defined by:

$$\text{Young's modulus} = \frac{\text{tensile stress}}{\text{tensile strain}} \qquad E = \frac{\sigma}{\varepsilon}$$

Units N m^{-2} or Pa.

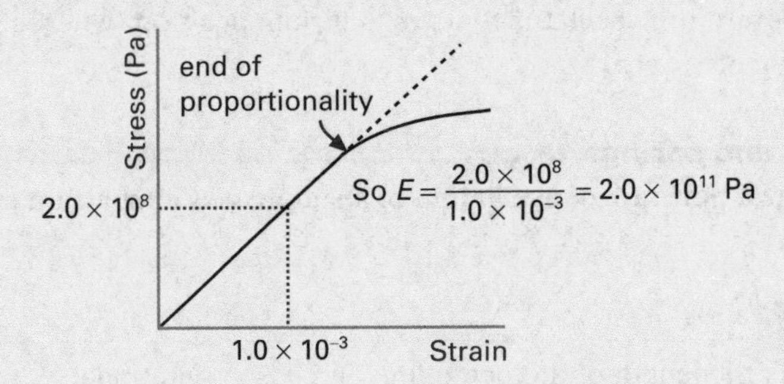

→ The larger the value of Young's modulus, the more rigid the material.

It is sometimes useful to expand the equation for Young's modulus to:

$$E = \frac{(F/A)}{(x/l)}$$

Rearranging, this becomes:

$$E = \frac{Fl}{xA}$$

Links

See *density and pressure*, pages 18–19. Compressive stress is a pressure transmitted through a solid.

Checkpoint 1

Remember extension $\boldsymbol{x}$ is the extra length gained when an object is stretched. The definition still stands under compression (negative extension). What is the strain for an elastic band stretched to double its original length?

Watch out!

Strain depends on stress, so you would normally plot strain on the y-axis. However, stress–strain graphs are usually plotted the 'wrong way around' with stress on the y-axis, so that E is the gradient.

Checkpoint 2

A material obeys Hooke's law if strain is directly proportional to stress. Check for yourself that this general statement of Hooke's law can be used to derive $\boldsymbol{F} = k\boldsymbol{x}$ for a spring.

Watch out!

Stay in shape – don't go beyond the elastic limit. There's no going back!

Exam questions answers: page 37

1 A steel cable of diameter 2.1 cm and length 12 m is used on a crane to lift a 1 500 kg load. The steel has a Young's modulus of 2.0×10^{11} Pa. (a) How much does it stretch when used to lift a 1 500 kg load? (b) Repeat the calculation (same length, diameter and load) for a copper cable. (Young's modulus for copper = 1.1×10^{11} Pa.) (15 min)

2 You are given a bag of identical springs, each with a spring constant of 15.0 N cm^{-1} and an elastic limit of 20 N. (a) Calculate the smallest number of springs required to lift a 500 N weight. (b) How are the springs arranged? (c) Calculate the extension of each spring. (d) Find the spring constant of the single spring that would behave in the same way. (10 min)

Vibrations and resonance

Links

See *waves and oscillations*, pages 97–130.

The jargon

Reciprocal This is simply 1 divided by the number. The reciprocal of 2 is $\frac{1}{2}$; the reciprocal of 0.1 is 10, etc. Frequency is the reciprocal of period (and period is the reciprocal of frequency).

The jargon

A *simple pendulum* has its mass concentrated in the bob. A *compound pendulum* has a bob attached to a (significantly) heavy oscillating bar. The distribution of mass in a compound pendulum complicates its behaviour slightly.

Checkpoint 1

What length would a simple pendulum have to be in order to have a period of 1 s?

Vibrating objects can generate waves. Waves carry energy and information. Resonance allows waves to transfer this energy and information to new objects – by making them vibrate or oscillate.

Frequency and period of oscillation

- **Frequency** f is the number of complete vibrations per unit time. Frequency is measured in *hertz* (Hz). 1 Hz = 1 vibration per second.
- **Period** T is the time taken for one *complete oscillation* (the time between one vibration and the next). Period is measured in seconds.

Provided we stick to SI units, frequency and period are linked by the following equations:

$$f = 1/T \qquad T = 1/f$$

Natural frequency

Every oscillator has a **natural frequency**. If you give a swing a single push, it will swing back and forth for some time at its natural frequency. The natural frequency is determined by the **restoring forces** which tend to return the oscillator to its equilibrium position (where its displacement is zero).

Swings and springs

The **natural period** T of oscillation of a simple pendulum (or a swing) is given by:

$$T = 2\pi\sqrt{(l/g)}$$

Where l is the length of the pendulum and g is gravitational acceleration.

The natural period of oscillation of a mass on a spring is given by:

$$T = 2\pi\sqrt{(m/k)}$$

Where m is the mass and k is the spring constant.

In both examples, we can see that the period is defined by a set of constants; it does not vary, so oscillating pendulums and weights on springs can mark time. All clocks feature oscillating systems.

Damping

Most oscillators gradually lose energy to their surroundings – perhaps through friction or air resistance. This loss of energy **damps** the oscillation, *reducing its amplitude*. There are degrees of damping:

- *Light damping* It takes many oscillations before the amplitude is reduced to zero.
- *Critical damping* The displaced body just returns to zero displacement without over-shooting. Critical damping is the minimum degree of damping required to prevent oscillation.
- *Heavy damping* There is no oscillation! The displaced body slowly returns towards zero displacement.

Forced vibrations and resonance

If you push a swing just once, it will oscillate gently at its natural frequency. Its amplitude will gradually diminish because of air resistance and friction (damping). If you push it again every time it returns to you, its amplitude of oscillation will increase. It will *resonate*! You will have succeeded in transferring energy (repeatedly) to the swing.

→ **Resonance** is the strong vibration that builds up when an oscillator is driven at its natural frequency.

To find an oscillator's natural frequency, you could try driving it at different frequencies. The oscillator will normally resonate (vibrate with greatest amplitude) when driving frequency matches natural frequency.

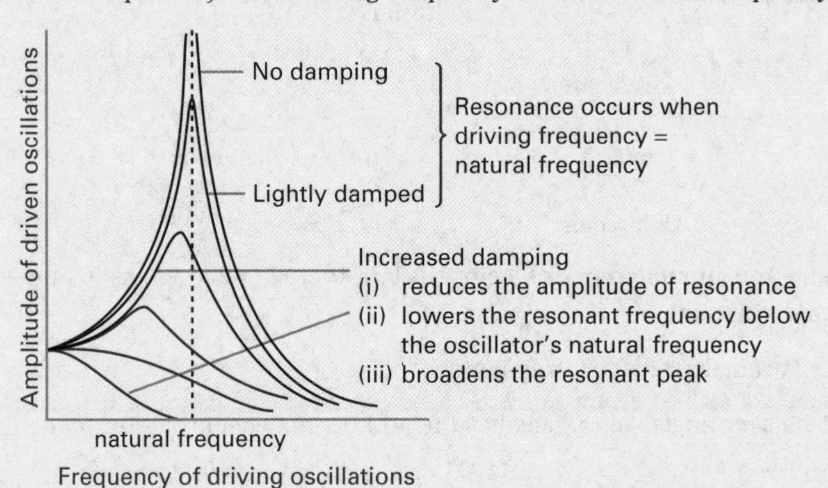

The importance of resonance

Wherever waves transfer energy, resonance plays a part. Here are just a few examples:

→ *Earthquakes* Buildings with natural frequencies close to the quake's frequency are most susceptible to damage.
→ *Microwave ovens* Interatomic bonds in water molecules have natural frequencies in the microwave region of the electromagnetic spectrum.
→ *Hearing* Different regions of the cochlea (a shell-shaped organ in the inner ear) resonate at different audible frequencies.

Car shock absorbers – applied damping

Shock absorbers are meant to cushion the car user from uncomfortable jolts and bumps, by soaking up vibrational energy peaks. The damping provided by shock absorbers should be just short of critical damping.

→ If damping is too heavy, the shock absorbers will not recover from one jolt in time to respond to the next.
→ If damping is too light, a single pot hole can set the car bouncing up and down uncomfortably.

Exam question answer: page 38

Sketch displacement–time curves to show light, moderate, critical and heavy damping. Which one would characterize a good car shock absorber? (15 min)

The jargon

Forced vibration Obvious forces drive the oscillator. These forces are typically repetitive and periodic. Truly *free vibrations* should be free of both cause and impediment. In practice, the term is used to describe any undamped or very lightly damped vibrations at natural frequencies.

Check the net

In 1940 the first Tacoma Narrows suspension bridge collapsed due to wind-induced vibrations. It had been open for traffic for only a few months. To see a film of the spectacular collapse, go to www.civeng.carleton.ca/Exhibits/Tacoma_Narrows/

Checkpoint 2

How can heavy damping change the resonant frequency of an oscillator? (What effect does damping have on the net restoring force acting on an oscillator?)

Checkpoint 3

Can you think of more examples of resonance which are:
(a) beneficial?
(b) harmful?

Links

See *medical and health physics 2*, pages 160–1.

Circular motion

"People travel to wonder at the height of the mountains, at the huge waves of the seas, at the long course of the rivers, at the vast compass of the ocean, at the circular motion of the stars, and yet they pass by themselves without wondering."

St Augustine, AD400

We are bound by gravity to a spinning planet orbiting a (spinning) star on a very nearly circular path. On a big scale, true linear motion must be a rarity in a universe governed by gravity!

Angles and angular velocities

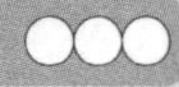

Radians

Radians are more fundamental units of angle than degrees. The definition of a degree is arbitrary: 1/360th of a full circle turn. In radians, an angle is defined by:

$$\text{angle (rad)} = \frac{\text{arc length (along a circle's circumference)}}{\text{radius}}$$

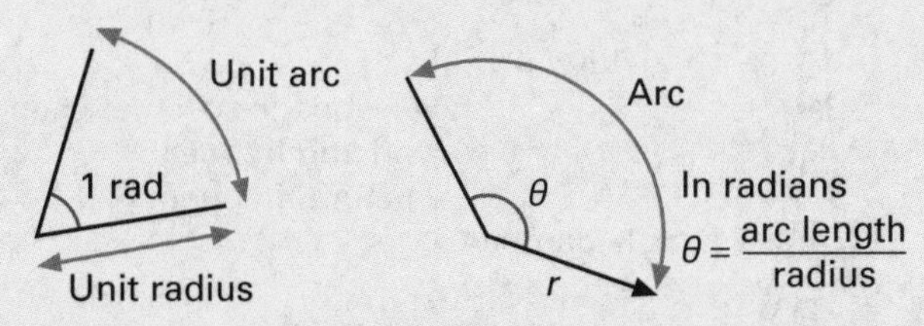

Since the circumference of a circle is $2\pi r$, there are 2π radians in one full revolution.

$$2\pi \text{ radians} = 360° = 1 \text{ revolution}$$

Measuring angles in radians has the additional benefit that *for small angles*:

$$\sin\theta \approx \tan\theta \approx \theta \text{ (in radians)}$$

Checkpoint 1

Convert 5° to radians. Find the sine and tangent of 5°. How close are the three values? (*Optional extra*: repeat the exercise, using greater angles to see how the three values diverge as the angle increases.)

Links

See *vibrations and resonance*, pages 28–9. Frequency and period are related by $f = 1/T$. If you know one of them, you can calculate the other.

Uniform circular motion and angular velocity

- **Uniform circular motion** is simply motion along a circular path, at a constant speed. If the period (time for one complete orbit) is T, speed v is given by:

  $$v = 2\pi r/T$$

- **Angular velocity** is rate of change of angular displacement – a measure of rate of rotation. Angular velocity is usually denoted by the Greek letter ω. Its units are rad s^{-1}.
- To convert from a frequency f (revolutions per second), to angular velocity ω (radians per second), simply multiply by 2π:

  $$\omega = 2\pi f$$

  Angular velocity ω and linear speed v are related by the equation:

  $$v = r\omega$$

Checkpoint 2

Write equations:
(a) giving speed of uniform circular motion v in terms of orbit radius and *frequency*
(b) giving angular velocity ω in terms of orbit *period*

Checkpoint 3

Calculate (a) the angular velocity ω and (b) the speed v of the Earth's orbit around the Sun. (The distance from the Sun to the Earth is 1.50×10^8 km.)

Centripetal acceleration and force

Nothing will follow a circular path unless it is forced to.

- **Centripetal force** is the force required to keep a body in uniform circular motion.

Links

See *Newton's laws of motion*, pages 14–15.

→ Centripetal force and acceleration are always directed towards the centre of the circular path.

If you need to convince yourself of the direction of the centripetal force, consider a conker swinging around on a string. The string can only pull the conker (you can't push something with a string!) and it pulls it inwards.

$$\boldsymbol{a} = (\boldsymbol{v} - \boldsymbol{u})/t$$

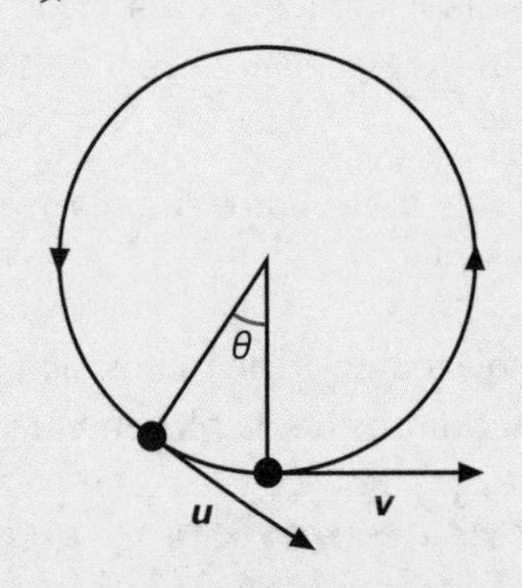

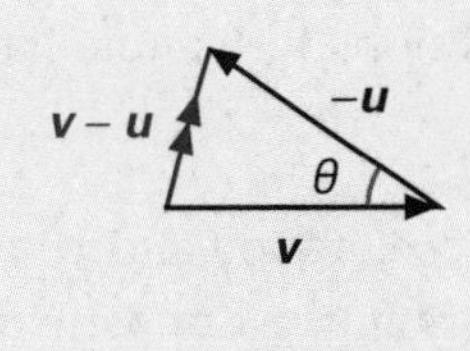

Provided the time interval (and therefore also the angle θ) is small, $\boldsymbol{v} - \boldsymbol{u}$ is a vector directed towards the centre of the circle. The links between θ, time, speed and angular velocity allow derivation of the equations below.

Equations for centripetal acceleration can be derived from the small change in velocity that occurs in a small time increment.

$$a = v^2/r$$

It is often more convenient to calculate centripetal acceleration from angular velocity, in which case:

$$a = r\omega^2$$

Centripetal force is the product of mass and centripetal acceleration:

$$F = mv^2/r$$

or

$$F = mr\omega^2$$

Banking

Aircraft steer by banking. A component of the lift force acting on the wings is directed horizontally into the turn. Banking questions test your ability to resolve forces *and* your understanding of circular motion – making them firm favourites with examiners.

Exam questions answers: page 38

1 A bobsleigh corners on a frictionless, horizontal banked track.
 (a) Draw a free-body force diagram for the bobsleigh.
 (b) By resolving forces, prove that the banking angle θ (the angle of the track relative to the horizontal) is given by $\tan\theta = v^2/(rg)$, where v is the bobsleigh's speed, r is the radius of its turning. (20 min)

2 The radius of a CD is 0.06 m and it rotates at 3.5 rev/s when playing music at the outer edge. Find the maximum tangential speed of the disk. (5 min)

Links

See *scalars and vectors*, pages 4–5 for rules of vector subtraction.

Examiner's secrets

To get you to really think, how would you explain that an object can be accelerating *and* moving at constant speed?

Checkpoint 4

(a) Calculate the centripetal acceleration of the Earth in its orbit around the Sun.
(b) Calculate the centripetal acceleration of the Moon around the Earth.
(Period of Moon's orbit = 27.3 days, Earth–Moon separation = 3.8 x 10^8 m.)

Checkpoint 5

Calculate the centripetal force needed to keep a person of mass 80 kg on the surface of the Earth from flying off into space (due to tangential motion):
(a) at the equator
(b) at the North pole.
(Assume a value of 6 400 km for the Earth's radius.)

Simple harmonic motion

Understanding simple harmonic motion is the first step towards understanding any mechanical oscillation, but be warned – SHM is not as simple as its title suggests.

Conditions for SHM

A body will oscillate with **simple harmonic motion** if the restoring force acting on it (pulling the body back towards a rest position) is directly proportional to the body's displacement. A restoring force results in an acceleration which is in the opposite direction to the body's displacement.

→ The conditions for simple harmonic motion are summarized by:

$$\boldsymbol{a} \propto -\boldsymbol{x}$$

Where $\boldsymbol{a}$ is acceleration and $\boldsymbol{x}$ is displacement. The minus sign shows that acceleration is in the opposite direction to the displacement vector.

Angular velocity and circular motion

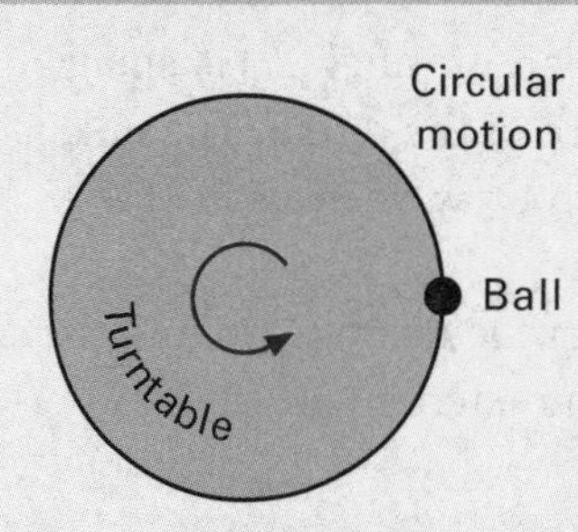

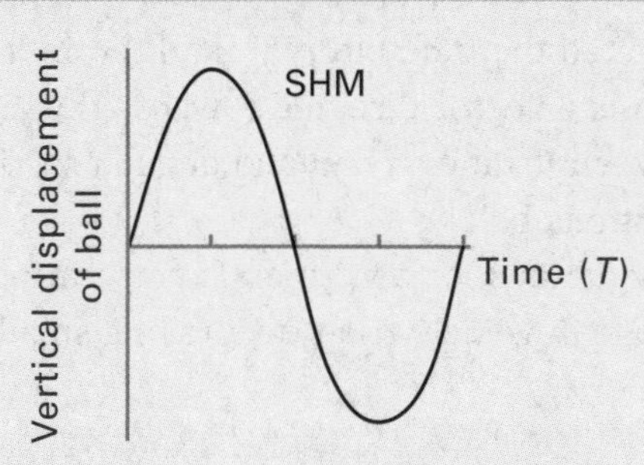

If you place an object on a turntable and view it from the side, you will see it oscillate from side to side as the turntable rotates. The object's angular velocity ω is equal to $2\pi/T$ (since there are 2π radians in one full circle), where T is the time for one complete revolution. The same definition of ω can be applied to the apparent lateral oscillation of the object viewed from the side, and to any regular oscillation:

$$\omega = 2\pi/T = 2\pi f$$

Phase and phase difference

Different points on the rim of a spinning wheel are said to be *out of phase*. The phase difference between any two such points is the angle between them (subtended at the wheel's centre). When the term phase is applied to waves and oscillations more generally, one complete oscillation is taken to be equivalent to one rotation or 2π radians and phase differences (usually denoted by the Greek letter epsilon, ε) are still measured in radians.

In phase
phase difference = 0

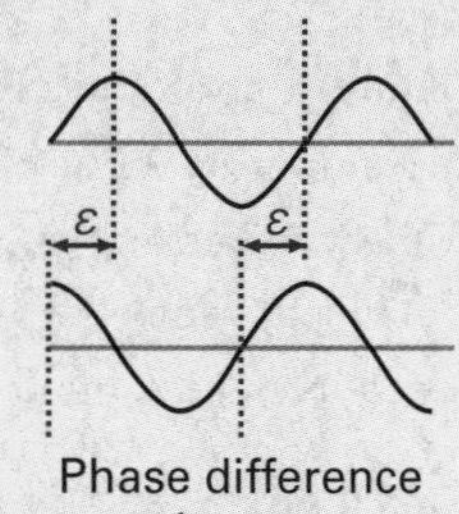

Phase difference
= $\frac{1}{4}$ cycle
= $\frac{\pi}{2}$ radians

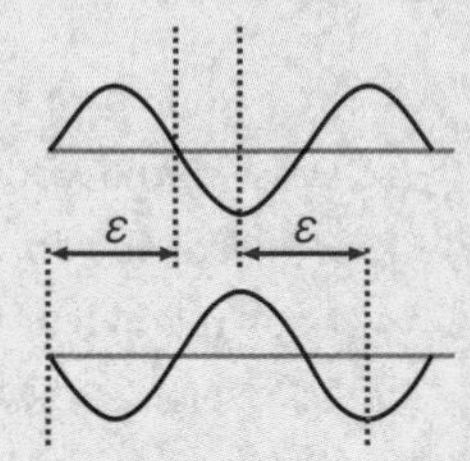

Completely out of phase
Phase difference
= $\frac{1}{2}$ cycle
= π radians

Links

Pendulums and weights on springs are both examples of simple harmonic oscillators that you should be aware of. See *vibrations and resonance*, pages 28–9 for details (equations for periodic motion etc.).

Examiner's secrets

Time period is independent of amplitude for objects that oscillate with SHM, but this does *not* define SHM.

The jargon

The term *angular velocity* is sometimes dropped in favour of *omega* when used to describe SHM. (Some books use the term *angular frequency*, which makes sense given that ω is measured in radians per second.)

Checkpoint 1

What are the angular velocities of:
(a) a spoke on a bicycle wheel of radius 20 cm if the bike is travelling at 12 m s^{-1}
(b) a loudspeaker cone vibrating at 300 Hz?

Don't forget

Time period, $\boldsymbol{T}$, is the reciprocal of frequency, $\boldsymbol{f}$.

Examiner's secrets

Many students try to use the wave equation ($\boldsymbol{v} = \boldsymbol{f\lambda}$) in questions about oscillations because there is a link with wave motion. Don't even go there!

Links

See *phase difference*, page 110. Waves that meet up in phase combine constructively, waves that meet up out of phase combine destructively.

Graphs and equations for SHM

The equations and graphs below work for continuous, undamped SHM.

→ Damping complicates things by reducing the amplitude A over time.
→ The oscillator must already be oscillating before time $t = 0$.

Displacement against time

The displacement $\boldsymbol{x}$ of a particle vibrating with SHM is given by:

$$\boldsymbol{x} = A \sin \omega t$$

Where A = amplitude (i.e. maximum displacement), ω is angular velocity and t is time. ωt is measured in radians.

Velocity against time

Velocity is rate of increase in displacement. At any instant it is the gradient of a displacement–time graph:

$$\boldsymbol{v} = A\omega \cos \omega t$$

Maximum velocity depends on both amplitude and angular velocity and since the maximum value a cosine can have is 1:

$$\boldsymbol{v}_{max} = A\omega$$

Maximum velocity is achieved when displacement $\boldsymbol{x} = 0$.

Acceleration against time

Acceleration is rate of increase in velocity. At any instant it is the gradient of a velocity–time graph:

$$\boldsymbol{a} = -A\omega^2 \sin \omega t, \text{ but } x = A \sin \omega t, \text{ so } \boldsymbol{a} = -\omega^2 x$$

Maximum acceleration is achieved at maximum displacement; i.e. when $|\boldsymbol{x}| = A$ (and $\boldsymbol{v} = 0$).

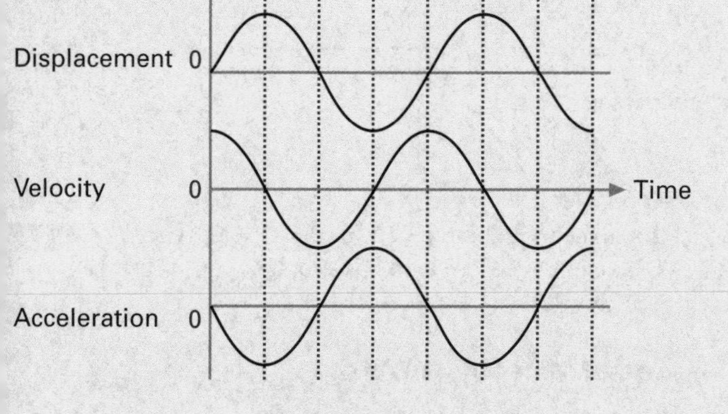

Velocity is the gradient of the displacement–time graph.

Acceleration is the gradient of the velocity–time graph

Exam questions answers: page 38

1 Define simple harmonic motion. (5 min)

2 A fairground attraction has a vibrating floor. It oscillates with simple harmonic motion, with an amplitude of 1 m. (a) Find the minimum frequency for customers to just lift off the ground. (b) What would happen to the minimum frequency if the amplitude were reduced? (15 min)

Watch out!

The displacement equation assumes that when $t = 0$, the oscillator's displacement is zero. This is a bit dodgy, because zero initial displacement implies zero initial restoring force and therefore no oscillation! Some exam boards prefer to use the more proper, but mathematically tricky: $\boldsymbol{x} = A\cos \omega t = A\cos 2\pi ft$ (same shape, but the oscillation starts at maximum displacement), in which case: $\boldsymbol{v} = -A\omega \sin \omega t = \pm 2\pi f\sqrt{(A^2 - \boldsymbol{x}^2)}$ and $\boldsymbol{a} = -A\omega^2 \cos \omega t = -(2\pi f)^2\boldsymbol{x}$.
Check your syllabus!

Watch out!

Take care with trigonometric functions. Make sure your calculator knows that the numbers you are typing in are radians, not degrees (a common mistake)!

The jargon

Two lines around a vector denote the *modulus of*. So $|\boldsymbol{x}|$ means *the modulus of* $\boldsymbol{x}$. The modulus of a vector is its magnitude (size!).

Action point

SHM involves the continuous transfer between kinetic energy and potential energy. The *total* amount of energy is constant. On the same axes, show how the KE and PE changes with displacement of a pendulum.

Examiner's secrets

SHM is a good topic to use to check your understanding of the links between displacement, velocity and acceleration. Given a graph of displacement against time, you must be able to say what is happening to velocity and acceleration.

Answers
Mechanics

Scalars and vectors

Checkpoints

1 *Scalar* mass, temperature, energy;
vector weight, acceleration.

2 If your ruler is accurate to the nearest mm, a 1 cm measurement has a 10% error whilst a 10 cm measurement has a 1% error. (If absolute errors are the same, increasing the scale reduces relative errors.)

3 $\sin\theta = \text{opp/hyp} = \boldsymbol{v}_y/\boldsymbol{v}$, so $\boldsymbol{v}_y = \boldsymbol{v}\sin\theta$
$\cos\theta = \text{adj/hyp} = \boldsymbol{v}_x/\boldsymbol{v}$, so $\boldsymbol{v}_x = \boldsymbol{v}\cos\theta$

Action point

7.55 kg or 7 550 g.

Exam questions

1 (a) Magnitude: $R^2 = (120)^2 + (50)^2 = 14\,400 + 2\,500 = 16\,900$
$R = 130$ N
(b) Direction: $\tan\theta = 50 \div 120 = 0.4167$
$\theta = \tan^{-1}(0.4167) = 22.6°$

2 (a) 10 m downstream (the boat travels 1 m downstream for every 2 m across).
(b)

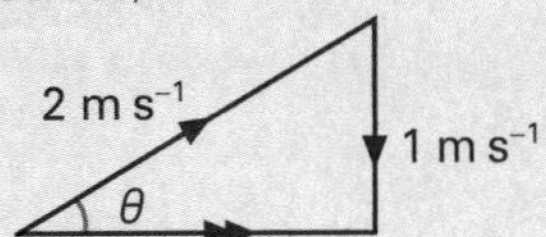

$\theta = \sin^{-1}0.5$ so $\theta = 30°$
(Heading is 60° E of north)

3

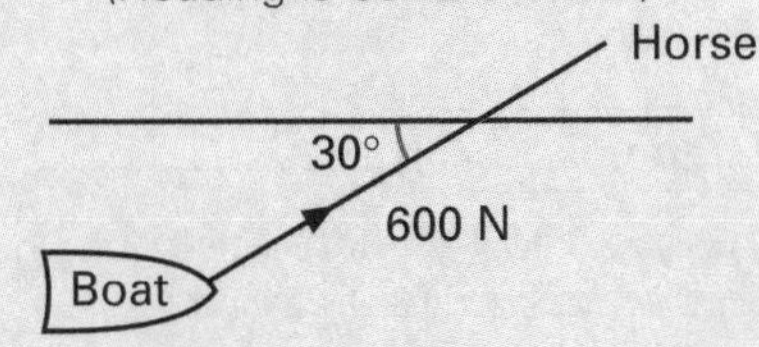

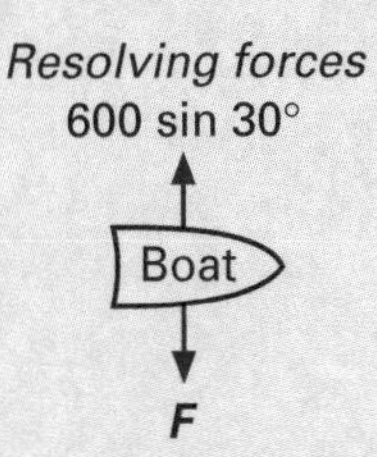

Resolving the tension in the rope into its components acting along and at a right angle to the boat (and bank). The component towards the bank must be balanced by an equal and opposite force (***F*** say) from the rudder etc.
$\boldsymbol{F} = 600 \sin 30° = 300$ N away from the bank.

Forces and moments in equilibrium

Checkpoints

1

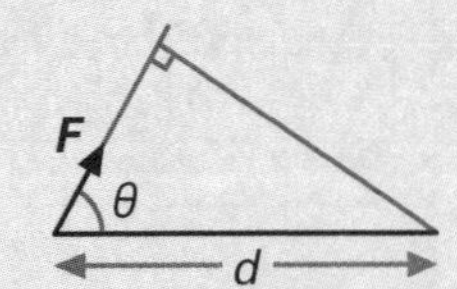

Using moment = perpendicular force × distance;
perpendicular force = $\boldsymbol{F}\sin\theta$, so moment = $\boldsymbol{F}d\sin\theta$
Using moment = force × perpendicular distance;
perpendicular distance = $d\sin\theta$, so moment = $\boldsymbol{F}d\sin\theta$

Exam questions

1 $\tan\theta = 7.5/0.6 = 12.5$ $\quad\theta = 85°$
Total upward force on walker = $2T\cos\theta$
Equilibrium ⇒ $2T\cos\theta = 500$ N
$T = 500/2\cos\theta = 3.1 \times 10^3$ N (to 2 sig. fig.)

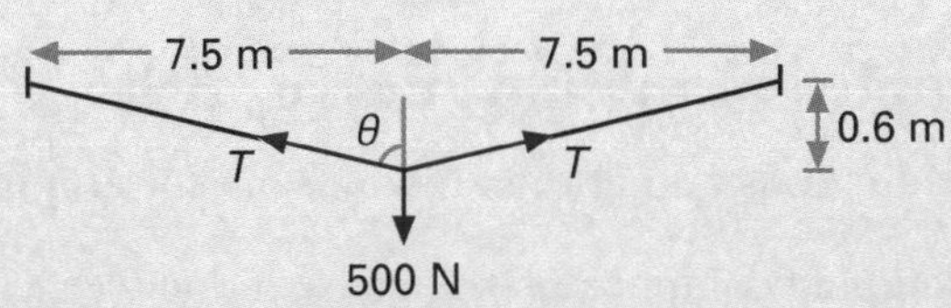

2 Vertical equilibrium ⇒ $\boldsymbol{F}_A + \boldsymbol{F}_B = 2.0 \times 10^5$ N
Rotational equilibrium ⇒ moments about any point are balanced.
Moments about point of support on pillar A
$2.0 \times 10^5 \times 100 = \boldsymbol{F}_B \times 140$
$\therefore \boldsymbol{F}_B = 1.43 \times 10^5$ N
Moments about point of support on pillar B
$\boldsymbol{F}_A \times 140 = 200\,000 \times 40$
$\therefore \boldsymbol{F}_A = 0.57 \times 10^5$ N
Check $\boldsymbol{F}_A + \boldsymbol{F}_B = 2.0 \times 10^5$ N

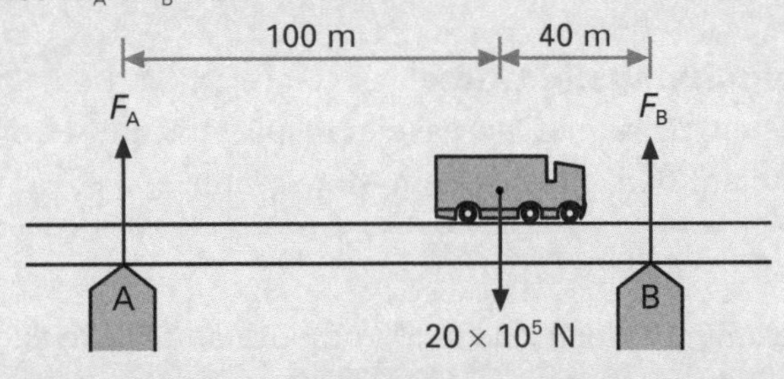

Ways of describing motion

Checkpoints

1 Average velocity = displacement/time taken. The start and finish of a lap are in the same place – zero displacement.

2

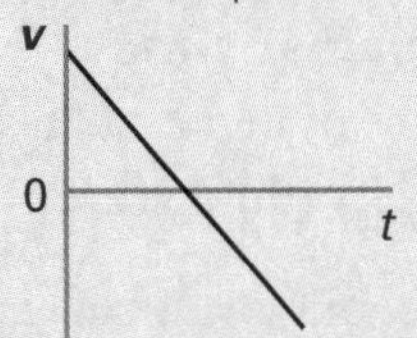

Displacement is the area under the graph.

Exam question

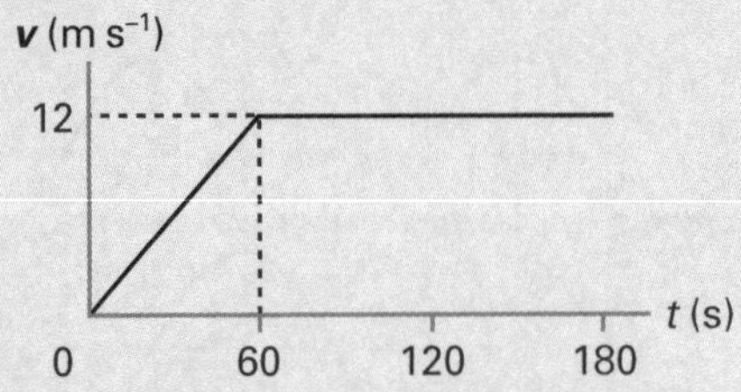

Top speed = $0.2 \times 60 = 12$ m s^{-1}
Distance travelled = area under graph = $(\frac{1}{2} \times 60 \times 12) + (120 \times 12) = 1\,800$ m

Equations of motion

Checkpoints

1 *Equation 4* $\quad\boldsymbol{v} = \boldsymbol{u} + \boldsymbol{a}t \Rightarrow \boldsymbol{u} = \boldsymbol{v} - \boldsymbol{a}t$
Substituting into $\boldsymbol{s} = \boldsymbol{u}t + \frac{1}{2}\boldsymbol{a}t^2$ we get $\boldsymbol{s} = (\boldsymbol{v} - \boldsymbol{a}t)t + \frac{1}{2}\boldsymbol{a}t^2$ which simplifies to $\boldsymbol{s} = \boldsymbol{v}t - \frac{1}{2}\boldsymbol{a}t^2$
Equation 5 Average velocity = $(\boldsymbol{v} + \boldsymbol{u})/2$;
Displacement $\boldsymbol{s}$ = average velocity × time taken = $(\boldsymbol{v} + \boldsymbol{u})t/2$

2 $\boldsymbol{a} = (\boldsymbol{v} - \boldsymbol{u})/t = (18 - 10)/4 = 2$ m s^{-2}; distance travelled = 56 m

3 $\boldsymbol{u} = 10$ m s^{-1}, $\boldsymbol{v} = 0$ m s^{-1}, $\boldsymbol{a} = -10$ m s^{-2}
Using $\boldsymbol{v}^2 = \boldsymbol{u}^2 + 2\boldsymbol{a}\boldsymbol{s}$, $0 = 100 - 20\,\boldsymbol{s} \Rightarrow \boldsymbol{s} = 5$ m

Exam questions

1 Use $v^2 = u^2 + 2as$
$100 = 0 + 2 \times a \times 20$
$\therefore a = 2.5 \text{ m s}^{-2}$

2 (a) Use $v = u + at$
$v = 0 + 2.5 \times 4$
$\therefore v = 10 \text{ m s}^{-1}$
(b) Average speed = distance/time
Time taken for accelerating phase = 4 s
Distance travelled in accelerating phase:
$s = ut + \frac{1}{2}at^2 = \frac{1}{2} \times 2.5 \times 16 = 20$ m
$\therefore$ Remaining 80 m were covered at 10 m s^{-1}
Time for constant speed phase = 8 s
Total time = 12 s
Average speed = $100/12 = 8.3 \text{ m s}^{-1}$

3 Use $v^2 = u^2 + 2as$
Top of flight, $v = 0$, so
$0 = u^2 + 2 \times (-10) \times 8$
$u = \sqrt{160} = 13 \text{ m s}^{-1}$
Time taken to reach peak:
$v = u + at$
$t = (v - u)/a = (0 - 13)/(-10) = 1.3$ s
Total time taken = 2.6 s

4 Thinking distance = $20 \times 0.5 = 10$ m
Braking time: $t = (v - u)/a = (0 - 20)/(-10)$
$t = 2$ s
Braking distance: $s = ut + \frac{1}{2}at^2$
$s = 20 \times 2 + \frac{1}{2} \times (-10) \times 2^2$
so $s = 20$ m
Total distance travelled by car = 30 m
It stops 15 m past the traffic lights!

Projectiles

Exam questions

1 (a) Horizontally, distance = $8 \times 3 = 24$ m
(b) Vertically, $s = ut + \frac{1}{2}at^2$
$a = 9.81 \text{ m s}^{-2}$
$s = 0 + \frac{1}{2} \times 9.81 \times 3^2 = 44.1$ m
Height of cliff = 44.1 m
(c)

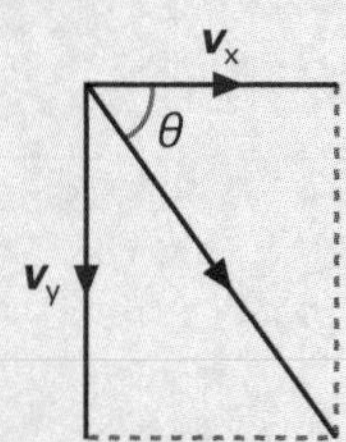

$\tan\theta = v_y/v_x$
Vertical velocity on impact, $v_y = u + at = 29.4 \text{ m s}^{-1}$
$\tan\theta = 29.4/8 = 3.675$ so $\theta = 74.8°$

2 Resolve initial velocity v into horizontal and vertical components v_x and v_y. $v_x = v\cos\theta$; $v_y = v\sin\theta$.
Range = $v_x \times$ time of flight.
Vertically, time taken to reach maximum height is given by $v = u + at$, which gives $0 = 20 \times \sin 45° - 9.81t$.
So $t = 1.44$ s.
Range = $20 \times \cos 45° \times 2.88 = 40.8$ m

3 (a) Vertically, $s = ut + \frac{1}{2}at^2$
$500 = 0 + \frac{1}{2} \times 9.81 \times t^2$
$t = \sqrt{(1\,000/9.81)} = 10.1$ s
(b) Horizontal distance travelled = $50 \times 10.1 = 505$ m
(c) Speed is the resultant of horizontal and vertical velocities. Vertical velocity as the package hits the ground = $9.81 \times 10.1 = 99.1 \text{ m s}^{-1}$. Using Pythagoras' theorem, speed $v = \sqrt{(50^2 + 99.1^2)} = 111 \text{ m s}^{-1}$

Newton's laws of motion

Checkpoints

1 (a) The pull of the Moon's gravity on the Earth.
(b) The gravitational attraction of the man on the Earth! (Weight is a gravitational force, so its reaction must be too.) The forces acting at surfaces are electrostatic and the action of a body on a surface has a reaction – the force the surface exerts on the body.
(c) The impact of the dartboard on the dart.
(*Note* Action and reaction are always the same type of force.)

2 (a)

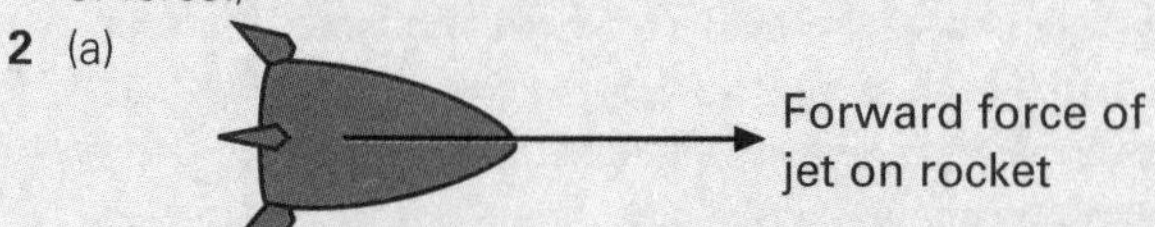

(b)
Backward force of rocket on exhaust jet

Exam questions

1 Use $v^2 = u^2 + 2as$ to find acceleration.
$50^2 = 0 + 2 \times a \times 50$ so $a = 25 \text{ m s}^{-2}$
$F = 12\,000 \times 25 = 300\,000$ N

2 This question is designed to make you think. It is not explained in enough detail for an answer that will satisfy everyone. You have to make some dodgy assumptions!

Doubling the force on a spring usually doubles its extension. Doubling the average force also doubles the average rate of acceleration. The arrow's top speed v is given by: $v^2 = 2as$ ($u = 0$). If both a and s are doubled, v^2 will be four times as big, and v will be twice as big.

Top speed is doubled by doubling the force. Try tackling the problem from an energy conservation viewpoint – see *work, energy and power*, pages 22–3, and *stress, strain and Hooke's law*, pages 26–7. (Real bows are not usually quite so simple in their behaviour!)

3 Treating truck and trailer as one unit:
Trailer and truck together

net force = $5\,000 \times 1.2 = 6\,000$ N
drive – drag = 6 000 N
drive force = 9 000 N

Trailer and truck separately

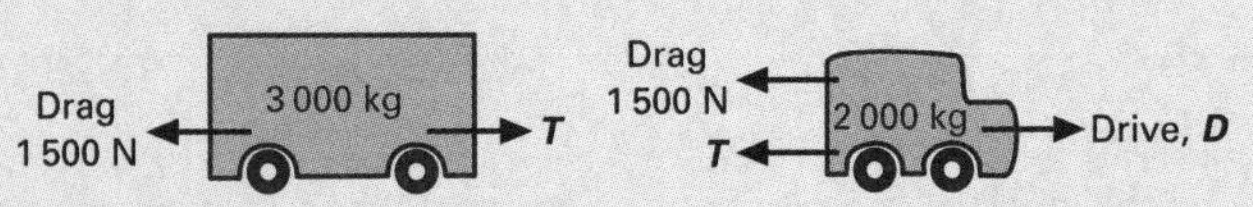

For the trailer:

$\boldsymbol{T}$ – drag = 3 000 × 1.2

$\boldsymbol{T}$ = 3 600 + 1 500 = 5 100 N

For the truck:

$\boldsymbol{D}$ – 1 500 – $\boldsymbol{T}$ = 2 000 × 1.2

$\boldsymbol{T}$ = 9 000 – 1 500 – 2 400 = 5 100 N

Note It is worth doing calculations 'from both sides' to check for mistakes. If you got a different answer for $\boldsymbol{T}$ from the truck calculation, you would know something was wrong.

Some important forces

Checkpoints

1 As a racing car speeds up, the downward force on the aerofoils increases (see *drag and lift – the Bernoulli equation*, page 21). This increases the contact force on the road. Friction is actually proportional to the normal reaction, and so grip is increased at high speed. For maximum acceleration, racing cars must be as light as possible. Aerodynamic lift (upside down!) compensates for the lack of weight and keeps the contact force high.

2

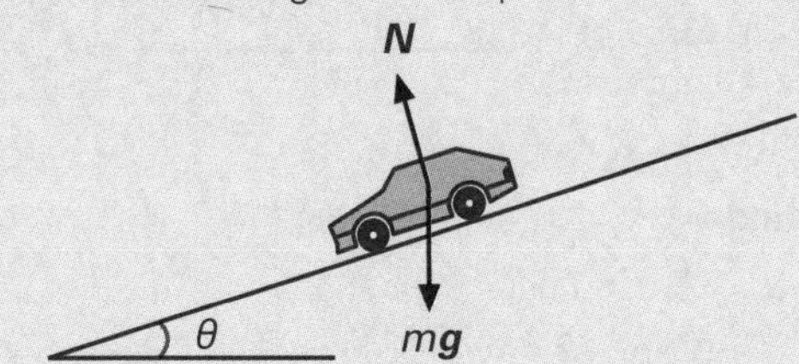

On a hill, the normal reaction is usually given by $\boldsymbol{mg}\cos\theta$, where θ is the slope of the hill; this has a maximum value (of mg) when $\theta = 0$ (horizontal road).

3 Considering the forces on the block: (a) up the slope (b) down the slope (c) up the slope.

Exam questions

1

N
Thrust = 8 000 N
Drag + friction = 800 N
15°
15°
mg

Resolve weight into components normal to the surface and along the surface.

Normal forces balance ($m\boldsymbol{g}\cos\theta = \boldsymbol{N}$).

Along the plane, forward force = 8 000 N;

Backward force = 800 + $m\boldsymbol{g}\sin\theta$ = 800 + 900 × 9.81 × sin 15° = 3 085 N

Resultant = 8 000 – 3 085 = 4 915 N

$\boldsymbol{a}$ = 4 915/900 = 5.46 m s^{-2}

2 Resolve weight into components along and normal to the slope (normal forces balance). Forces along slope balance ∴ $\boldsymbol{T} = m\boldsymbol{g}\sin\theta$ = 50 × 9.81 × sin 25° so $\boldsymbol{T}$ = 207 N

3 (a) If $\boldsymbol{g}$ = 10 m s^{-1}, $\boldsymbol{W}$ = 7 000 N

(b) When the lift falls at 10 m s^{-2}, relative acceleration = 0 and the scales register zero weight (Force on scales = $m\boldsymbol{a}$; $\boldsymbol{a}$ is the relative acceleration of man and scales.)

(c) Scales will register 1 400 N (same reasoning).

(*Note* This kind of reasoning helped Einstein to develop his theory of general relativity.)

Density and pressure

Checkpoints

1 Sea water is more dense than pure water. Warm water is *generally* less dense than cold water, but strange things happen close to freezing point. Ice is less dense than water. Water's density hits a maximum at 4 °C. As a pond cools, the cold water sinks. This continues until 4 °C is reached, then the pond freezes from the top down. Fishes etc. are usually safe in the dense 4 °C water below!

2 No effect! Floating ice is already displacing its weight in water; it continues to do so when it melts. Melting of the Arctic ice caps will have no effect on sea levels. The ice is already displacing its weight in water. (The melting of the Antarctic ice cap may have quite devastating effects, since it is mainly supported by land and will flow into the sea.)

3 *Crab* weight > upthrust; *bubble* weight < upthrust; *submarine* weight = upthrust.

4 Pressure decreases as they rise (Boyle's law).

5 Liquids transmit pressure. Force = pressure × area. Big pistons on the load side extract a big force.

6 Their lungs are filled with high pressure air. As they rise, the external pressure decreases and the air expands. If they don't release the air fast enough, it enters the blood stream, forming bubbles and causing 'the bends'.

7 Air is compressible. The work of pressing on the brake is wasted compressing air bubbles and the rise in pressure in the brake fluid is dangerously reduced and delayed.

Exam questions

1 (a) $V = m/\rho$

$V = 0.5/920 = 5.4 \times 10^{-4}$ m^3

(b) $V = 0.5/1\,000 = 5.0 \times 10^{-4}$ m^3

0.5 kg of ice occupies a greater volume than the same mass of water. So, the pipes expand (and may burst).

2 Atmospheric pressure = $\dfrac{\text{weight of air}}{\text{area it sits on}}$

weight = $mg = \rho Vg$

Since $V = Ah$, $p = \rho gh$

Pressure exerted by depth h of (constant density) air = pressure exerted by 76 cm of mercury.

$\rho_a gh = \rho_{Hg} g \times 0.76$

$h = 0.76 \times 13\,600/1.2 = 8\,600$ m (to 2 sig. fig.)

3 $V = 0.75$ m^3 $m = \rho V = 750$ kg $W = 7\,400$ N

$p = 7\,400/3 = 2\,500$ Pa (to 2 sig. fig.)

Drag and lift

Checkpoints

1 Drag ∝ v^2. When v is doubled, drag is quadrupled. A disproportionate increase in drag with increased speed will lower fuel efficiency. It's cheaper to drive slowly.

2

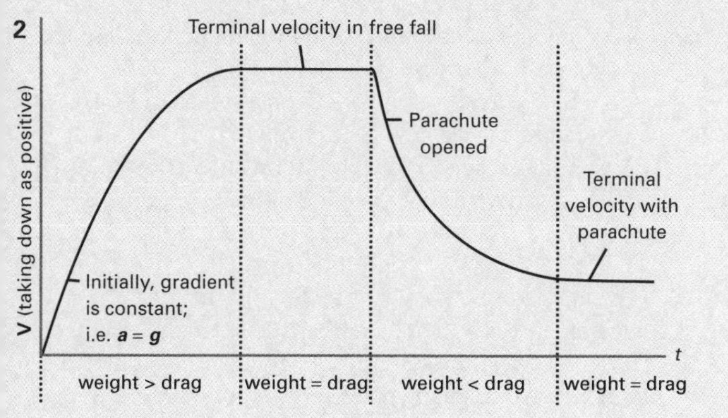

3 $kg\,m^{-3} \times m^2 \times m\,s^{-1} = kg\,s^{-1}$
Volume flow rate is measured in $m^3\,s^{-1}$.

4 (a) Zero viscosity (all liquids are viscous).
(b) Incompressibility (all gases are compressible).

5 E_k per unit volume is given by the $\frac{1}{2}\rho v^2$ terms. E_p per unit volume is given by the ρgh terms.

6 An aerofoil is shaped to make the air travel further and faster over the top than the bottom. (For a constant mass flow rate, constant density and reduced streamline cross-sectional area, the air travelling further within a streamline must travel faster.) For horizontal movement, the lift is proportional to the difference in relative airspeed above and below the aerofoil.

Exam question

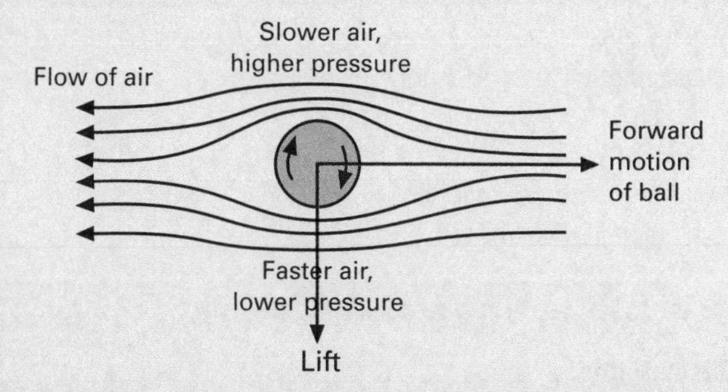

Work, energy and power

Checkpoints

1 Because it keeps the sums easy. We are only interested in *changes* in gravitational potential energy, not absolute quantities.

2 A comet has its greatest gravitational potential energy when it is furthest from the Sun.

Exam question

If the engine supplies the only significant force slowing the boat down, the work it must do is equal to the kinetic energy it must lose.

$E_k = \frac{1}{2}mv^2 = \frac{1}{2} \times 8000 \times 1.2^2 = 5760$ J

work done $= F \times 6.0 \quad \therefore F = 960$ N

power $= F \times v$, where v = average speed

$\therefore$ power $= 960 \times 0.6 = 580$ W (to 2 sig. fig.)

Note You could also tackle this using equations of motion to find (acceleration and) time taken (10 s), which gives exactly the same answer.

Momentum and impulse

Checkpoints

1 Momentum is always conserved, but to demonstrate it, you need to be able to measure the masses and velocities of all bodies involved. Friction links trolleys to the Earth. We can't detect any changes in the Earth's velocity, and so it is hard to prove momentum conservation. Better bearings, use of an air-track, etc., improve the situation by reducing the transfer of momentum to the Earth. A friction-compensated track could also be used.

Exam questions

1 (a) (i) $\boldsymbol{p} = m\boldsymbol{v} = 0.04 \times 500 = 20$ kg m s^{-1}.
(ii) $E_k = \frac{1}{2}m\boldsymbol{v}^2 = 5\,000$ J. (iii) recoil speed = 20/2.5 = 8.0 m s^{-1}. (iv) rifle's $E_k = 80$ J.
(b) No. *Collisions* can sometimes be perfectly elastic (same total amount of kinetic energy before and after), but in an *explosion*, energy is converted from some other form into kinetic energy. If you don't have any kinetic energy before the explosion, and you do after, there has to have been an increase in the system's kinetic energy!

2 $\boldsymbol{a} = 35/20 = 1.75$ m s^{-2}
$\boldsymbol{F} = 1\,400 \times 1.75 = 2\,450$ N

3 Initial momentum of the two skaters is 120×8 = 960 kg m s^{-1}. Final momentum of 50 kg skater is 500 kg m s^{-1}, so final momentum of 80 kg skater = 460 kg m s^{-1}. Final velocity = 460/80 = 5.75 m s^{-1}.

4 The hard snowball has a greater change in momentum than the soft snowball. The impulse on it is greater and so the impulse on the person is greater (Newton's third law). The hard snowball could impart an impulse of up to twice that imparted by the soft snowball. If the acceleration period is the same for both cases, the force exerted by the hard snowball will be correspondingly greater. In fact, the contact time is generally shorter when hard objects collide than when soft objects collide. This increases the impact.

Stress, strain and Hooke's law

Checkpoints

1 Strain = $\boldsymbol{x}/l$. If $\boldsymbol{x} + l = 2l$, $\boldsymbol{x} = l$ and strain = 1

2 Stress $\propto$ strain $\Rightarrow \boldsymbol{F}/A \propto \boldsymbol{x}/l \Rightarrow \boldsymbol{F} \propto x$ if A and l are constants.

Exam questions

1 $d = 2.1 \times 10^{-2}$ m, so cross-sectional area
$A = 3.46 \times 10^{-4}$ m^2
weight of load = 15 000 N (to 2 sig. fig.)
$\therefore$ stress $= 4.3 \times 10^7$ N m^{-2}
E = stress/strain
$\therefore$ strain $= 4.25 \times 10^7/2.0 \times 10^{11} = 2.13 \times 10^{-4}$
$\therefore x = 12 \times 2.13 \times 10^{-4} = 2.5 \times 10^{-3}$ m (2.5 mm)

2 (a) 25 springs (500/20)
(b) In parallel
(c) For each spring, $\boldsymbol{F}$ is exactly 20 N so x is 15 cm.
(d) The total load = 500 N, x = 15 cm, so $k = 33.3$ N cm^{-1}.

Vibrations and resonance

Checkpoints

1 $T = 2\pi\sqrt{(l/g)} \Rightarrow T^2 = 4\pi^2 l/g \Rightarrow l = gT^2/4\pi^2$
For $T = 1$ s, $l = 0.248$ m

2 Damping reduces the net restoring force acting on an oscillator, increasing its oscillation period and reducing its resonant frequency.

3 (a) *Beneficial* Grills and radiant heaters cause surface molecules to resonate and absorb heat rays. Lasers at certain frequencies can destroy the colour in certain tattoo dyes by resonance. Radio receivers are tuned to resonate at certain frequencies. Musical instruments make use of resonance.

(b) *Harmful* Too much energy can always be harmful. Regular gusts of wind can cause havoc at sea – to sailing boats with resonant frequencies equal to the gust frequency. Car wheels can develop a dangerous wobble at certain speeds due to resonance (generally cured by balancing the wheel).

Exam question

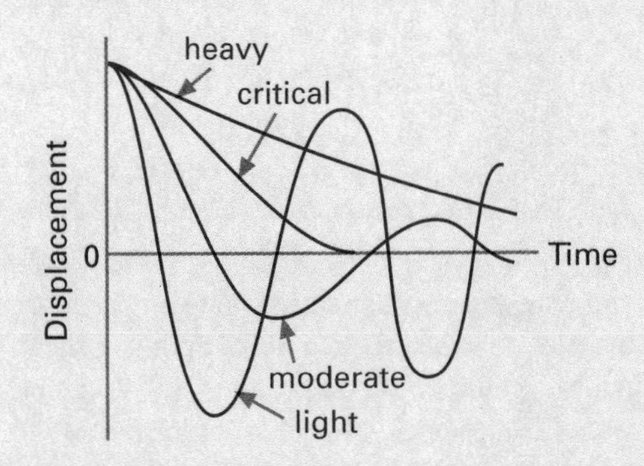

Moderate damping. Critical damping is slightly too stiff; the shock absorber doesn't fully recover in time to absorb the next shock.

Circular motion

Checkpoints

1 5° = 0.087 3 rad; sin 5° = 0.087 2; tan 5° = 0.087 5 (very little difference between θ in radians, $\sin\theta$ and $\tan\theta$).
10° = 0.174 5 rad; sin 10° = 0.173 6; tan 10° = 0.176 3 (some divergence, but still *approximately* equal – OK for rough calculations).
20° = 0.349 1 rad; sin 20° = 0.342 0; tan 20° = 0.364 0 (significant errors would arise if you tried assuming $\theta = \sin\theta = \tan\theta$).

2 (a) $v = r\omega$; $\omega = 2\pi f$ $\therefore$ $v = 2\pi r f$

(b) $\omega = 2\pi f$; $f = 1/T$ $\therefore$ $\omega = 2\pi/T$

3 (a) The Earth's period of orbit is 1 year. The easy answer is $\omega = 2\pi$ radians per year. To convert to radians per second: 1 year = 365 × 24 × 60 × 60 = 31 536 000 s
$\therefore$ $\omega = 2\pi/31\,536\,000 = 1.992 \times 10^{-7}$ rad s^{-1}

(b) Earth's speed in orbit around the Sun, $v = r\omega$ $= 1.50 \times 10^8 \times 1.992 \times 10^{-7}$. $v = 29.9$ km s^{-1}

4 (a) Centripetal acceleration of Earth around the Sun; use either $\mathbf{a} = \mathbf{v}^2/r$ or $\mathbf{a} = r\omega^2$.
$\mathbf{a} = 5.95 \times 10^{-3}$ m s^2. (*Note* convert $\mathbf{v}$ to m s^{-1} and r to m before calculating.)

(b) Centripetal acceleration of Moon around Earth = $r\omega^2$.
$T = 27.3$ days $= 27.3 \times 24 \times 3\,600$ s. $T = 2.3 \times 10^6$ s
$\omega = 2\pi/T = 2.664 \times 10^{-6}$ rad s^{-1}; $r = 3.8 \times 10^8$
$\therefore$ $\mathbf{a} = 2.69 \times 10^{-3}$ m s^{-2} (to 3 sig. fig.)

5 $\mathbf{F} = mr\omega^2$. $m = 80$ kg, $\omega = 1.992 \times 10^{-7}$ rad s^{-1}. At the equator, $r = 6.4 \times 10^6$ m, so $\mathbf{F} = 2.03 \times 10^{-5}$ N (not much danger of being thrown off!). At the poles, $r = 0$ and the force needed to keep you from flying off = 0.

Exam questions

1

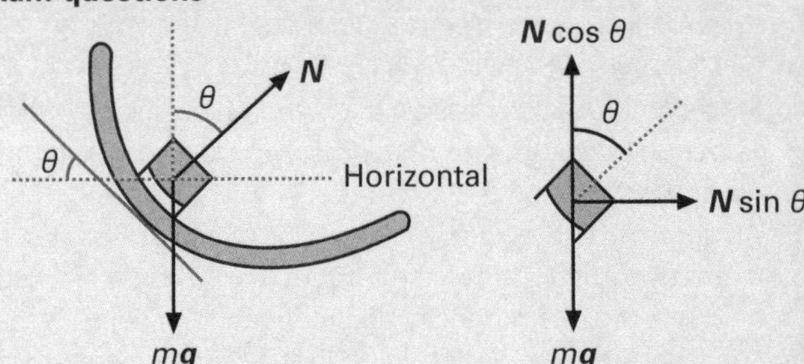

θ is the angle of the bobsleigh relative to the normal.
N is the normal reaction of the surface on the bobsleigh.
Resolving forces:

horizontally $N\sin\theta = \dfrac{mv^2}{r}$ (the centripetal force)

vertically $N\cos\theta = m\mathbf{g}$

Dividing the two equations gives

$$\frac{\sin\theta}{\cos\theta} = \frac{v^2}{\mathbf{g}r} = \tan\theta$$

2 $\omega = 2\pi \times 3.5 = 22$ rad s^{-1}
$v = r\omega = 0.06 \times 22 = 1.3$ m s^{-1}

Simple harmonic motion

Checkpoints

1 (a) $\boldsymbol{\omega} = v/r = 12/0.2 = 60$ rad s^{-1}

(b) $\boldsymbol{\omega} = 2\pi f = 1\,880$ rad s^{-1}

Exam questions

1 SHM is oscillatory motion where acceleration is proportional in size, but opposite in direction, to displacement; the simplest definition is $\mathbf{a} \propto -\mathbf{x}$.
$\mathbf{a} = -A\omega^2 \sin\omega t$ would do!

2 (a) The problem is this: What is the frequency when the floor's maximum downward acceleration just equals $\mathbf{g}$?
$\mathbf{a} = -A\omega^2 \sin\omega t$
When the floor is at its greatest positive displacement, $\sin\omega t = +1$ and $\mathbf{a} = -A\omega^2$ (i.e. maximum downward acceleration $= A\omega^2$)
($\mathbf{a} \propto -\mathbf{x}$. When $\mathbf{x} = +A$, $\omega t = \pi/2$)
$\boldsymbol{\omega} = 2\pi f$, so when the maximum downward acceleration is $\mathbf{g}$, $A(2\pi f)^2 = \mathbf{g}$ $f = \sqrt{(9.81/4\pi^2)}$ $f = 0.498$ Hz
At any greater frequency, the customers' feet will leave the ground.

(b) If you reduce A, the necessary frequency increases ($f \propto 1/\sqrt{A}$).

Radioactivity and the structure of the atom

The joint discoveries of radioactivity and atomic structure heralded a new era for physics – an era dominated by the newly discovered laws of quantum physics and relativity. You will get a flavour of both of these pillars of modern physics in this chapter.

Exam themes

- *Scientific evidence* Your ability to remember, analyse and interpret the results from key experiments can (and will) be tested in this section of the course. What is the evidence for:
  - the nuclear structure of the atom
  - the size and spacing of atoms
  - the size of nuclei?

How do α-, β- and γ-rays differ? (What experiments could you do to tell them apart?) You need to be clear about the clues and implications of all key experiments.

- *Applications* How can radioactivity be put to use? To apply knowledge requires a good understanding. Examiners like to see if you can see the benefits and limitations of new techniques. Nuclear power, medicine, measurement techniques and radioactive dating are such important applications that examiners find them difficult to ignore!
- *Knowledge* of nuclear structure, notation and the rules governing nuclear reactions.
- *Radioactive decay* Awareness of randomness and the need for a statistical approach.
- *Ability to interpret graphical data* N v Z for stable nuclides; mass defect per nucleon.

Topic checklist

○ AS ● A2	AQA/A	AQA/B	CCEA	EDEXCEL	OCR/A	OCR/B	WJEC
The nuclear atom	○	○	○	○	●	●	○
Elements and isotopes	○	○	○	○	●	●	○●
Nuclear instability	●		○	○	●	●	○●
Properties of ionizing radiation	●	○	○	○	●	●	○●
Radioactive decay	●	○	○	○	●	●	○●
Binding energy and mass defect	●	●	○	○	●	●	●
Applications of radioactivity	●	●	○	○●	●	●	●
Probing matter	●	●	○	○●	●	●	●

The nuclear atom

"When it comes to atoms, language can be used only as in poetry."

Niels Bohr

Checkpoint 1

Why did Thomson think the positive and negative charges should be evenly distributed within the atom?

Checkpoint 2

Alpha particles have about 1/50th the mass of a gold atom and they are positively charged. If a gold atom was a pudding of evenly distributed mass and charge, what do you think would happen to the alpha particles?

- → Would they be slowed down?
- → Would they be deflected?

Links

A modern equivalent of Rutherford scattering uses fast moving electrons to probe deep into protons. So-called *deep inelastic scattering* resulted in the discovery of quarks and other exotic particles. See *nuclear and particle physics*, Part 1, pp. 148–9, Part 2, pp. 150–1.

Examiner's secrets

You should be able to draw the paths of alpha particles scattered by a nucleus. The closer the path to the nucleus, the greater the deflection.

Examiner's secrets

This is a classic experiment. Learn the set-up and the reasons for:

- → vacuum
- → collimated source and detector
- → thin foil of gold

The idea that everything is made of tiny particles is fundamental to physics. Until the start of the 20th century, atoms were believed to be the tiny fundamental particles everything else was made from, but then the electron was discovered and it became clear that atoms were themselves made of still smaller parts.

Thomson's 'plum-pudding' model of the atom

Atoms are not charged. Since electrons are negatively charged constituents of an atom, the rest of the atom must be positively charged. Thomson's model of the atom shown below consists of a positively charged 'pudding', with negatively charged 'plums' or 'raisins' (electrons) embedded in it.

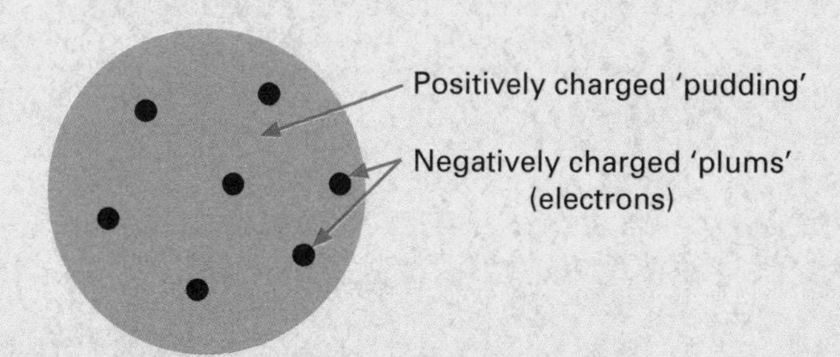

Discovery of the nucleus

In 1899, Ernest *Rutherford* discriminated between three types of ionizing radiation emitted by unstable nuclides: alpha, beta and gamma radiation. In 1909, researchers in his group were using *alpha particles* to probe gold atoms. *Geiger* and *Marsden* fired alpha particles at a thin gold foil. They used a zinc sulphide screen (a simple *scintillation detector*) to see where the alpha particles ended up.

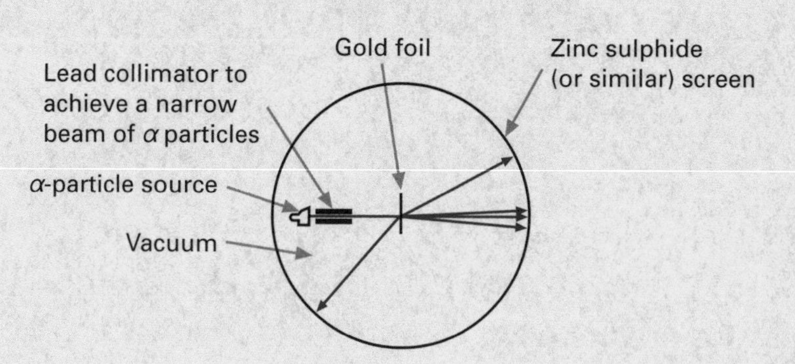

Geiger and Marsden discovered that:

- → Most of the alpha particles went straight through the gold foil without any change in direction and without any loss of energy.
- → Some alpha particles were deflected and some even rebounded!

Rutherford set about interpreting the data. The most obvious conclusions to be drawn from these experiments were:

- → most of an atom is empty space
- → the positive charge is concentrated in the nucleus
- → the mass is concentrated in the nucleus

Rutherford was able to determine the size of the nucleus by working out the force needed to give the necessary alpha deflections, using Coulomb's law (an inverse-square law for the force between two charges). He found that while an *atom's diameter* might be around 10^{-10} m, the *diameter of the nucleus* is typically around 10^{-15} m – 100 000 times smaller!

Problems for 'classical physics'

Rutherford's new model posed serious problems that classical physics could not answer. The electrons had to be in orbit, since static electrons would be captured by the nucleus, but

- → orbiting electrons are being constantly accelerated
- → according to classical physics, accelerated charges should emit radiation (they don't!)

The evidence for the nuclear structure of the atom was so strong that it was classical physics that had to give way. Niels Bohr used atomic line spectra to show that electron orbits had specific energy levels and Louis de Broglie provided a theoretical mechanism for these states by linking particle and wave properties!

The nuclear atom

Moseley's suggestion that positive charge should be quantized and Chadwick's discovery of the neutron completed the basic picture. The approximate radius of an atom = 10^{-10} m, and the approximate radius of the nucleus = 10^{-15} m.

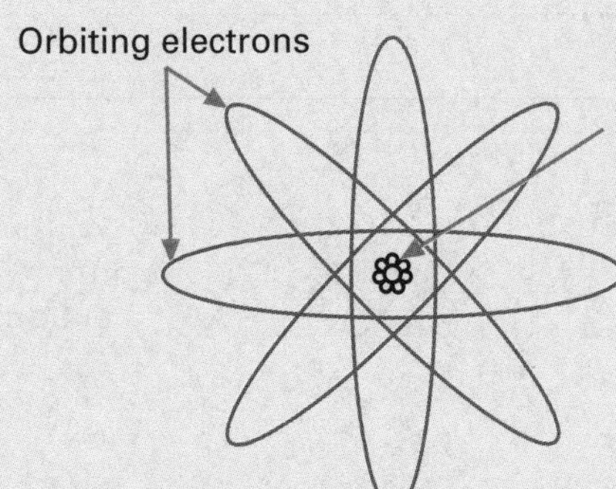

Exam questions answers: page 56

1. What were the key results of Geiger and Marsden's α-particle scattering experiments and what conclusions were drawn about the structure of the atom? (15 min)

2. Experiments suggest that the radius r of a nucleus (in metres) is given by the equation:

   $r = A^{1/3} \times 1.2 \times 10^{-15}$

   where A is the nucleon number of the atom.

   Calculate the radius of a gold atom ($A = 197$) and estimate the distance between neighbouring protons. (30 min)

"It was almost as incredible as if you had fired a fifteen inch shell at a piece of tissue and it came back and hit you."

Ernest Rutherford

Links

See *Coulomb's law*, page 136.
A version is $F = k\,q_1q_2/r^2$
Where k is a constant, q_1 and q_2 are charges and r represents the separation of the charges. Check for yourself that halving r quadruples the force between the charges.

The jargon

Classical physics is the old stuff. Galileo, Newton and even Maxwell were classical physicists. Classical physicists believed in determinism: perfect laws of nature from which everything could ultimately be predicted. Modern physicists accept uncertainty!

Checkpoint 3

What would happen if electrons constantly lost energy by electromagnetic radiation?

Links

See *wave–particle duality*, page 54.
According to quantum theory, particles moving at speed behave like waves. Standing waves don't transfer energy. In certain orbits, electrons behave as standing waves!

Links

A modern quantum interpretation of electrons describes their orbits in terms of *probabilities* – see *quantum behaviour*, pages 118–19.

Elements and isotopes

"'Excellent!' Dr Watson cried. 'Elementary', said Holmes."

Sir Arthur Conan Doyle

Watch out!

Isotopes of an element have the same number of electrons, so they have the same *chemical* properties. They have different *physical* properties due to their different masses.

Don't forget

Nucleon number (A) is sometimes called *atomic mass number* and the proton number (Z) is also known as the *atomic number*.

Links

In 1919 Rutherford was the first to detect protons, see pages 122–3.

Checkpoint 1

Notice the distinction between *atomic mass* and A. The mass number A is always a whole number (an integer). Atomic mass is not. Explain.

Examiner's secrets

Generally, physics questions are about physical processes. Try not to wander off into chemistry – easily done when thinking about electrons.

Elements are the substances that everything else is made of. They were first identified by their chemical behaviour. Measurements of atomic mass were then used to put the elements in order and some underlying chemical patterns began to emerge, but it was hard to make much sense of the periodic table until the discovery of the basic nuclear structure of the atom.

Proton number or atomic number Z

→ Z completely defines the *element*.

All carbon atoms have six protons, all nitrogen atoms have seven, etc. The number of electrons in an atom equals the number of protons (protons and electrons have equal and opposite charge), which is why Z determines an atom's chemical behaviour.

Nucleon number, or mass number, A

→ A defines the particular *isotope*.

A is the total number of *nucleons* (protons and neutrons) in the atom. Different **isotopes** of an element have different numbers of neutrons, but the same number of protons. All carbon atoms have six protons; most also have six neutrons (so for normal carbon, $A = 12$), but there is an isotope of carbon which has eight neutrons ($A = 14$). C-14 atoms are heavier than C-12 atoms, but they react in exactly the same way as normal carbon. (This isotope also happens to be radioactive.) The three isotopes of hydrogen are shown below.

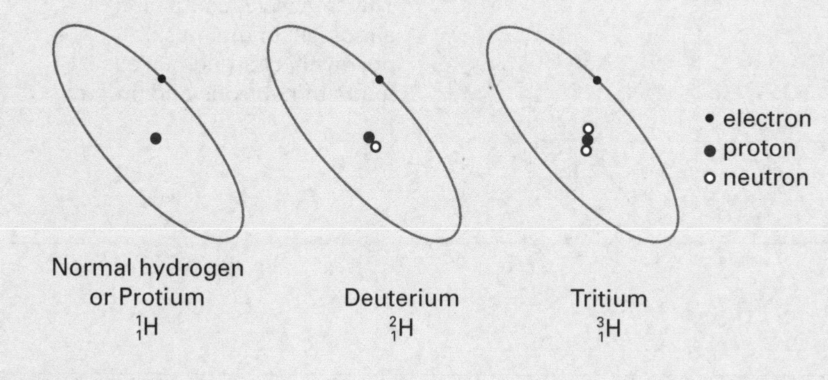

Nuclear notation

Nucleon number → A

Proton number → Z

X ← Chemical symbol

Atomic mass units u

Weighing atoms in kilograms yields difficult, tiny numbers. A suitably small alternative unit is needed. The mass of a proton or neutron would be a sensible choice, and this is actually very close to what an **atomic**

mass unit (symbol u) is, but just to keep things tricky, u is defined as one-twelfth of the mass of a normal carbon atom, i.e. carbon-12.

$1\ \text{u} = 1.660\ 566 \times 10^{-27}$ kg

Particle	*Z*	*A*	*Charge (C)*	*Mass (u)*
Electron	−1	0	-1.6×10^{-19}	0.000 55
Proton	1	1	$+1.6 \times 10^{-19}$	1.007 28
Neutron	0	1	0	1.008 67

Nuclear reactions

In *nuclear reactions* (and *only* in nuclear reactions), elements can **transmute**, i.e. change into new elements.

The rules

- Proton number Z is conserved.
- Nucleon number A is conserved.

Some examples of nuclear reactions

- In 1919 *Rutherford* bombarded nitrogen gas with alpha particles

  $${}^{4}_{2}\text{He} + {}^{14}_{7}\text{N} \Rightarrow {}^{17}_{8}\text{O} + {}^{1}_{1}\text{H}$$

  (α-particle) (nitrogen) (oxygen) (proton)

  and observed transmutation for the first time. Notice that an alpha particle is identical to a helium nucleus and a proton is identical to a hydrogen nucleus.
- In 1932 *Cockcroft and Walton* were first to use an artificially accelerated beam of protons to induce a nuclear reaction:

  $${}^{7}_{3}\text{Li} + {}^{1}_{1}\text{H} \Rightarrow {}^{4}_{2}\text{He} + {}^{4}_{2}\text{He}$$

  Careful measurement of particle masses showed the energy releasing potential of nuclear reactions. The products weigh less than the reactants. The lost mass is converted to energy according to Einstein's famous equation $E = mc^2$.

Exam questions answers: page 56

1 A caesium atom has an atomic mass number of 137 and an atomic number of 55.
(a) Give the number of:
(i) neutrons
(ii) protons
(iii) electrons.
(b) Explain why all atoms are neutral.

2 A material is known to be an isotope of tin. Given only this information, can you specify:
(a) its proton number
(b) its neutron number
(c) its nucleon number?
Explain. (10 min)

3 Uranium-238 has a proton number of 92. Write a balanced equation for its decay by alpha-particle emission (to an isotope of thorium, Th). (5 min)

Examiner's secrets

Generally your answers should be quoted to 2–4 significant figures. In questions involving atomic mass units, you are allowed to use 6, but don't make a habit of it!

Checkpoint 2

Values of atomic mass for most elements are close to whole number multiples of the mass of one hydrogen atom, but some are not (e.g. chlorine has an atomic mass of 35.5). Why is this?

Checkpoint 3

Chlorine's proton number is 17. How many neutrons are there in each of its isotopes: Cl-35 and Cl-37? Can you work out which isotope is most abundant? By how much?

Links

See *properties of ionizing radiation*, pages 46–7 to find out more about alpha particles.

Examiner's secrets

Nuclear equations appear with great regularity in exam papers. They generally aim to test whether you:
(a) can balance a nuclear equation
(b) understand how Z and A define the nuclide.

Nuclear instability

Whether a particular nuclide will be stable or not depends upon the balance (or imbalance) that exists between the forces tending to rip it apart and the forces tending to glue it together.

The jargon

β-decay posed problems for nuclear physicists. β-decay seemed to break the law of conservation of energy (and that won't do!). Wolfgang Pauli solved the problem by proposing the existence of a new, difficult to detect, particle – the *neutrino* – which escapes with all the missing energy. (The new particle needed a new force to eject it: the weak nuclear force.) *Neutrinos* have since been detected and are thought to exist in huge numbers throughout the universe.

Fundamental forces

There are only four known fundamental forces: **gravity** and the **electromagnetic force** account for every push or pull encountered outside the nucleus, but inside the nucleus, two more forces exist. These are called, rather unimaginatively, the **strong nuclear force** and the **weak nuclear force** (or interaction). The strong force is the main focus of this section; it binds the nuclides together. The weak nuclear force plays a role in β-decay. The two forces that battle it out in the nucleus and determine whether or not it will be stable are the electrostatic repulsion between protons and the strong nuclear force.

The strong nuclear force

- Is a force of attraction between neighbouring nucleons (which turns to repulsion when the nucleons try to squeeze too close!).
- Does not depend on charge (the attraction between two protons is about the same as that between two neutrons, or between a proton and a neutron).
- Has an *extremely* short range.

Checkpoint 1

How do we know that:
(a) The nuclear force is stronger than the electromagnetic force at close range?
(b) The strong force acts on both protons and neutrons?

The strong nuclear force can be thought of as a contact force. Each nucleon is strongly attached to its immediate neighbours (only). *The binding effect of each nucleon is extremely localized.* If it helps you, think of the nucleons as tiny balls with Velcro surfaces!

The electrostatic force

Although every nucleon is not necessarily attracted to every other nucleon by the strong nuclear force, *every proton in the nucleus is repelled by every other proton.* The electrostatic force:

- is a force of repulsion within the nucleus (between protons, due to their similar charge)
- obeys an inverse-square law and acts over a far greater range than the strong nuclear force

Links

See *Coulomb's inverse-square law*, page 136.

The strong nuclear force dominates at short distances. This explains why atoms of low atomic mass tend to be stable (and why the helium nucleus is a particularly stable unit). In larger atoms, extra protons tend to exert a destabilizing influence. Bismuth ($Z = 83$) is the stable nuclide with the highest proton number.

The stable nuclides

Nuclide means nuclear combination. Any particular nuclide is defined by its mix of protons and neutrons. The stable nuclides are shown below. There are about 270 of them. More than 2 500 unstable nuclides (not shown) have also been discovered (or artificially induced). They tend to decay, leaving more stable 'daughters' (i.e. daughters lying closer to the trend line).

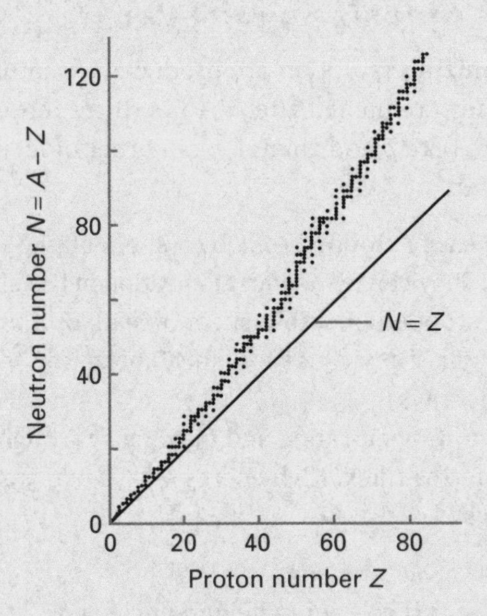

Notice that stable nuclides with the highest atomic numbers tend to have the greatest excess of neutrons over protons. No stable nucleus has more than 83 protons.

Important

→ *$Z_{max} = 83$* There is a limit to the number of protons a nucleus can hold before their mutual repulsion makes the nucleus unstable.

→ *For $Z < 20$, $Z \approx N$* At low atomic number, the strong nuclear force dominates – to such an extent that nucleon charge makes very little difference to stability.

→ *For $Z > 20$, $N > Z$* At high atomic numbers, extra neutrons increase stability. Extra neutrons add a bit of extra nuclear binding (at least locally) and increase the average proton separation (which reduces their inclination to blow the nucleus apart).

The jargon

Get used to words like *nuclide, radionuclide, isotope, radioisotope,* but don't get scared by them. The context often gives away the meaning. Physicists tend to talk about *nuclides,* because the nucleus is the physicist's domain! The definition of a nuclide ignores the existence of electrons (and therefore of chemistry!) completely, but the words nuclide and isotope are usually interchangeable. The *radio* prefix is used as in *radioactive* and really means unstable.

Links

A radionuclide above the stability line decays by beta emission – see *properties of ionizing radiation,* pages 46–7.

Examiner's secrets

When you are asked a question concerning nuclear stability, always mention *both* forces involved.

Exam questions

answers: page 56

1 'The discovery of the nuclear structure of the atom was also the discovery of the third fundamental force – the strong nuclear force, which binds the nucleus.' True or false? Give your reasons. (15 min)

2 (a) Krypton's atomic number is 36.
   (i) Plot the position of Kr-89 on the figure above.
   (ii) It decays by β-emission. Plot its daughter's position.
 (b) Radon has an atomic number of 86.
   (i) Plot the position of Rn-222 on the figure above.
   (ii) It decays by α-emission. Plot the position of its daughter.
 (c) What in general terms is the effect of each type of decay on stability? (20 min)

Properties of ionizing radiation

> **The jargon**
> The word *ray* can be used to describe radiated particles as well as photons of electromagnetic radiation. (A ray is just something which has been radiated!)

> **Checkpoint 1**
> Beta particles are fast moving electrons that come from the nucleus, but the nucleus does *not* contain any electrons! Can you explain this? See *nuclear and particle physics 2*, pages 148–9.

> **Checkpoint 2**
> What is the link between the degree of ionization each type of ray causes and its range?

> **Examiner's secrets**
> Never write just *radiation* (which includes light and anything else that is radiated) when you really mean *ionizing radiation*. The ability of α-, β- and γ-rays to cause ionization is the thing that makes them special!

> **The jargon**
> *X-rays* are high-energy photons, like γ-rays. The only difference between the two is the way in which they are generated. X-rays are emitted by excited atoms when electrons make big jumps in energy level, but γ-rays are emitted by the nuclear quantum energy leaps associated with radioactive decay!

> **Checkpoint 3**
> Geiger counters are best at detecting α- and β-rays. They are less efficient at detecting γ-rays. Why is this?
> *Hint* Think about the range, penetrating power and degree of local ionization caused by each type of ray.

Radioactive decay releases ionizing radiation, i.e. radiation capable of knocking electrons out of atoms and molecules.

Three types of ionizing radiation

Three types of **ionizing radiation** are produced by naturally occurring radioisotopes. They are named after the first three letters of the Greek alphabet (alpha α, beta β and gamma γ), in order of increasing ability to penetrate matter.

- α-*particles* are easily stopped (e.g. by a sheet of paper, or a few centimetres of air). They cause the most ionization in the shortest distance.
- β-*particles* are stopped by a thin sheet of lead, or a few tens of centimetres of air. They cause less local ionization than α-particles, but more than γ-rays.
- γ-*rays* are far more penetrating and may pass through a layer of lead several centimetres thick. High-energy γ-rays may pass through entire buildings!

Radiation	*Description*	*Z*	*A*
α	High-energy helium nucleus	2	4
β	High-energy electron	–1	0
γ	High-energy photon	0	0

Detecting radiation

Photographic film was the first detector of ionizing radiation ever used (by Henri Becquerel in 1896). There are now better detectors for most applications, though film still has its uses. Detectors that count individual events include:

- Geiger–Müller tubes
- scintillation detectors
- semiconductor devices

You need to understand the workings of the Geiger–Müller tube.

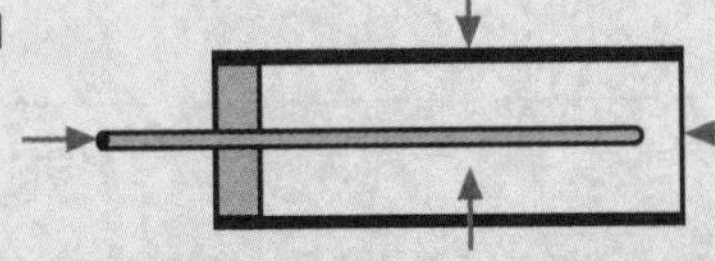

α-, β- or γ-rays ionize the argon gas. The released electrons accelerate towards the central electrode and a brief pulse of current is recorded by an electronic timer. (The accelerated electrons cause further ionization on the way, amplifying the electrical pulse.)

Cloud chambers

Cloud chambers show the actual paths taken by ionizing particles. Again, they rely on ionization. Clouds condense readily on charged particles. A typical cloud chamber is cooled (with dry ice) and supersaturated with ethanol. The ethanol vapour is desperate to condense. Wherever an ion is found, a droplet of condensation will appear. α-particles cause the most ionization in the least space and show up as straight lines in a cloud chamber. The length of the line gives a measure of the α-particle's energy. β-particles may show up as fainter traces. γ-rays are difficult to trace because they cause less local ionization.

Note that all tracks have the same length, indicating that α-particles are emitted with specific energies and not in a continuous spectrum.

Deflection by electric and magnetic fields

Moving charged particles are deflected by electric and magnetic fields. The degree of deflection depends on charge, mass and speed; the direction of deflection depends on whether the charge is positive or negative. The drawing below shows the effects of a perpendicular magnetic field on α-, β- and γ-rays.

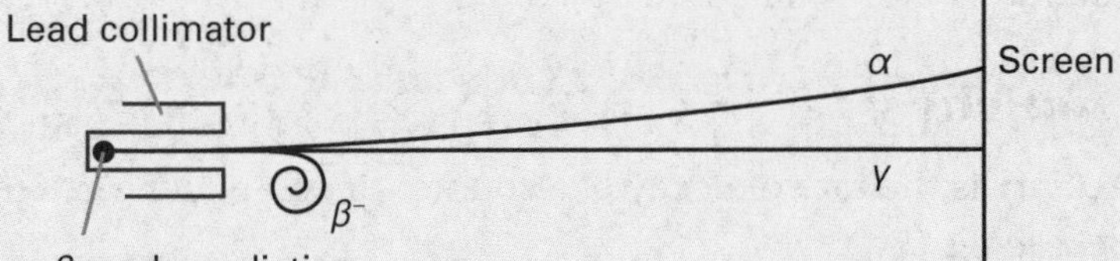

γ-rays are not deflected by a magnetic field, proving they are uncharged. α- and β-rays are deflected in opposite directions by the field, proving they are oppositely charged. β-rays are deflected very much more than α-rays, proving they have a significantly higher *charge/mass ratio*.

Exam questions answers: page 57

1 Which type of radiation could be used to detect: (a) cracks in steel girders; (b) smoke; (c) under-filled cereal packs? Give reasons. (10 min)

2 (a) Describe experiments which could be used to determine the types of ionizing radiation emitted by an unknown source.
(b) How could you compare the energies of: (i) α-particles and (ii) β-particles from different sources? Explain your method(s). (20 min)

The jargon

Bubble chambers do much the same job as cloud chambers, but they are bigger and more impressive. The chamber is filled with a liquid (liquid hydrogen) which is on the verge of boiling. Tiny bubbles form around ionization tracks. Magnetic fields are often used to deflect charged particles.

Watch out!

Many people think that γ-rays are the most dangerous because they are the most penetrating. In fact, α-particles can do you the most damage. Can you explain why?

Examiner's secrets

You may be expected to explain the differences between tracks made by α and β particles.

Links

See *radiation and risk*, pages 162–3.

Checkpoint 3

Use Fleming's left-hand rule to work out the direction of the magnetic field used in the figure opposite. Does it go into or out of the page?

Checkpoint 4

Explain why β-particles have a higher charge-to-mass ratio than α-particles. How much higher is it?

Radioactive decay

Unstable nuclei have too much energy locked up inside them for comfort. Sooner or later, they have to release some of this energy – by radioactive decay – in order to achieve a more stable (less energetic) state. Nuclear parents are less stable than their offspring!

Checkpoint 1

Why do we have to measure background radiation levels *every time* we carry out quantitative experiments with ionizing radiation?

Background radiation

If you turn on a Geiger counter well away from any known radioactive sources, it will start to click away gently and randomly. It is recording **background radiation**. Sources of background radiation include:

- the *Sun* (particularly during periods of sun-spot activity)
- *cosmic rays* (from far off stellar explosions etc.)
- the *Earth* (radioactive rocks and gases)

Background radiation has always existed; you can't escape it. It contributes to our radiation dose and it provides a constant background of radiation noise, which must be subtracted from any other radiation measurements we might be interested in. Background radiation levels vary with location and they vary over time.

Checkpoint 2

If the *absolute error* in a radiation count N is $\pm\sqrt{N}$, what is the *relative error* in the same count? Calculate absolute and percentage errors in radiation counts of 100 and 10 000.
Note $\sqrt{N}/N$ simplifies to $1/\sqrt{N}$.

Randomness

- Radioactive decay is unaffected by temperature and pressure (at least within the normal earthly range of conditions!).
- Radioactive decay is a *random* process. You may know that a particular nucleus is unstable, but you cannot know exactly when it will decay.

You can hear this randomness in the uneven clicking of a Geiger counter measuring low-level radiation. You can see it in the fluctuating count rate from any radioactive source. We have to use large radiation counts if we want to minimize the effects of these random fluctuations. (Statisticians tell us that the expected random variation in any radiation count N is plus or minus $\sqrt{N}$.)

Examiner's secrets

Calculus notation You should be familiar with the notation used in calculus even though you are not required to understand how to *do* calculus (integration and differentiation of graphs). In calculus notation, activity A is written 'dN/dt'; which means the rate of change in N (the gradient of a graph of N against t).

Activity

Activity is the rate of decay of a source – its output, measured in *Becquerels* (Bq).

1 Bq = 1 decay event per second

The activity of a source depends on the number of unstable atoms it contains. If a source contains N unstable atoms, then we can write:

$$A \propto -N$$

Where A is the activity of the source. The minus sign is necessary because each decay event *reduces* N (by one). This becomes:

$$A = -\lambda N \qquad [1]$$

Where λ, the proportionality constant which links activity to number of unstable atoms remaining, is called the **decay constant**. λ is closely related to (but not equal to) the probability that an atom will decay in any given interval. The most stable radionuclides have the smallest decay constants (and vice versa).

The exponential decay equation 

If we integrate equation 1, we get the *exponential decay equation*:

$$N = N_0 \, e^{-\lambda t} \qquad [2]$$

Where N_0 is the number of unstable parent atoms at the start (when $t = 0$), N is the number of unstable parent atoms remaining after time t, λ is the decay constant and e is the special number, loved by mathematicians, which is the base of natural logarithms ($e \approx 2.72$).

- ➜ Since a source's activity is directly proportional to the number of unstable atoms remaining, we can substitute initial and final activity or decay rates for N_0 and N and the equation will still hold.

The graph on the left shows raw data for exponential decay and the right-hand graph shows the best-fit curve from this data used to determine the half-life.

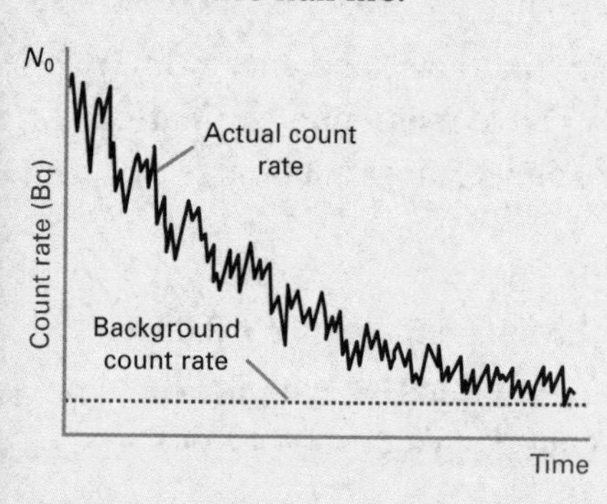

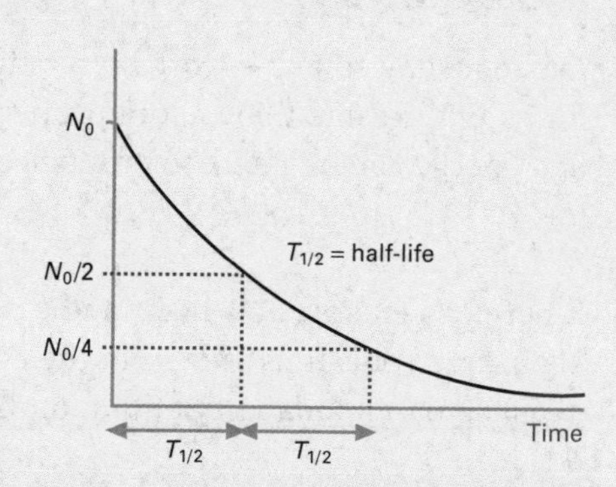

Half-life $T_{1/2}$ 

The radioactive **half-life** of a source is the time it takes for the source's activity to drop by a half. Half-life can normally be found from the decay curve. Remember to:

- ➜ subtract background radiation
- ➜ draw a best-fit curve to smooth out the random variations
- ➜ work out half-life from at least two starting points and take an average

Half-life and the decay constant

The radioactive half-life of a source is linked to its decay constant by the equation:

$$\lambda T_{1/2} = \ln 2$$

ln 2 is the natural logarithm of 2, which has the value 0.693. This equation can be derived from the exponential decay equation by substituting $T_{1/2}$ for t and $N_0/2$ for N (see action point).

Exam questions answers: page 57

1. A radionuclide contains 400 000 nuclei. Its decay constant is 0.30 s^{-1}. What is the initial activity? (2 min)
2. Two samples A and B of the same radionuclide have different activities. Give the reason for this. (5 min)
3. The half-life of thorium-228 is 1.91 years. Calculate its decay constant and hence find the time needed for its activity to fall by a factor of ten. (*Note* If $e^x = y$, then $x = \ln(y)$.) (15 min)

The jargon

Mathematically speaking, an *exponent* is a power, and so an equation with powers in can be called an *exponential equation.*

Checkpoint 3

Rearrange equation 2 to make $e^{-\lambda t}$ the subject. Write down in your own words what the new version of the equation means.

Checkpoint 4

If a source has a half-life of 1 day, how many days will it take for its activity to fall below 10% of its initial activity?

Examiner's secrets

Make sure you learn that N equals the number of unstable nuclei *remaining.* It is easy to forget this because the number remaining is equal to the number that have decayed after the first half-life. This is the *only* time this is true.

Watch out!

The units used for time and for decay constant must cancel. If time is measured in seconds, λ must be given in s^{-1}; if time is measured in years, λ must be in y^{-1}.

Action point

You can work out for yourself that $e^{-\lambda t}$ represents the fraction of unstable parent atoms (N/N_0) remaining after time t. So when $t = T_{1/2}$, this fraction must equal 1/2. So:

$e^{-\lambda T_{1/2}} = 1/2$

or $e^{\lambda T_{1/2}} = 2$

Taking natural logarithms gives us:

$\lambda T_{1/2} = \ln 2$

Binding energy and mass defect

With the exception of hydrogen, all atoms weigh slightly less than the sum of their constituents! They have a mass defect. The law of conservation of mass is broken and must be replaced with the more general law of conservation of mass–energy discovered by Albert Einstein. In this section, we get to use the most famous equation in physics: $E = mc^2$!

The jargon

Special relativity is based on the premise that the speed of light is constant for all observers (regardless of their motion).

Binding energy

Nucleons are bound together by the *strong nuclear force*. You would have to do work against this binding force to pull the nucleons apart. The amount of work you would have to do to separate all the constituent nucleons from a nucleus is called the **binding energy**.

Mass–energy

Einstein showed in his work on relativity that mass and energy are two forms of the same thing, more properly called **mass–energy**. Matter and energy can be interconverted according to the equation:

$$E = mc^2$$

Where E is energy, m is mass and c is the speed of light ($c = 3.00 \times 10^8$ m s^{-1}). You should see from the equation that the annihilation of a small amount of matter yields a lot of energy. (1 kg $\rightarrow 9 \times 10^{16}$ J!)

Checkpoint 1

How long could you run a 1 000 MW power station on 1 kg of matter (assuming *all* of it is converted into useful energy)?

Mass defect

The **mass defect**, Δm, of a nucleus is the difference between the summed mass of its constituent nucleons and electrons and its actual mass.

$$\Delta m = (m_{\text{protons}} + m_{\text{neutrons}} + m_{\text{electrons}}) - m_{\text{atom}}$$

Links

Refer to the *big bang theory* – the formation and annihilation of matter and antimatter, see page 157.

It is the mass equivalent of the atom's binding energy (the work you would have to do to separate each nucleon from the nucleus increases its mass!)

$$\underset{\text{binding energy}}{\Delta E} = \underset{\text{mass defect} \times \text{speed of light squared}}{\Delta m\, c^2}$$

Total binding energy increases with *nucleon number*, but this does *not* mean that high nucleon number nuclides are the most stable. (With more nucleons, the glue is spread more thinly.)

Links

See the section on the strong nuclear force in *nuclear instability*, pages 44–5.

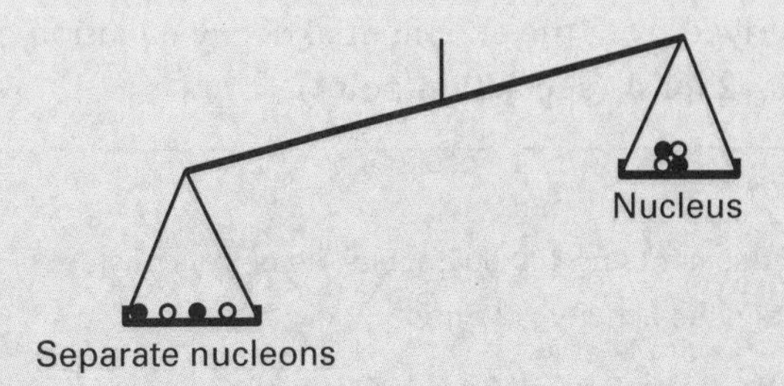

An atom or nucleus always weighs less than the particles it is made of.

Mass defect and binding energy per nucleon

If you divide mass defect or binding energy by the number of nucleons, you get a useful measure of nuclear stability. The greater the mass defect (and binding energy) per nucleon, the more stable the nucleus.

→ $1 \text{ eV} = 1.6 \times 10^{-19}$ J
→ $1 \text{ u} = 1.661 \times 10^{-27} \text{ kg} = 931.5$ MeV

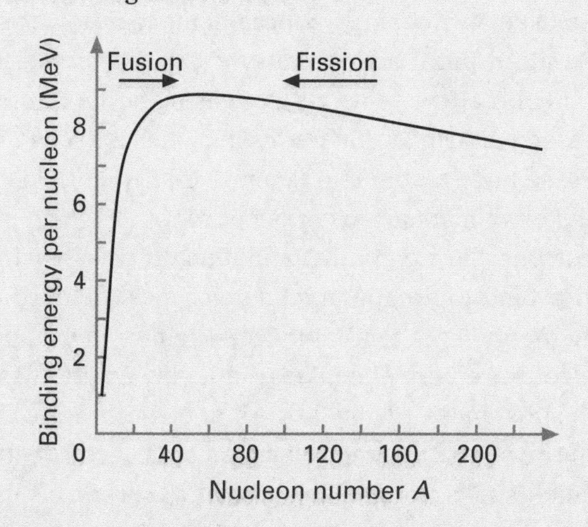

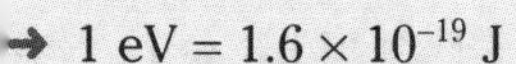

Some features of the graph

→ Mass defect per nucleon of normal hydrogen is zero, because there is only one nucleon, so the strong nuclear force is not involved, and so no work need be done to overcome it!
→ Mass defect per nucleon rises rapidly with A, peaking between $A = 50$ and $A = 80$. Iron ($A = 56$) is the most stable atom.
→ Mass defect per nucleon falls as A increases beyond this point.

Energy release from fusion and fission

→ **Fusion** is sticking together, **fission** is breaking apart.
→ Fusion powers the Sun and the stars. Fission powers nuclear power stations.
→ Both processes release energy. (How is this possible?)

The graph above holds the key to understanding energy release by fusion and by fission. Any process which increases mass defect *per nucleon* releases energy. Looking at the graph:

→ fusion can be interpreted as a shift to the right (increasing A)
→ fission is a shift to the left (decreasing A)
→ fusion of low atomic mass nuclei (only) increases mass defect per nucleon and therefore releases energy
→ fission of high atomic mass nuclei (only) also increases mass defect per nucleon and therefore releases energy

Exam question

answer: pages 57–8

Two deuterium (^{2}H) nuclei (mass = 2.0141 u) fuse to produce one ^{3}He nucleus (mass = 3.0160 u) and a neutron.

(i) Write the nuclear equation for this reaction.
(ii) Calculate the total mass defects before and after the fusion.
(iii) Find the energy released in MeV. (20 min)

Examiner's secrets

Make sure you are happy with the definitions of mass defect and binding energy and with the idea of mass–energy conversions.

The jargon

Binding energy per nucleon can be thought of as an energy *well*. The deeper the well, the more tightly bound the nucleus.

Examiner's secrets

This is a key graph. Study it closely. Energy release leads to greater stability. Fusion of low A nuclei and fission of high A nuclei release energy.

Watch out!

Binding energy is sometimes considered a negative quantity, since it is an energy deficit – so you might find the figure on the right in an inverted form (a reflection in the x-axis), in which case iron sits at the bottom of the binding energy trough.

Action point

Notice from the graph that the binding energy per nucleon is typically 8 MeV. Find out which element has the most stable nucleus.

Watch out!

Once you reach iron in nuclear reactions, it's all over. No more energy available. Iron is cold and dead in terms of available nuclear energy.

Applications of radioactivity

The applications of radioactivity and nuclear physics range from treating cancer to the development of the neutron bomb. Nuclear physics has changed the world!

Nuclear power

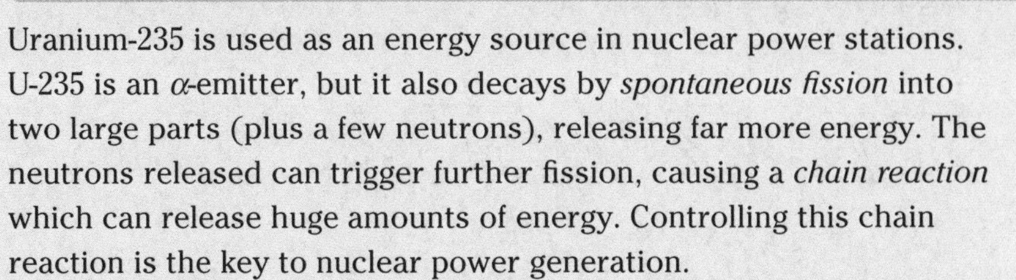

Uranium-235 is used as an energy source in nuclear power stations. U-235 is an α-emitter, but it also decays by *spontaneous fission* into two large parts (plus a few neutrons), releasing far more energy. The neutrons released can trigger further fission, causing a *chain reaction* which can release huge amounts of energy. Controlling this chain reaction is the key to nuclear power generation.

Ejected neutrons travel very fast and interact weakly with most matter – so they tend to escape. A *moderator* must be used to slow them down (to *thermalize* them) and keep them from escaping. (Thermal neutrons are very effective at inducing further fission.) Hydrogen-rich substances readily absorb neutron energy, but they tend to also absorb the neutrons themselves (as hydrogen is converted to deuterium) which spoils things. Pure graphite (carbon) is usually used as a moderator.

If more than one neutron per fission goes on to cause further fission, the reaction rate will rapidly rise. *Control rods* made of cadmium or boron (good neutron absorbers) are used to make sure the chain reaction does not get out of hand.

Checkpoint 1

99.3% of uranium is of the isotope U-238; only 0.7% of mined uranium is of the isotope U-235 which provides the power in a normal fission reactor. Why can't U-235 be separated chemically from U-238?

Watch out!

Be sure you understand the distinction between absorbing neutrons and absorbing their energy. The moderator should absorb only the energy; the control rods must mop up the particles themselves.

Nuclear bombs

The simplest nuclear bomb consists of two lumps of nuclear fuel, each of which is too small by itself to sustain a nuclear chain reaction (too many neutrons escape). The bomb is detonated by pushing the lumps together. This takes the combined lump beyond the fuel's *critical mass*, starting an explosive chain reaction.

Radioactive dating

Radioactive decay takes time. If you know a source's initial activity and its half-life, its present activity tells you the time (give or take all the uncertainties involved). Radioactive dating techniques have to rely on certain assumptions about initial conditions.

Checkpoint 2

Carbon-14 dating is only accurate(ish) for organic objects between 200 and 10 000 years old. Why?

Carbon dating

Atmospheric levels of carbon-14 are thought to have been fairly steady for thousands of years. C-14 is constantly being produced in the atmosphere by the action of cosmic rays. It decays by β-emission, with a half-life of 5 730 years. Over the millennia, an equilibrium has been set up so that the rate of decay is balanced by the rate of formation. Every living thing contains C-14 in the same proportion as occurs in the atmosphere. When organisms die, the proportion of C-14 begins to fall, so the specific activity (activity per kg) of C-14 in any organic matter tells us how long ago it died.

Checkpoint 3

The age of the Earth All uranium on Earth is older than the Earth itself. U-238 has a half-life of around 4.5 billion years; U-235 has a (much shorter) half-life of 7.1×10^8 years. Is it any surprise that U-238 is the more abundant isotope? (If you assume they were equally abundant when formed, you can calculate how long ago that was! The answer comes to around 6 billion years, which is okay as an upper limit.)

Non-destructive measurement and detection

- *α-particles* are used to detect smoke (open up your smoke detector and look for the radiation danger sticker).
- *β-particles* are used to check the thickness of paper and card.
- *γ-rays* are used to check aircraft wings and oil rigs for cracks. They are used in the coal mining industry to monitor the rate of production automatically (on the output conveyor belt). γ-rays have found an enormous range of uses. They come in a range of energies, making them suitable for measuring a wide range of objects! Dual-energy beams can be used to distinguish between different materials (e.g. coal and ash or steel and plastic).

Links

Penetration (see *properties of ionizing radiation*, pages 46–7) and half-life (see *radioactive decay*, page 49) are key properties in any application.

Nuclear medicine

Tracer techniques

Radioactive tracers are used to aid diagnosis. A (fairly) short half-life gamma emitter is injected and a gamma-camera is used to map its movement and distribution within the body. The tracer can be attached to sugars to highlight sugar-greedy cancer growth, or to a pharmaceutical which will accumulate in the site of interest.

Links

See *medical and health physics 2*, pages 160–1, for a fuller discussion.

Treatment of cancer

Radiation has proved a most effective treatment for many types of cancer. Cancer cells divide and grow more rapidly and are more susceptible to radiation damage than normal cells.

- *External treatment* makes use of gamma-ray beams which target the tumour from different directions so that the tumour (and only the tumour) receives a fatal radiation dose.
- *Internal treatment* can involve inserting a sealed β-source for short time periods or injecting a radiopharmaceutical which is designed to be as site specific as possible.

Checkpoint 4

The range and penetration of γ-rays depends upon their energy. How would the ideal source energy for treating a child differ from that used for an adult?

Killing microbes

- γ-rays are used to sterilize surgical equipment (γ-rays can penetrate packaging without damaging it, which offers obvious benefits). The process is quick, clean and simple.
- γ-rays can also be used to sterilize tinned foods and (more controversially) to increase the shelf life of fresh fruit and vegetables.

Checkpoint 5

What other methods could be used for sterilizing surgical equipment? What are their disadvantages (compared to γ-ray sterilization)? Which method would you support?

Exam question answer: page 58

The specific activity of carbon in living organisms is always 0.23 Bq per gram. A fossilized bone is discovered to have an activity of 0.0052 ± 0.0006 Bq per gram of carbon.

(a) Calculate the range of values the bone's age lies within.

(b) Explain the source of uncertainty and state any measures which could be taken to reduce it. (15 min)

Probing matter

Atoms were supposed to be the fundamental particles everything is made of, but it turns out that it's not quite that simple! Nuclear physicists continue to probe and smash atoms and their nuclei in the hopes of finding evidence for a grand unified theory – a theory of everything.

Links

See *diffraction and resolution*, pages 106–7.

Diffraction patterns

Light *diffraction patterns* can be used to measure the spacing of slits in a *diffraction grating*. X-ray diffraction patterns can be used to measure the spacing of atoms in a crystal.

→ You can't distinguish between two points closer than one wavelength apart. Fine *resolution* requires short wavelengths.

The Bragg equation gives the link between wavelength λ, spacing s and displacement angle θ for the nth constructive fringe.

$$n\lambda = s \sin\theta$$

Note The wavelength must be smaller than the spacing being measured or you get no *interference fringes* (see checkpoint). The number of *constructive interference* fringes you get is equal to s/λ, rounded down to the nearest whole number.

Checkpoint 1

If $\lambda = s$, the Bragg equation becomes $n = \sin\theta$. For $n = 1$, $\sin\theta = 1$ and θ is 90°. The first order constructive fringe grazes off at 90° to the target. What is the smallest spacing that can be measured by diffraction of light (wavelength 0.4 to 0.7 μm)?

Particle diffraction

Particles sometimes behave as waves (and vice versa). The wavelength of a particle is related to its momentum by the de Broglie equation:

$$p = h/\lambda$$

Where p is momentum (the product of mass and velocity), h is Planck's constant and λ is wavelength. The greater the particle's momentum, the shorter its wavelength.

Examiner's secrets

Many examinations now include questions that ask you to recall details of standard laboratory experiments. You should keep concise and up-to-date records of class experiments and demonstrations.

Low-energy electron diffraction

Louis de Broglie's theory of wave–particle duality was first confirmed in 1927, when George Thomson (son of J. J. Thomson, who discovered electrons) noticed diffraction patterns produced by a beam of electrons he had fired at gold foil. The spacing of the fringes allowed calculation of atom separations.

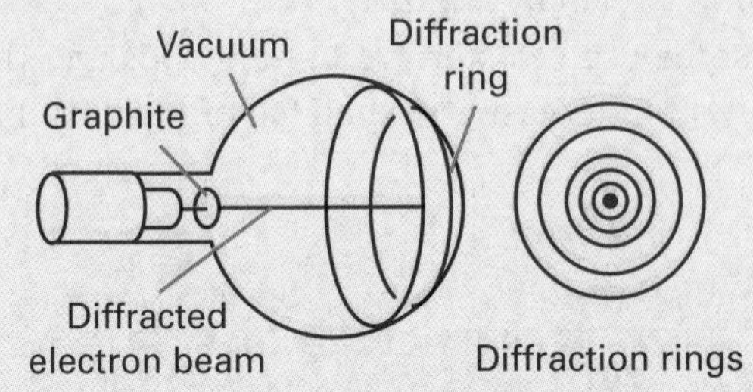

Diffraction patterns produced by a graphite sheet in a cathode-ray tube are shown above. Typical values: tube voltage ≈ 5 000 V, electron speed ≈ 4.2×10^7 m s^{-1}, wavelength ≈ 1.7×10^{-11} m, carbon atom separation ≈ 1.2×10^{-10} m.

Checkpoint 2

What wavelength (roughly) would you choose to measure:
(a) atom sizes and separations?
(b) nucleon sizes and separations?
Give your reasons.
(c) High energy X-rays may have wavelengths as short as 10^{-11} m. What are the implications for probing matter?

High-energy electron diffraction

With a high enough voltage in a good enough vacuum, you can give electrons enough momentum to produce diffraction patterns when they are scattered elastically by nuclei. Being *leptons*, electrons have the advantage that they are not subject to the strong nuclear force.

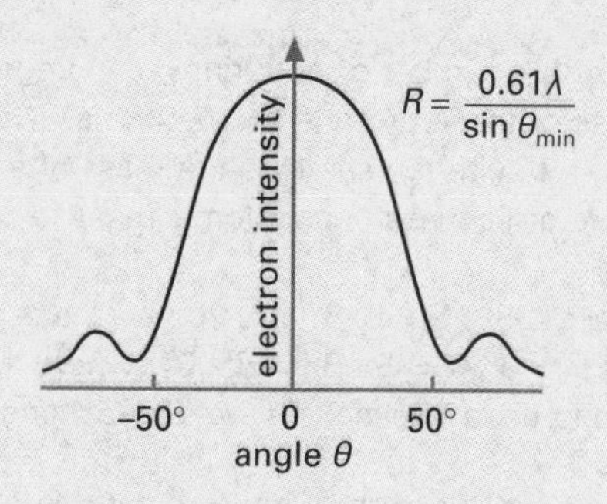

The high-energy (420 MeV) electron diffraction by carbon nuclei is shown above. Note that the geometry is different here. The nuclei are treated as individual tiny balls and the angular displacement of the first minimum is used to calculate the size.

Neutron diffraction

Just about every type of particle has been used to probe the atom. Charged particles such as electrons, protons and alpha particles interact with the charge held within the nucleus. Neutrons have the advantage of no charge, and so neutron diffraction can be used to give information on the distribution of the strong nuclear force.

Variation in nuclear size with nucleon number

Variation in nuclear volumes and radii with nucleon number (data from high-energy electron diffraction experiments) are shown below.

Nucleus	*A*	*r* ($\times 10^{-15}$ m)	*V* ($\times 10^{-45}$ m^3)
H	1	1.00	4.2
C	12	3.04	117
Si	28	3.92	252
Ca	40	4.54	392
Co	59	4.94	505
Au	197	6.87	1358

These results show that volume is roughly proportional to nucleon number *A*. This tells us that nucleons are incompressible; nuclear density is constant. The radius, *r*, of a nucleus is given by:

$$r = r_0 A^{1/3}$$

Where r_0, the constant of proportionality, equals 1.2×10^{-15} m. r_0 is the radius of a single nucleon.

Exam question answer: page 58

(a) Use the data in the table above to plot a graph of *r* against *A*.

(b) Now plot a graph capable of testing the equation $r = r_0A^{1/3}$.

(c) Comment on your results (to what extent do they support the equation?). (30 min)

Watch out!

Don't try to work out the momentum of a high-speed particle using its rest mass. Mass increases significantly as you approach light speed. (One of the effects explained by Einstein's theory of special relativity.)

The jargon

Relativistic is a term used to describe particles moving so fast that relativity must be accounted for. You can generally ignore relativity below speeds of 10^8 m s^{-1}.

Checkpoint 3

The main drawback with neutrons is that because they have no charge, there is no way to artificially accelerate them. Calculate the wavelength of a neutron ejected from a nucleus at a speed of 10^7 m s^{-1} ($h = 6.63 \times 10^{-34}$ J s).

Jargon

The liquid drop model of the nucleus: like a drop of water – a nucleus has a limited size; a nucleus cannot be compressed; a nucleus can split into smaller parts.

Action point

Can you show that the density of all nuclei is the same? (*Hint* you'll need to know the formula for the volume of a sphere.)

Action point

Make sure this equation makes sense to you. If $V \propto A$ and $V \propto r^3$, r^3 should be proportional to *A*.

Answers

Radioactivity and the structure of the atom

The nuclear atom

Checkpoints

1 Because opposite charges attract.
2 They would be slowed down, but they would not be deflected (imagine firing pellets through a real pudding).
3 They would slow down and fall into the nucleus.

Exam questions

1 Key results of Geiger and Marsden's experiments and their implications:
(i) Most of the α-particles passed straight through, without loss of energy, showing that most of the atom is empty space.
(ii) Some of the α-particles are deflected and a few even rebounded, showing that mass and charge must be concentrated in the nucleus (in a tiny space).

Examiner's secrets

Hints and tips The basic implications are easily learned. A-grade students should explain that the law of conservation of momentum requires that the things the α-particles rebounded off must have greater mass than the α-particles themselves and that electrostatic forces obey an inverse-square law; every doubling of separation quarters the size of the force. The force needed to reverse the α-particle's momentum can be calculated and from this, you can work out how close the α-particle must get to the repulsive nucleus!

2 This is a tough question! First the easy bit.
$R = 1.2 \times 10^{-15} \times 197^{1/3} = 6.98 \times 10^{-15}$ m
Now the difficult part:
Average spacing means centre-to-centre separation.
The average spacing of nucleons = average diameter of nucleons. If we call the volume of one nucleon v, and the volume of the entire nucleus V, then we know:
$V = Av \quad v = (4/3)\pi r^3$ and $V = (4/3)\pi R^3 \quad r = R/\sqrt[3]{A}$
The average nucleon separation is twice this value (diameter = 2 × radius!).
The final twist is to realize that the average separation of protons is found by substituting Z for A in the analysis above.
average separation of protons = $2 \times R/\sqrt[3]{Z}$
The question doesn't give Z for gold, so you have to make an educated guess:
sensible value of Z (75 – 98)
answer: $2 \times 6.98 \times 10^{-15}/79^{0.33} = 1.63 \times 10^{-15}$ m

Examiner's secrets

Questions like this require you to make a few assumptions. Write them down. (Very few people would see a route to the answer straight away!) If you work out the right formula for proton separation, you get full marks – regardless of the route taken – provided the examiner can follow your workings.

Elements and isotopes

Checkpoints

1 A = number of protons + number of neutrons = an integer. Atomic mass = mass of these protons and neutrons. The mass of a proton is not quite the same as the mass of a neutron, so even in atomic mass units, atomic mass is not a whole number.
2 Cl-35 has 18 neutrons (35 – 17). Cl-37 has 20 neutrons. Average atomic mass = 35.5 = $A \times 35 + B \times 37$, where A and B are the proportions of each isotope (as fractions). $A + B = 1$.
3 $35.5 = 35A + (1 - A) \times 37$; $A = 0.75$. So 75% Cl-35 and 25% Cl-37.

Exam questions

1 (a) (i) 137 – 55 = 82 (ii) 55 (iii) 82
(b) There is an equal number of electrons and protons.
2 (a) Yes. All isotopes of any particular element have the same proton number.
(b) No. Different isotopes have different numbers of neutrons.
(c) No. Nucleon number incorporates the (unspecified) neutron number.
3 $^{238}_{92}U \rightarrow ^{234}_{90}Th + ^{4}_{2}He$

Nuclear instability

Checkpoints

1 (a) Electrostatic repulsion would tear the nucleus apart.
(b) There has to be a force to bind all nucleons. Neutrons are not repelled electrostatically, but they would tend to drift about everywhere if not held tightly in the nucleus by the strong force.

Exam questions

1 True. The nucleus is made of protons and neutrons. Protons are all positively charged and are therefore mutually repulsive. There has to be a binding force strong enough (at least at short range) to overcome this repulsion. (The new strong force must also bind the uncharged neutrons.)

Examiner's secrets

You might argue that the discovery of the nuclear structure was *not* also the discovery of the strong force; you can still get credit if your case is good enough! (Rutherford's group discovered the nucleus before anyone knew it was made of protons and neutrons. If you treat each nucleus as a tiny blob of concentrated mass and charge, there may be no need for a new force, but this approach explains nothing!)

2 (a) (i) Kr-89: $Z = 36$, $N = 53$ plotted correctly.
(ii) Daughter after β-decay: $Z = 37$, $N = 52$.
(b) (i) Rn-222: $Z = 86$, $N = 136$ plotted correctly.
(ii) Daughter after α-decay: $Z = 84$, $N = 134$.
(c) In each case, the daughter lies closer to the line of stability; decay increases stability.

Properties of ionizing radiation

Checkpoints

1 A neutron converts into a proton and an electron. The electron cannot be held in the nucleus and so it is ejected as a beta particle. In beta+ decay, a proton converts into a neutron and an anti-electron (positron).

2 Most γ-rays will pass straight through a Geiger counter without causing ionization and therefore not being detected. (Note that for inverse-square law experiments, you normally have the GM tube held perpendicular to the γ-ray beam. X-rays can penetrate the metal walls and this arrangement makes it easier to measure the appropriate distance.)

3 Into the page

4 α-particles: $m = 4$ u, $q = +2e$; $q/m = 0.5e\,u^{-1}$; β-particles: $m \approx 1/2\,000$ u, $q = -e$; $q/m \approx -2\,000e\,u^{-1}$. Charge-to-mass ratio of a β-particle is about 4 000 times that of an α-particle!

Exam questions

1 (a) γ-rays. Neither α- nor β-radiation has sufficient penetration to probe for internal cracks. The size of steel girders rules out α-particles (the source and detector would have to be more than 10 cm apart, so α-particles wouldn't ever make the journey). β-particles could penetrate the air, so they could be used to detect holes (but so could light etc.). Only γ-rays have sufficient penetration to get through steel to probe for internal faults.

(b) α-particles. Only α-particles will be significantly attenuated by smoke.

(c) β-particles. α-particles would never get through the cardboard pack – whatever its contents. The intensity of a beam of γ-rays would not be sufficiently affected by the presence or absence of cereal in a small box. β-particles should be able to get through an empty cardboard box, and would be significantly attenuated by any cereal contained.

Examiner's secrets

Most A-level students would get the right type of radiation for each job, but only the best students give full explanations of their choices. For any application, two key questions are:
(i) Is the radiation sufficiently penetrating for the job?
(ii) Is the radiation going to be measurably affected by the change it is trying to detect?

2 (a) Credit is given for any methods that work. You could exploit differences in charge, charge-to mass ratio, penetration or ionizing ability.

To test for γ-radiation, you could place a lead shield (say 1 cm thick) in front of the source. If you still detect significant radiation, then the source must be a γ-emitter. α- and β-emitters could be identified from their cloud chamber traces. α-particles cause the most ionization and the clearest tracks; α-tracks are straight and end abruptly (because α-particles have high momentum and are mono-energetic). β-particle tracks are fainter and wander from the straight and narrow. You could use an arrangement similar to that given in the lower drawing on page 47 to exploit differences in charge (and charge-to-mass ratio).

Essentials

(i) Methods show knowledge of distinct characteristics of α-, β- and γ-radiation.

(ii) Methods would work. Suitable detectors and clear explanations of outcomes.

(b) (i) *α-particle energies* Use cloud chamber track. The longer the track, the greater the energy.

(ii) *β-particle energies* Compare ability to penetrate a suitable absorber ('suitable' means sufficiently thick and dense to absorb some, but not all, β-particles).

More penetrating ⇒ greater energy. Greater energy ⇒ more work must be done to stop the particles ⇒ greater range.

Note cloud chamber track-lengths do not provide an easy method of comparing β-particle energies. Many types of detector can measure energy directly. Some credit can be given for suggesting their use. For full credit, details of *how* they discriminate between energies is needed.

Radioactive decay

Checkpoints

1 Background levels vary – over time as well as with location. Surface activity on the Sun and stellar explosions in outer space all affect background levels of ionizing radiation.

2 If $N = 100$, absolute error = 10 and percentage error = 10%. If $N = 10\,000$, absolute error = 100 and percentage error = 1%

3 $e^{-\lambda t} = N/N_0$, i.e. the fraction of the original activity that still remains after time $t = e^{-\lambda t}$.

4 4 days.

Exam questions

1 $A = \lambda N = 0.30 \times 400\,000 = 120\,000\ s^{-1}$ (Bq)

2 They must have different masses (different number of nuclei)

3 $\lambda = \ln(2)/T_{1/2} = 0.693/1.91 = 0.363\ y^{-1}$

$A = A_0 e^{-\lambda t}$

Problem is to find t when $A = A_0/10$

$A_0/10 = A_0 e^{-\lambda t}$

$e^{\lambda t} = 10$

$\lambda t = \ln(10)$

$t = \ln(10)/\lambda = 2.303/0.363 = 6.35$ y

Binding energy and mass defect

Checkpoints

1 Energy equivalence of 1 kg of matter $= 1 \times (3.00 \times 10^8)^2 = 9.00 \times 10^{16}$ J. $E = P \times t$, so $t = 9.00 \times 10^{16}/10^9 = 9.00 \times 10^7$ s = 25 hours (≈ 1 day)

Action point

iron-56 at 8.79 MeV.

Exam question

(i) ${}^2_1H + {}^2_1H \rightarrow {}^3_2He + {}^1_0n$

(ii) Mass of nucleons = $(2 \times 1.0078 + 2 \times 1.0087)$ u
= 4.0330 u
Mass of nuclei before reaction = (2×2.0141) u
= 4.0282 u
Total mass defect before = (4.0330 – 4.0282) u
= 0.0048 u
Mass of nuclei after reaction = (3.0161 + 1.0087) u = 4.0248 u
Total mass defect after = (4.0330 – 4.0248) u = 0.0082 u
Change in mass defect = 0.0034 u

(iii) Change in mass defect in kg
$= 0.0034 \times 1.661 \times 10^{-27} = 5.65 \times 10^{-30}$ kg
$\Delta E = \Delta mc^2$
$= 5.65 \times 10^{-30} \times (3.00 \times 10^8)^2 = 5.08 \times 10^{-13}$ J
1 eV = 1.6×10^{-19} J, so 1 MeV = 1.6×10^{-13} J
1 J = $1/(1.60 \times 10^{-13})$ MeV
∴ energy released = 3.17 MeV.

Applications of radioactivity

Checkpoints

1 U-235 and U-238 are both uranium – same proton number, same electron structure, same chemistry!
2 Radioactive dating works best if the time period being measured is roughly the same as the half-life of the radioisotope being used.
(i) Activity of C-14 in living matter is low (0.23 Bq g^{-1}).
(ii) Enough time must have passed for a significant change in activity to have occurred (after 200 y the activity of C-14 only falls by about 2%; shorter time scales are impossible to measure without very big samples).
(iii) If too much time has passed, the residual activity will be hard to measure (after 10 000 y, residual activity is around 30% initial activity; because initial activity is low, it is hard to get accurate C-14 datings beyond this sort of time scale).
3 No! U-238's longer half-life means the more time that passes, the greater its abundance relative to U-235's etc. will be.
4 Ideal source for use on a child would have a lower energy and therefore a higher attenuation coefficient.
5 Steam-sterilization in an autoclave; chemical sterilization (followed by rinsing).

Exam question

$A_0 = 0.23$ Bq g^{-1}
(a) For minimum age, $A = 0.0058$; for maximum age, $A = 0.0046$ Bq g^{-1}
$A = A_0 e^{-\lambda t}$
$\lambda = \ln 2/t_{1/2} = 0.693/5\,730\ y^{-1} = 1.21 \times 10^{-4}\ y^{-1}$
Min age, t_{min} = $[\ln(0.23/0.00058)]/1.29 \times 10^{-4} = 28\,500$ y
Max age, t_{max} = $[\ln(0.23/0.0046)]/1.29 \times 10^{-4} = 30\,300$ y
(b) The main source of uncertainty is the randomness of radioactive decay.
To minimize uncertainty, you need the biggest counts possible, so
(i) count over a long period
(ii) use as large a sample as possible
(iii) take background readings over a long period, under similar conditions
(iv) use a detector with a high capture efficiency

Probing matter

Checkpoints

1 For $\lambda = 0.4$ μm, spacing must be > 0.4 μm for any constructive fringes to appear. For $\lambda = 0.7$ μm, s must be > 0.7 μm.
2 (a) and (b) The ideal wavelength must be at least as small as the particle being measured, but not too much smaller. $10^{-11} - 10^{-10}$ m for an atom, $10^{-15} - 10^{-14}$ m for a nucleus. You don't get diffraction patterns with longer wavelengths and the fringes become too closely spaced with much shorter wavelengths. (c) 10^{-11} m is not even nearly short enough for probing the nucleus, but it's ideal for measuring atom separations.
3 Momentum, $p = mv$. $m = 1.0087 \times 1.6651 \times 10^{-27}$ kg (see *elements and isotopes*, pages 42–3)
∴ $m = 1.6796 \times 10^{-27}$ kg ⇒ $p = 1.6796 \times 10^{-20}$ kg m s^{-1}.
$\lambda = h/p = 6.63 \times 10^{-34}/1.68 \times 10^{-27} = 3.95 \times 10^{-7}$ m.

Exam question

(a) See below. Labelled axes, units, clear points, best-fit curve.

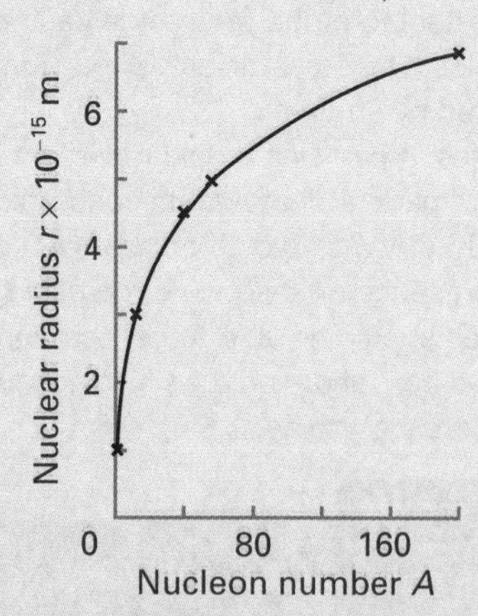

(b) A plot of either r against $A^{1/3}$ or r^3 against A. See below. Labelled axes, units, clear points, best-fit straight line.
(c) Data supports the equation well.
All data lies close to a straight line which goes through the origin.

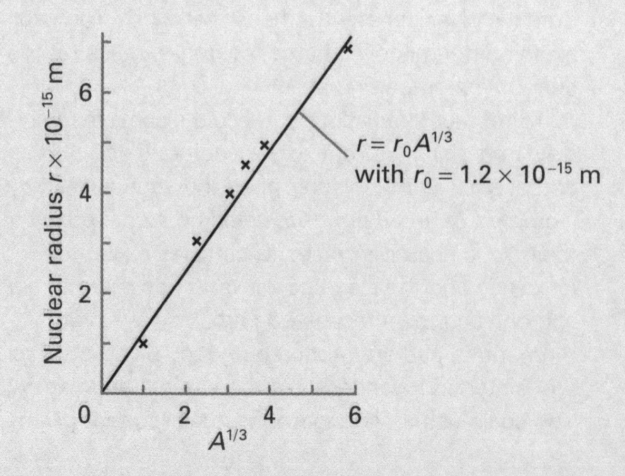

Electricity and electromagnetism

Electricity comes from the Greek word *elektron*, meaning amber. A Greek philosopher, Thales, discovered that when he rubbed amber (a resin) with a cloth, it attracted feathers. That was around 600 BC. 2 300 years passed before electricity really captured the interest of the scientific world. One particularly bizarre experiment was performed by Luigi Galvani in 1786. He hung the legs of a dead frog from a railing during a thunderstorm to discover whether lightning would make them twitch. The frog's legs twitched even before the lightning arrived. This was the inspiration for Mary Shelley's horror story *Frankenstein.*

Exam themes

- → *Understanding* For instance, why does a capacitor initially discharge quickly?
- → *Modelling* Use of formulae such as $C = Q/V$ to model physical phenomena.
- → *Mathematical competence* Such as using equations to solve circuit problems.
- → *Recall* Remembering, for example, the distinction that is made between EMF and PD.
- → *Applications* Such as why do car lights dim if a driver tries to start a car with the lights on?
- → *Graphical analysis* For example, interpreting a graph of terminal PD against current to find E and r.
- → *Links* Building an overview of physics, e.g. the recurring appearance of exponential changes.
- → *Practical work* For instance, obtaining IV characteristics.

Topic checklist

○ AS ● A2	AQA/A	AQA/B	CCEA	EDEXCEL	OCR/A	OCR/B	WJEC
Current as a flow of charge	○	○	○	○	○	○	○
Current, PD and resistance	○	○	○	○	○	○	○
Resistors and resistivity	○	○	○	○	○	○	○
Electrical energy and power	○	○	○	○	○	○	○
Kirchhoff's laws	○	○	○	○	○	○	○
Potential dividers and their uses		○	○	○	○	○	○
EMF and internal resistance	○	○	○●	○	○	○	○
Alternating currents	○		●			●	●
Capacitors	●	●	○	●	●	●	●
Electromagnetism	●	●	●	●	○●	●	●
Electromagnetic induction	●	●	●	●	●	●	●

Current as a flow of charge

Why are birds sitting on *live* electrical wires not electrocuted? It seems that to get an electric shock, electrons have to flow through the victim and go into the ground. There is an old saying that 'volts jolt but mils kills'. Just 26 milliamps (0.026 A) flowing through your heart may well kill you, but not many volunteers have tried to find out!

Checkpoint 1

Electricity can be subdivided into two smaller topics – static and current electricity. Explain the difference(s) between them.

Electric current

When you switch on a light bulb, it lights up almost instantly. This does not mean that electrons carried energy from the power station at the speed of light. The electrons were already in the metal filament of the bulb just waiting!

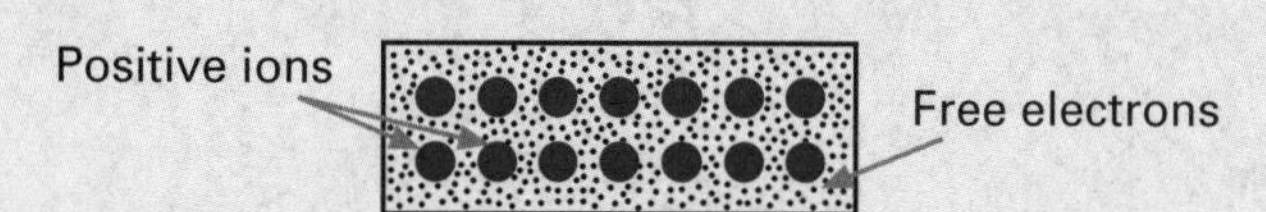

This diagram represents the internal structure of a metal. A regular lattice (framework) of atoms that have lost some electrons (positive ions) is surrounded by a sea of free electrons.

An **electric current** is a net (overall) movement of charged particles in a certain direction. In metals, the charged particles that can be persuaded to move by completing the circuit with an energy source (e.g. a battery) are electrons.

The jargon

A *cell* is a component that converts chemical energy into electrical energy. More than one cell is referred to as a battery. Remember a cell means one, like one room in a prison. A battery means more than one, like a battery of soldiers.

The unit of charge

Charge is measured in coulombs (C). Each electron carries just -1.6×10^{-19} C so it takes nearly 10^{19} electrons to carry 1 C of charge. A current of 1 ampere (A) flows when 1 C of charge flows past a point in a circuit in 1 s. Therefore:

$$I = \frac{Q}{t} \quad \text{or} \quad Q = It$$

Action point

Can you show that 1 ampere is a flow of 6.25×10^{18} electrons every second?

Checkpoint 2

Calculate how much charge passes a point if 12 A flows for 3 hours.

Consider a graph showing a steady current of 10 A flowing for 10 s.

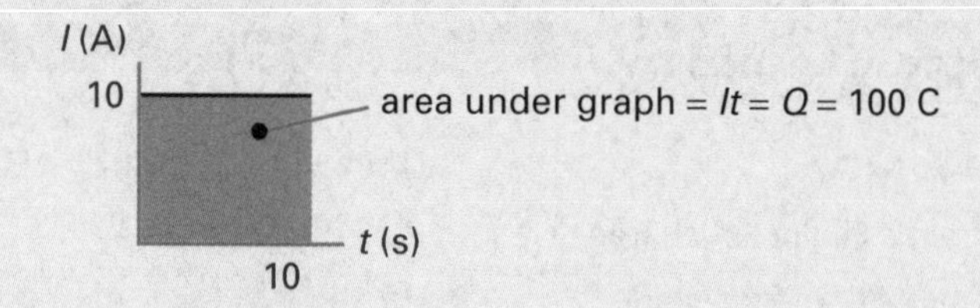

As a rule, charge carried equals the area under an *I–t* (current against time) graph. The rule applies whether current is steady or varying.

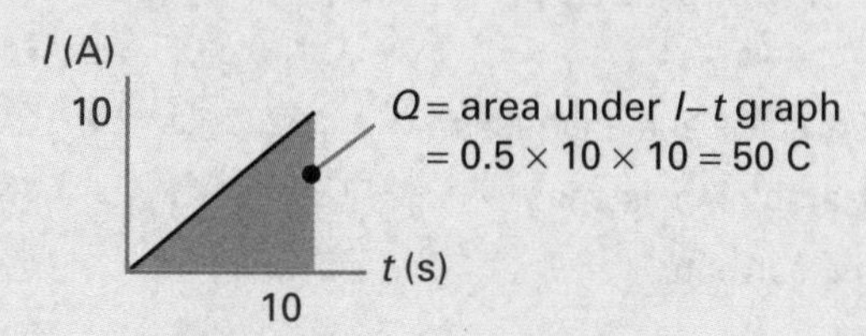

In a short time Δt, the charge that passed a point is $\Delta Q = I\Delta t$. This can be rearranged to read $I = \Delta Q/\Delta t$, or in words: current equals the rate of flow of charge.

Measuring current

Points to remember when using ammeters to measure current include:

- ➜ connect your ammeter in series
- ➜ ammeters can be delicate – start on the least sensitive setting, then move on to a more sensitive scale when you know that the current will not exceed the upper limit of that scale
- ➜ an ideal ammeter would have zero resistance

Conventional current and electron flow

Think of a number! You almost certainly did not think of a minus number. When scientists were trying to decide what would move when current flowed, they assumed that it would be a positive quantity. (The electron had not been discovered then!) As like charges repel and opposite charges attract one another, a positive charge would be repelled from the positive terminal of a battery and attracted to the negative. This is the direction of conventional current.

+ charges would be repelled from the + terminal of the cell and attracted towards the – terminal

This is the direction of conventional current.

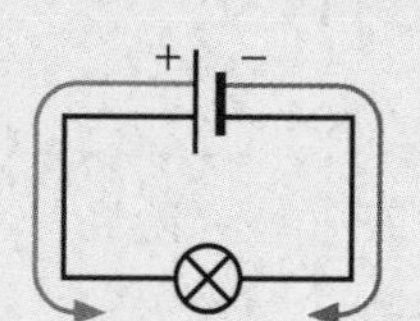

– charges are repelled from the – terminal of the cell and attracted to the + terminal

This is the actual direction of electron flow.

An equation for current

The current I through a metal, e.g. a wire, depends upon:

- ➜ the number of free electrons per unit volume n (metals are good conductors because they have many free electrons)
- ➜ the cross-sectional area of the wire A (thick wires allow the electrons to move more easily than thin wires)
- ➜ the charge carried by each free electron e
- ➜ the drift speed with which the electrons move v (typically about $1\ \text{mm s}^{-1}$)

Current is therefore given by:

$$I = nAev$$

But current can flow through materials other than metals; e.g. ions can flow through liquids in electrolysis experiments. We need a more general equation:

$$I = nAqv$$

Where n is the number of charge carriers per unit volume and q is the charge on the charge carriers.

Exam question answer: page 82

(a) If 26 mA flow through your heart, how long would it take for 4.68 C to pass?

(b) Describe the difference between semiconductors and conductors in terms of the number of charge carriers they contain.

(c) Calculate the drift speed of free electrons in a copper wire of diameter 1 mm, carrying a current of 7 A. Take n to be 1.0×10^{29} and the charge on each electron to be 1.6×10^{-19} C. (15 min)

Checkpoint 3

Why would an ideal ammeter have zero resistance?

Examiner's secrets

All arrows representing current should point in the direction of conventional current. This rule extends to include all circuit symbols, e.g. in transistors.

The jargon

Electrons normally move around constantly in a haphazard way with no overall movement in any one direction. But they can be persuaded to drift in a certain direction by an electric cell. A gradual drift towards the positive plate is superimposed on their normal haphazard motion. This is their *drift speed* (or *drift velocity* if we mention direction too).

Examiner's secrets

Many students use clumsy expressions when describing physical processes. Current flows *through* components. It can help you visualize what is happening if you learn this kind of phrase.

Current, p.d. and resistance

> *"Truth is ever to be found in the simplicity, and not in the multiplicity and confusion of things."*
>
> Sir Isaac Newton

Checkpoint 1

State and explain three assumptions that have been made in this section.

Examiner's secrets

It is often helpful to visualize electrons picking up and dropping off energy. However, remember that they are not living things with minds of their own!

Check the net

For brief introductory notes on electricity go to http://www.abdn.ac.uk/physics/bio/fi20www/index.htm

Examiner's secrets

Don't annoy the examiner! Voltage does *not* go through things. Voltage, or p.d., is measured *across* components.

Checkpoint 2

What would be the resistance of an ideal voltmeter?

Current is a flow of charged particles. Electrons pass along the electrical circuits in our houses, entering appliances via the live wire and leaving along the neutral wire. So what do we buy when we pay our electricity bills? We are not paying for electrons as they enter and leave. We pay for the energy they deliver.

Energy transfers

Energy is stored as chemical energy in cells. When it is transferred to electrons it is known as electrical energy, or even kinetic energy as the electrons are moving. In the circuit below, electrons 'pick up' energy at the cell and deliver it to the bulbs. If the electrons were still carrying any energy when they returned to the cell, the bulbs in this circuit would get brighter and brighter. As this does not happen, we know that all the energy picked up at the cell is delivered to the circuit.

For **series circuits** $V = V_1 + V_2 + \ldots$

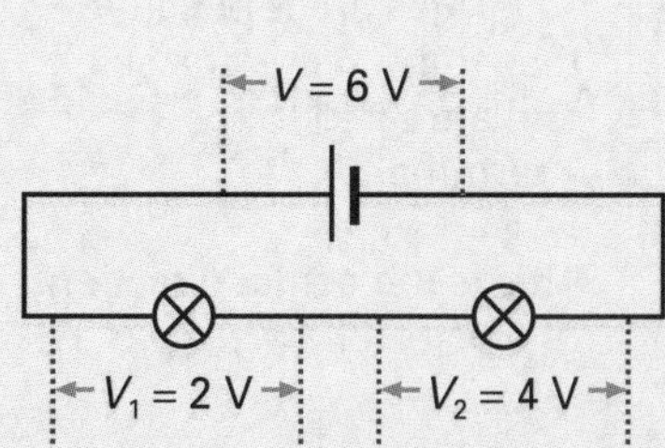

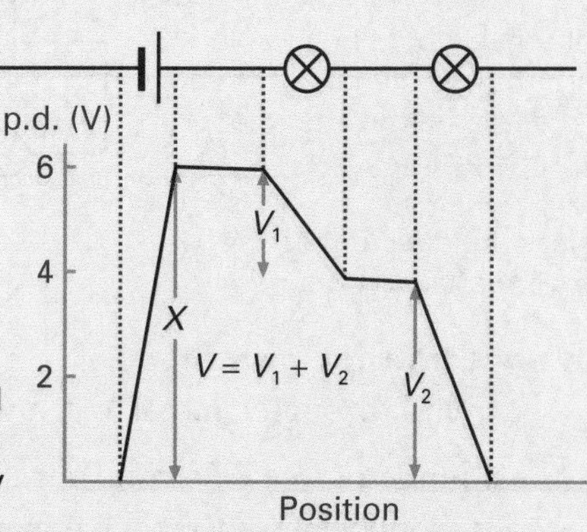

Imagine if the circuit above could still work if it was 'opened up' as shown on the right. A graph showing energy changes would look like this

Electrons give up all of their energy before they return to the cell. There are many electrons distributed all the way around the circuit, all carrying the same (negative) charge, all repelling one another. Electrons leaving the second bulb may have given up all their energy but there are many more coming behind to push them forward. Remember that:

- the potential at a point is defined as the electrical potential energy (PE) of a unit positive charge (i.e. + 1 C) at that point
- the **potential difference** (p.d.) between any two points in a circuit is the difference in electrical PE between the two points
- p.d. between two points is also the work done (or energy transferred) in moving +1 C of charge from one point to another
- $\text{p.d.} = \dfrac{\text{energy (work done)}}{\text{charge}}$ or $V = \dfrac{W}{Q}$
- 1 volt = 1 joule per coulomb ($1\text{ V} = 1\text{ J C}^{-1}$)

Measuring potential difference

As potential difference is used to compare two points in a circuit, say before and after a resistor, voltmeters are connected in parallel with the component. To avoid changing what we set out to measure, i.e. the electrical PE before and after the electrons enter the resistor, the voltmeter should have a very high resistance to avoid diverting electrons away from their original route through the resistor.

Resistance

As free electrons move through a wire, they collide with positive ions and with one another, slowing down as they do so. Some of their (kinetic) energy is transferred to the ions making the ions vibrate more so that the temperature of the wire increases. This is **resistance**.

→ resistance = $\frac{\text{potential difference}}{\text{current}}$ or $R = \frac{V}{I}$

→ If a p.d. of 1 V causes a current of 1 A to flow through a conductor, the resistance equals 1 ohm (Ω).

Series and parallel circuits

Series circuits have only one path for the electrons to follow.

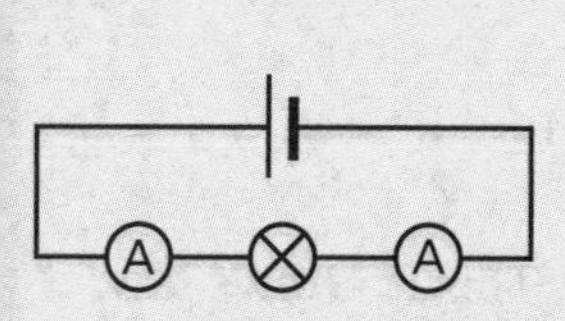

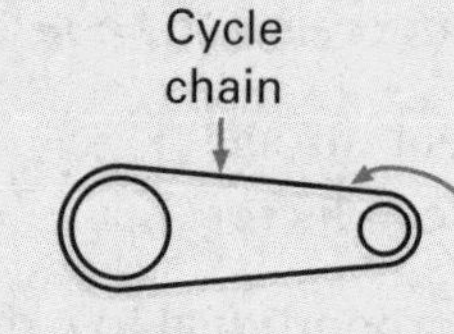

→ The current is the same at every point in a series circuit.

→ 1 A is defined as 1 C of charge flowing past a point every second.

→ As each electron carries a charge of -1.6×10^{-19} C, when 1 A flows, 6.25×10^{18} electrons go past every second!

→ The links in a bicycle chain are like electrons. There is only one path for them to take, and as they are all kept a certain distance apart from one another, they all travel around at the same speed.

→ If you change gear, it is as if more resistance has been included in an electrical circuit. All the chain links (electrons) slow down. All the electrons in a circuit are affected by the amount of resistance present, not just the electrons before or after the resistors.

Parallel or branching circuits offer electrons more than one route on their journey from the negative to the positive terminal of the cell.

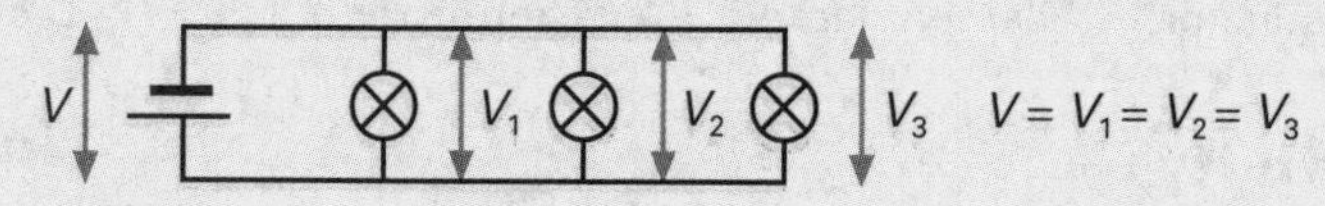

As with series circuits, the bulbs in a parallel circuit do not get brighter as time goes on. So each time electrons go to pick up another packet of energy from the cell, they must have already given up all of their previous load. As each electron has only passed through one arm of the circuit (in this case, just one of the bulbs), it must have given up all its energy in that arm (to that bulb).

For **parallel circuits** $V = V_1 = V_2 = V_3 = \ldots$

Exam question answer: page 82

(a) Describe one model used to simplify an aspect of electrical theory.

(b) An ideal way to measure the number of people using the London Underground would involve not slowing the travellers down. Why is this? Describe, with a reason, the characteristics of an ideal ammeter. (20 min)

Action point

If you get the opportunity, use a multimeter to measure the electrical resistance in your body. Hold one probe between the thumb and forefinger of one hand and the other probe between the thumb and forefinger of your other hand.

Action point

Now use the multimeter as a voltmeter. Hold one probe against the temple of a friend's head; hold the other probe against his/her other temple. Now ask the volunteer a question and watch what happens to the voltmeter's display!

Checkpoint 3

What group of materials have very high electrical resistances?

The jargon

The phrase *constancy of current* refers to the fact that current is the same all the way around a series circuit. The electrons maintain a constant speed.

Checkpoint 4

Why are parallel circuits used in household wiring?

Examiner's secrets

Marks are given for good use of English and spelling. Note the spellings of amps, amperes and ammeter.

Resistors and resistivity

Georg Simon Ohm found the relationship between p.d. and current in 1826. His father, Johann, had received no formal education yet he gave Georg an excellent introduction to science and mathematics that contrasted with the school-based teaching of that time. Georg's brilliance was recognized belatedly and the unit of resistance now bears his name.

Check the net

To find out more about Georg Ohm go to http://www-gap.dcs.stand.ac.uk/~history/Mathematicians/Ohm.html

Ohm's law

The resistance of a wire tells us how hard it is for electrons to flow along it. An ideal insulator would have infinite resistance whereas a superconductor has zero resistance. Ohm changed the p.d. across wires and then measured the current that flowed through them. He found that, provided the temperature remained constant,

$$\begin{pmatrix}\text{p.d. across the}\\ \text{conductor}\end{pmatrix} = \begin{pmatrix}\text{current through}\\ \text{the conductor}\end{pmatrix} \times \begin{pmatrix}\text{a constant}\end{pmatrix}$$

In other words, current I is proportional to p.d. V. This is **Ohm's law**. The constant in the equation above is resistance R, so:

$$V = IR$$

Links

See page 63 to remind yourself how resistance is defined by rearranging $V = IR$ into $R = V/I$.

Examiner's secrets

The resistance of any component can be found as the ratio of V and I. The equation V IR is not in itself a statement of Ohm's law. To obey Ohm's law, the resistance must be constant.

I–V graphs

Fingerprints are used to identify criminals. *I–V* graphs are used to identify electrical components. Ohmic conductors are identified by *I–V* graphs that have a straight line going through the origin.

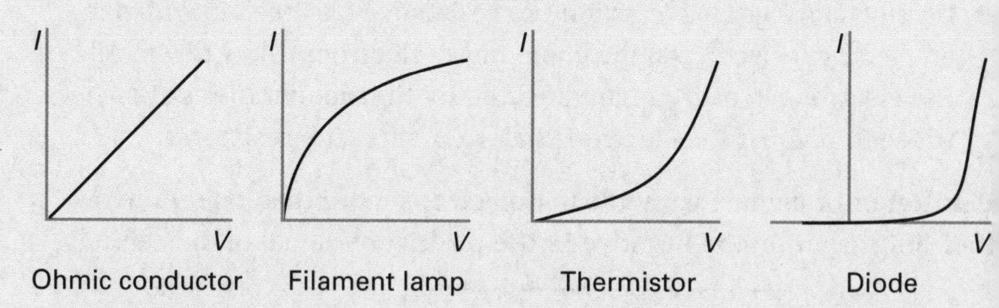

Ohmic conductor — Filament lamp — Thermistor — Diode

The jargon

Ohmic conductors are those that obey Ohm's law. A conductor that does not obey this law is called a *non-ohmic conductor*.

Watch out!

Some books plot graphs of V against I, rather than I against V as shown here.

Resistivity

Ohm also investigated the factors that affect resistance. It was found that length l, cross-sectional area A and the material the conductor was made of affected its resistance. These results are summarized by:

$R = \rho l/A$ ρ is a property of the material: **resistivity**

- Properties ending in *-ance*, e.g. resist*ance*, vary between samples under investigation. A long, thin sample of Nichrome wire would have a larger resistance than a short, thick sample. Therefore you cannot look up the resistance of Nichrome in a book.
- Words that end in *-ity*, e.g. resistiv*ity*, refer to a material rather than an individual sample. Silver has a lower resistivity than iron. If we have two identically sized samples held at the same temperature, the silver sample would provide less resistance than the piece of iron.
- The resistivity of materials tends to vary with temperature and changes if impurities are present in the material.

Checkpoint 1

Rearrange $R = \rho l/A$ to make ρ the subject of the equation, i.e. ρ = ? Then use this new equation to find the units for ρ.

Checkpoint 2

Conductivity $\sigma = 1/\rho$. What are the units of conductivity? Explain whether you would expect to find a table of conductivities quoted in a table of physical constants.

Using resistors

Imagine that you want to watch TV. You might glance at the TV to see if it is on standby (the red LED on–off indicator is protected against high currents by a fixed value resistor). To adjust the lighting in the room you could use a dimmer switch (a variable resistor that controls current). Finally you lower the volume of the TV (using a variable resistor again, this time to change the voltage).

Resistors in series and parallel

The total resistance R_T of three resistors R_1, R_2 and R_3 **in series** is:

$$R_T = R_1 + R_2 + R_3$$

The total resistance R_T if the resistors are **in parallel** is given by:

$$\frac{1}{R_T} = \frac{1}{R_1} + \frac{1}{R_2} + \frac{1}{R_3}$$

Light-dependent resistors (LDRs) and thermistors

LDRs and thermistors are really variable resistors. As more light falls on an LDR, or as the temperature of a thermistor increases, resistance falls. LDRs and thermistors are made from semiconductor materials. As more energy is supplied to them (either by shining light on LDRs or heating thermistors) more electrons are released to become free electrons so that more current can flow; i.e. resistance decreases.

What controls the conveyor belt carrying groceries at a supermarket checkout? When the goods reach the end of the belt, they block out a beam of light that had been falling on an LDR. The LDR is connected to the motor that moves the belt. Less light falls on the LDR, its resistance goes up, the current through the LDR and motor decreases, and so the belt stops!

Superconductors

- The resistance of most metals falls with temperature.
- If cooled sufficiently, some materials have no resistance at all.
- Some materials, based on oxides of bismuth, become superconductors (i.e. they have zero resistance) at −196 °C.
- This is their transition temperature.
- This technology is used in superconducting electromagnets.
- Their unusually stable magnetic fields are used in MRI scanners.

Exam question answer: page 82

(a) Sketch I–V characteristic graphs of: a metallic conductor at constant temperature, the filament of a light bulb, a negative temperature coefficient thermistor and a diode. Explain the shape of each graph.
(b) Describe an experiment to find the resistivity of a metallic conductor.
(c) State three uses of superconductors. (25 min)

"Thus the task is, not so much to see what no one has yet seen; but to think what nobody has thought, about that which everybody sees."

Erwin Schrödinger

Checkpoint 3

Why is the total resistance of three 2 Ω resistors connected in series more than if they were connected in parallel?

Watch out!

$1/R$ is called conductance G and has the unit Ω^{-1} or *siemens* (S). The total conductance for a parallel combination is $G_1 + G_2 + G_3$.

Action point

If you have access to an LDR and a multimeter, perhaps at school, try this. Switch the multimeter to its ohmmeter setting and connect it to the LDR. Now cover the LDR with your thumb and watch its resistance increase!

Links

For more on thermistors, see pages 86–7.

Check the net

To visit the UK's national centre for research into the origins and applications of superconductivity go to http://www.phy.cam.ac.uk/research/sucon/index.html

Links

See *materials 3*, on pages 174–5 to find out more about conductivity.

Examiner's secrets

When asked to sketch a graph, remember that scales are not required but labelled axes are!

Electrical energy and power

Energy is transferred to electrons in power stations. The electrons carry it around the country in a clean and convenient way. Unfortunately the electricity wires and pylons are ugly. If the wires were buried underground, our environment would benefit and less energy would be wasted before reaching its intended destination.

Energy and power

Power tells us how quickly energy is changed from one form to another.

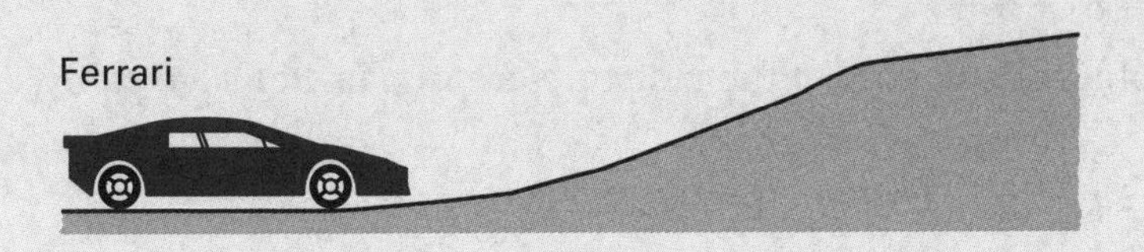

This Ferrari is very powerful – it can change chemical energy in its petrol into kinetic energy *very* quickly. So it goes uphill *very* quickly!

Energy and power in electrical circuits

The energy transferred to a component, e.g. a bulb, depends on:

- the p.d. across it (how much energy is dropped off at the component by the electrons)
- the current through it (how quickly the electrons are delivering the packets of energy)

So,

power = potential difference × current

$$P = VI$$

(watts) = (volts) × (amps)

Power and energy equations

There are several equations for power:

$$P = VI$$

As $V = IR$, this becomes:

$$P = IR \times I = I^2R$$

Alternatively, as $V = IR$, $I = V/R$. Using this expression for I we get:

$$P = V \times V/R = V^2/R$$

There are also several equations for energy. As power is how quickly energy can be converted:

$$P = E/t$$

$\therefore E = Pt$ where t is time in seconds

Using our expressions for power this becomes:

$$E = VIt = I^2Rt = V^2t/R$$

The jargon

Energy is a promise of work to be done in the future. It is the stored ability to do work. *Energy* is measured in joules (J).

Links

To compare this work with the use of the terms energy and power in mechanics, see pages 22–3.

Checkpoint 1

If all other factors were equal, why would you choose a 2 kW hairdryer rather than a 1500 W model?

Watch out!

In all these equations, *V* refers to the p.d. across the resistor and *I* refers to the current flowing through the resistor.

Examiner's secrets

Remember to express power in W, p.d. in V, current in A, resistance in Ω, and time in s before using any of these expressions. Fewer mistakes are made this way.

Checkpoint 2

What is the power produced by a power station that feeds a current of 100 A into the national grid at a potential of 440 kV? The resistance per unit length of a 20 km section of the power lines is 0.2 $\Omega\,m^{-1}$. Calculate the power loss in this section.

Fuses

The word fuse means *to melt*. We have already seen that when a current flows through a wire, the temperature of the wire increases. If the temperature of the wire exceeds the wire's melting point then the wire will obviously melt. Fuses are short lengths of wire, often copper covered with tin, that melt when the current flowing through them exceeds a predetermined maximum. Fuses are used to protect circuits from dangerously high currents. They are deliberate weak links in a circuit, included for safety reasons. Fuses are labelled with the maximum current they can carry without melting. Only certain values are commercially available; e.g. 3 A, 5 A, 13 A.

Paying for electricity

We have already made the point that we pay for the energy delivered by electrons rather than the electrons themselves. It is fairly easy to visualize this with a simple *direct current* (DC) circuit. The electrons pick up a packet of energy at the cell and deliver it to, let us say, a bulb before returning to the cell for another packet. This can continue until the cell 'runs flat'; i.e. until there are no more packets of energy left there. It is a little harder to imagine what happens with *alternating current* (AC) from power stations:

→ the number of electrons entering and leaving devices should always be the same (unless a fault develops)
→ with AC, electrons oscillate as if connected in a long chain
→ the power station provides the energy to vibrate the chain
→ this energy is dissipated in the device
→ as with DC, no electrons are used up!

If you check an electricity bill, the number of joules (J) of energy that you have used will not be quoted. 1 J is a very small amount of energy. Just lifting an apple from the floor up to table height involves a transfer of about 5 J! A kilowatt-hour (kW h) is a much bigger unit of energy (3.6 MJ) and so it is more convenient. Study this example.

A 3 kW heater is left on for 4 hours. Each kilowatt-hour costs 10 p. What was the cost of the energy used?

$$\begin{aligned}\text{energy cost} &= (\text{number of kW h}) \times (\text{cost of 1 kW h})\\ &= (\text{number of kW}) \times (\text{number of hours}) \times (10\text{ p})\\ &= 3\text{ kW} \times 4\text{ hours} \times 10\text{ p}\\ &= 120\text{ p}\end{aligned}$$

Exam question — answer: page 83

(a) Calculate the energy costs of using a 100 W lamp for 1 day. (1 kW h = 10 p)

(b) How much heat is produced by a 20 Ω resistor carrying 5 A for 60 s?

(c) Calculate the heat produced in 10 min by a pair of 15 Ω resistors connected in parallel with a PD of 3 V across the combination. (15 min)

The jargon

A *fuse rating* is the maximum current that can flow through the fuse without causing it to melt.

Checkpoint 3

A hairdryer, labelled 1 000 W, is connected to the 240 V mains. What fuse should be used to protect it?

The jargon

Direct current means that the electrons keep moving in the same direction all the time. *Alternating current* means that the size and direction of the current keep changing.

The jargon

Dissipate: to scatter or spread out.

Check the net

To view a gallery of energy pioneers go to www.energy.ca.gov/education/scientists/index.html

Checkpoint 4

What are the electrical energy costs of using a 5 kW oven for 2.5 hours if 1 kW h costs 10 p?

Examiner's secrets

Hybrid cars are a new generation of vehicle that combine the benefits of electrical energy and liquid-gas fuels. You can expect to use your knowledge and understanding of different areas of physics in the same question.

Kirchhoff's laws

Kirchhoff's laws for current and voltage built on th
work of Georg Ohm. The two fundamental concept
explained in this spread allow voltage and current t
be calculated at any point in a circuit. Kirchhoff's wo
was not confined to electricity. For example, his wo
on black-body radiation laid the foundations for th
quantum theory.

Check the net

For a more detailed analysis of Kirchhoff's laws for current and voltage, go to w3.scale.uiuc.edu/ECE110/lessons/kirchhoff/klhome.html

Kirchhoff's first law

This law is based on the fact that current (or charge) cannot build up a wire. A formal statement of this law is that:

- The (algebraic) sum of the currents into a point equals the sum of the currents out of that point.

The jargon

Current is a vector quantity. *Algebraic* sum means that you take the direction and therefore *sign* of the current into account.

This law can be expressed in a mathematical form as $\sum I = 0$.

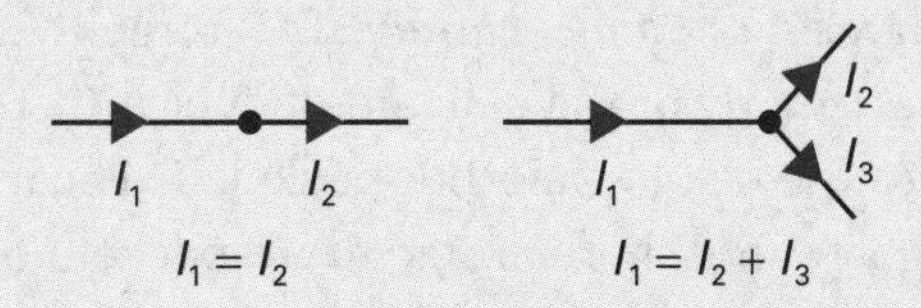

Checkpoint 1

Calculate the unknown current in the diagram below.

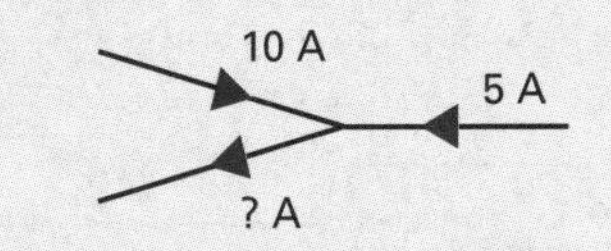

Kirchhoff's first law is a consequence of conservation of charge. For example, when electrons go into a point in a circuit they do not disappear, rather they come out of the other side. So the current that enters a point equals the current that leaves that point.

Don't forget

Conventional current flows from positive to negative.

Kirchhoff's second law

This law is based on the fact that all the energy that electrons pick up in the cell or battery is dropped off as they travel around the circuit. This means that:

- The sum of the EMFs in any closed loop in a circuit is equal to the sum of the p.d.s around that loop.

The jargon

An EMF (in volts) is the energy *gained* by 1 C of charge as it passes through an energy source. A p.d. is taken to be the energy *lost* by 1 C of charge as it moves through a resistor. ($1\ V = 1\ J\,C^{-1}$)

Mathematically this can be written as $\sum E = \sum IR$; i.e. in a closed loop, the sum of the EMFs equals the sum of the IR products.

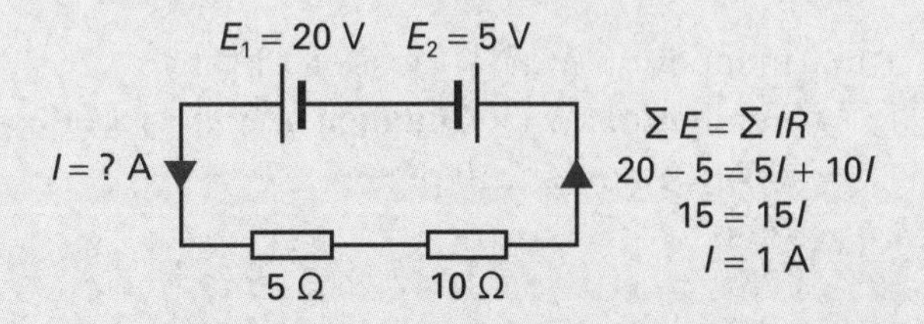

Checkpoint 2

In the example shown on the right, why do we assume that the direction of the unknown current is anticlockwise?

Kirchhoff's second law is a consequence of conservation of energy. Remember these important points about this law:

- energy gained by electrons moving through the cells equals the energy lost by the electrons as they move through the resistors
- in all the calculations involving Kirchhoff's second law, it is assume that the amount of energy lost in the wires is so small that it can be ignored

Examiner's secrets

This topic really lends itself to synoptic questions that assess several areas of your course. A question might be 'describe a practical application of two conservation laws'.

Using Kirchhoff's laws

Work your way through this example to find I_1, I_2 and I_3.

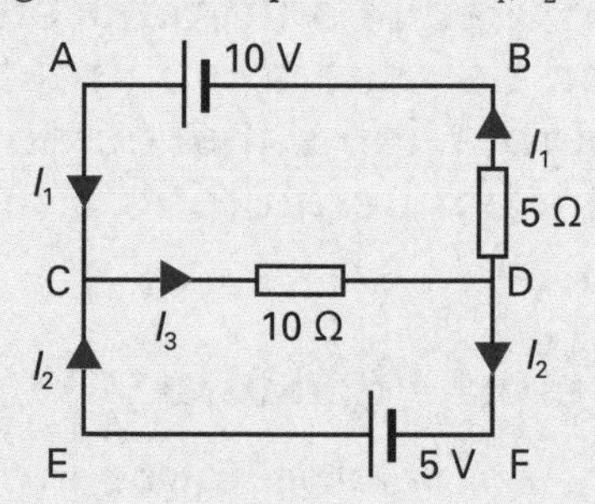

- Indicate the direction of the currents with arrows (see above). (Make a best guess regarding the directions! You will know if you have made a mistake, negative numbers will appear as the answers.)
- Apply Kirchhoff's first law:

  at C $\quad I_1 + I_2 = I_3$ [1]

- Select a closed loop and apply Kirchhoff's second law:

  for ABCD $\quad 10 = 10I_3 + 5I_1$ [2]

- Apply Kirchhoff's second law to another closed loop:

  For CDEF $\quad 5 = 10I_3$

  $I_3 = 5/10 = 0.5$ A

- Substitute this value for I_3 into equation 2:

  $10 = 5 + 5I_1$

  $I_1 = 1$ A

- Substitute these values for I_1 and I_3 into equation 1:

  $1 + I_2 = 0.5$

  $I_2 = -0.5$ A

- This means that I_2 flows in the opposite direction to that shown in the diagram and has a magnitude of 0.5 A.

Exam question answer: page 83

(a) Use Kirchhoff's second law to find the value of *R* in the circuit below.

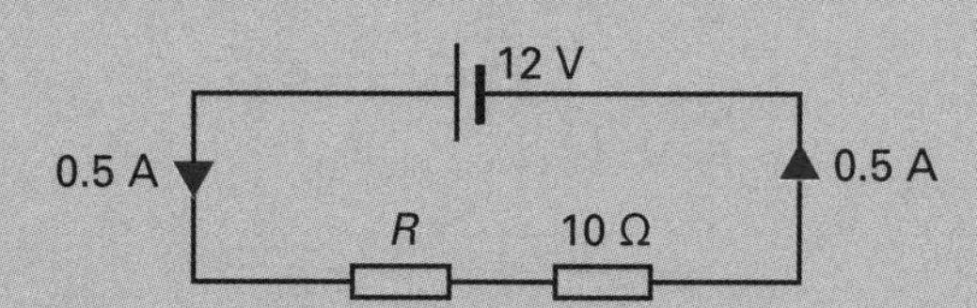

(b) Use Kirchhoff's laws to find the value of the unknown quantities in the circuit below.

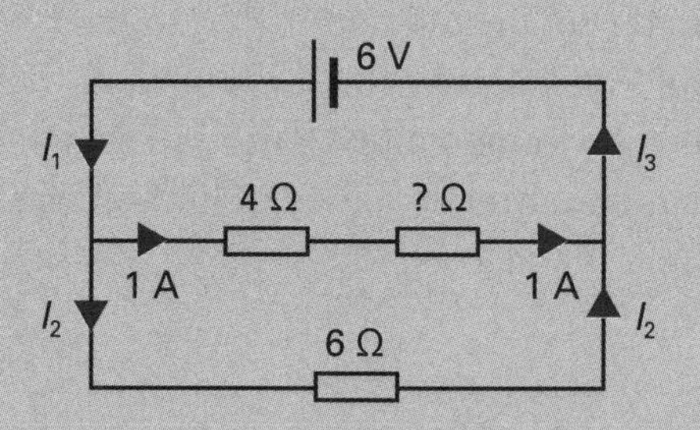

(10 min)

Examiner's secrets

Check your syllabus carefully. Not all expect you to be able to use simultaneous equations, as shown here.

The jargon

Simultaneous equations are a set of equations that are all satisfied by the same set of variables.

Examiner's secrets

Always set out your work in a neat, logical order (see the example on the right).

Checkpoint 3

Kirchhoff's second law, for closed loops, is rather like a circular walk around a hill if the walker returns to his/her starting point on completion of one circuit. Explain this analogy.

Examiner's secrets

It is unlikely that you will need to state Kirchhoff's laws, but you will certainly need to be able to use them.

Potential dividers and their uses

Potential dividers use resistors to divide a battery's voltage so that a portion of it can be used. Potential dividers are often used in automatic electronic circuits. These circuits can be used to control the temperature in a fish tank, operate lights that come on when it gets dark, maintain the temperature in a premature baby's incubator, and so on.

The jargon

It is usually better to avoid the term *voltage*. But electronic engineers often use it for convenience and so we will use it in this spread.

Principle of potential dividers

This circuit shows how a small voltage can be produced from a larger one. The equation for the smaller (output) voltage V_{out} is important.

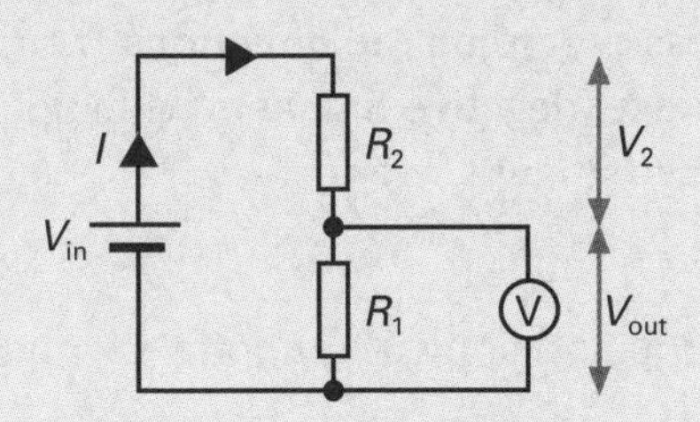

Checkpoint 1

Calculate the output voltage from this circuit if V_{in} = 12 V, R_1 = 200 Ω and R_2 = 100 Ω.

The voltmeter has such a high resistance that very little current goes through it. Therefore, we can consider this circuit as a simple series circuit comprising a cell in series with two resistors R_1 and R_2.

$$V_{in} = V_{out} + V_2$$
$$= IR_1 + IR_2 \qquad \text{(Kirchhoff's second law)}$$
$$= I(R_1 + R_2)$$

As $V_{out} = IR_1$:

$$\frac{V_{out}}{V_{in}} = \frac{IR_1}{I(R_1 + R_2)} = \frac{R_1}{R_1 + R_2}$$

$$V_{out} = \frac{V_{in}R_1}{R_1 + R_2}$$

Links

For more on Kirchhoff's laws for current and voltage see pages 68–9.

Links

If you need more detail on electronics see the *electronics option*, pages 164–9.

Action point

Can you show $\frac{V_1}{V_2} = \frac{R_1}{R_2}$? This is a really useful relationship when solving problems involving potential dividers. Learn it!

A more useful circuit

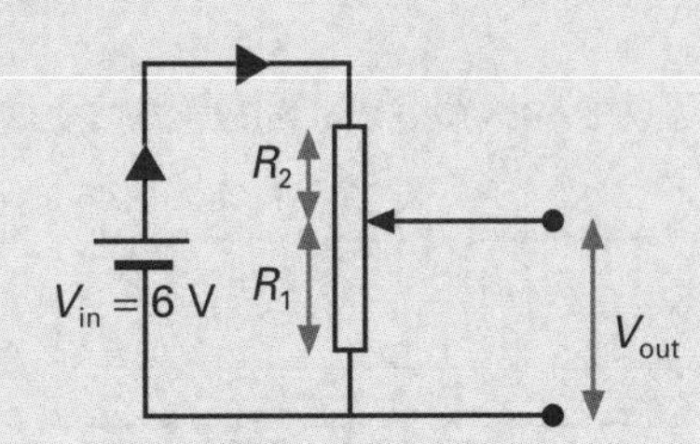

- By moving the sliding contact, any value for V_{out} between 0 V (slider at the bottom) and 6 V (slider at the top) can be obtained.
- A transistor can be connected across V_{out} to be used as a switch.
- A transistor could be selected that switches on when V_{out} = 0.6 V.
- This circuit relies on human intervention to move the sliding contact, so it is not automatic and is of limited use. (The next circuit is truly automatic.)

The jargon

Transistors are automatic switches. They have three terminals – the collector, base and emitter. If there is enough PD between the base and emitter arms a small current flows into the base, switching the transistor on. Now a larger current can flow into the transistor via the collector, leaving through the emitter.

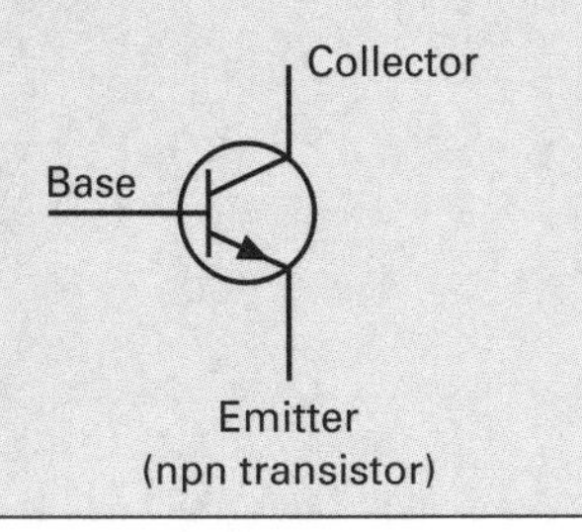

Real uses of potential dividers

The introduction to this spread mentioned some applications of potential-divider circuits. In each example, temperature or light intensity has to be monitored automatically. Thermistors and light-dependent resistors are the sensors used to carry out this job.

Circuits using thermistors

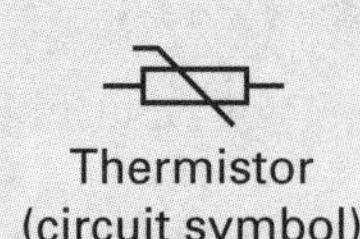

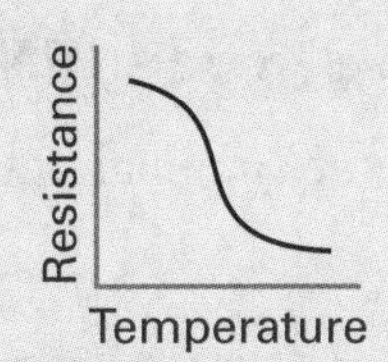

The resistance of thermistors varies with temperature. They are made from semiconductor materials. When hot, more free electrons are released inside a thermistor so its resistance falls. Therefore thermistors automatically sense temperature changes. If a thermistor is used, a potential divider produces a p.d. that depends on temperature.

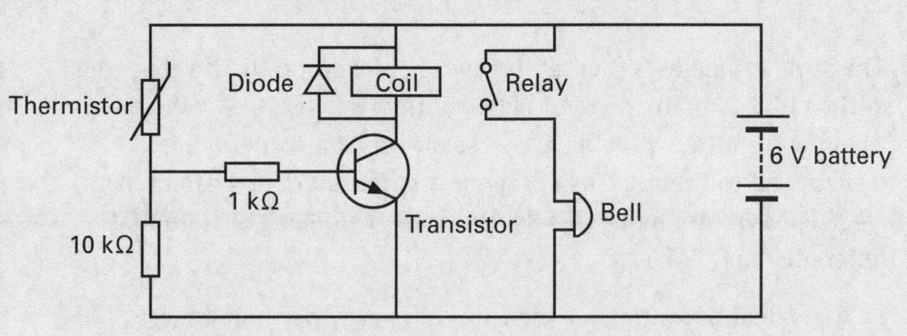

This circuit shows an automatic fire alarm. When hot, the resistance of the thermistor falls so that the p.d. across it falls and the PD V_{out} across the 10 kΩ resistor increases. The transistor switches on and current flows through the coil, magnetizing it so that the relay switches on. Current can now flow through the bell and the alarm will be raised!

Circuits involving light-dependent resistors (LDRs)

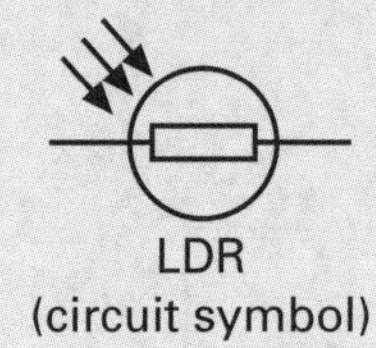

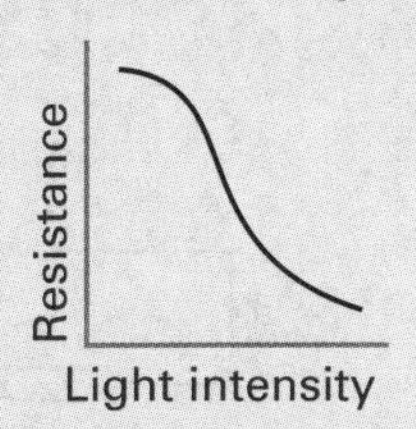

The resistance of an LDR depends on light intensity. More light energy falling on an LDR can release more electrons so that the resistance of this semiconductor-based component falls. If an LDR is used, a potential divider produces a p.d. that depends on light intensity.

Exam questions answers: page 83

1 Draw, and explain, how a potential-divider circuit could be used to automatically control a light so that it turned on in dark conditions. (15 min)

2 Two 10 kΩ resistors are connected as a potential divider with a supply p.d. of 10 mV. What is the current flowing and the p.d. across each resistor? (5 min)

Checkpoint 2

What other word does *therm*istor remind you of? What is the connection between these two words?

Checkpoint 3

What are the three essential components of an automatic decision-making circuit?

Checkpoint 4

State two advantages of this circuit.

Action point

Make lists of all the uses of potential-divider circuits mentioned in this spread that use:
(a) thermistors
(b) LDRs.
Do some research to add to these lists.

Examiner's secrets

If you get the first part of a question wrong, you may still get marks in the second part. Examiners mark numerical questions using an *error carried forward*. So don't give up if you struggle with the early calculations in questions.

EMF and internal resistance

"Prediction is very difficult, especially about the future."

Niels Bohr

'If it moves it's biology, if it smells it's chemistry and i[f] it doesn't work it's physics!' Why is it that if you selec[t] 6 V on a power pack and then check it with a volt[-]meter, there always seems to be a discrepancy? Th[e] power pack may not be very precise. But you will als[o] discover that the p.d. across it depends on the circui[t] it is connected to.

Electromotive force (EMF) *E*

You are probably very familiar with this type of circuit.

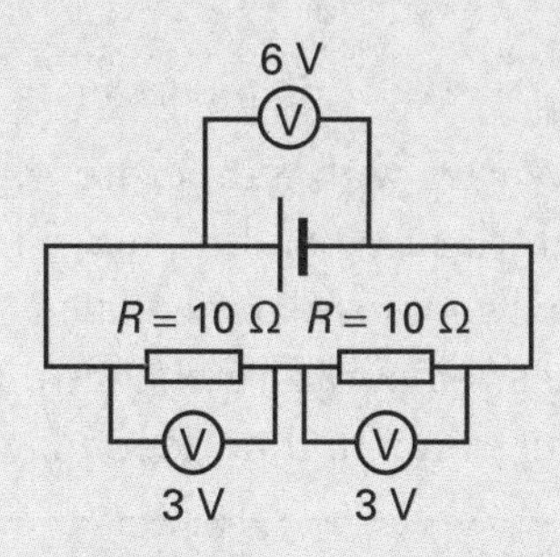

This circuit suggests that all the energy picked up by the electrons at the cell is equally divided between the two resistors. (We have assumed that the wires have no resistance so no energy is transferred to them.) This spread examines energy transfers in electrical circuits more closely. We start by looking at potential difference and EMF.

- **Electrical potential** is a measure of energy per unit charge.
- **Potential difference** (p.d.) is the difference in potential between electrons entering and leaving a component. That could mean electrons gaining energy as they cross a power source or losing energy as they cross, e.g. a resistor. To distinguish between energy gains and losses, PD conventionally refers to energy losses.
- **EMF** is the p.d. across a cell or other power supply. The electrons are gaining energy.

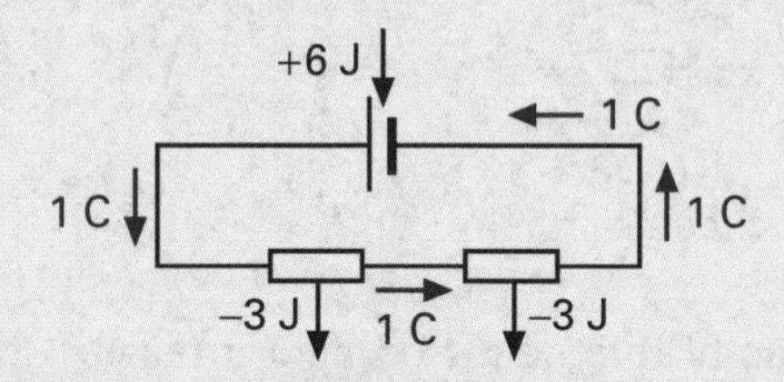

1 volt means that 1 coulomb of charge (carried by 6.24×10^{18} electrons) gains or loses 1 joule of energy. This diagram shows that the 6 V cell gives 6 J of energy to each coulomb of charge that passes through it. The p.d. of 3 V across each of the resistors means that in each case 3 J of electrical energy is converted into other forms (e.g. heating the resistor) as 1 C of charge (6.24×10^{18} electrons) passes by.

This leads to the equation $V = E/Q$ (1 volt = 1 joule per coulomb).

- p.d. = the energy transferred by 1 C of charge between two points
- p.d. = work done by 1 C of charge moving between two points.

Checkpoint 1

If the 10 Ω resistors were replaced by a 4 Ω and a 2 Ω resistor, what would the new PD across each resistor be now?

Watch out!

EMF is a misnomer! This phrase is very misleading. It suggests that the cell pushes along the electrons. Instead, think of energy changes. Chemical energy is converted into electrical energy (i.e. the KE of the electrons).

Action point

Study this spread then read pages 66–7. Now produce a concept map for both spreads. Put the word *energy* in a box in the centre of a piece of paper. Write down as many important, relevant words as you can. Link up the words that are connected in some way and then jot down their connection on the line you have drawn to join them.

Checkpoint 2

Calculate the amount of energy transferred to 15 C of charge as it passes through a cell of 4 V.

Internal resistance *r*

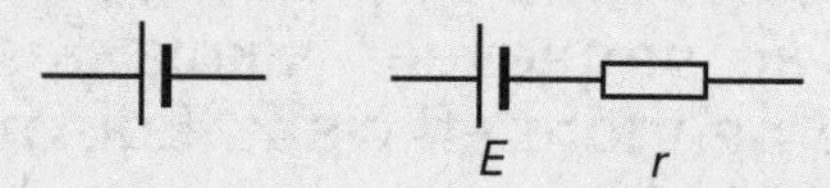

From now on we will replace our old cell symbol with the more accurate one shown on the right. As electrons go through a cell, energy is transferred to them from chemicals in that cell. The chemicals also provide some resistance to the movement of the electrons. This is called **internal resistance**. Power packs (transformers) also have internal resistance as their wires have electrical resistance.

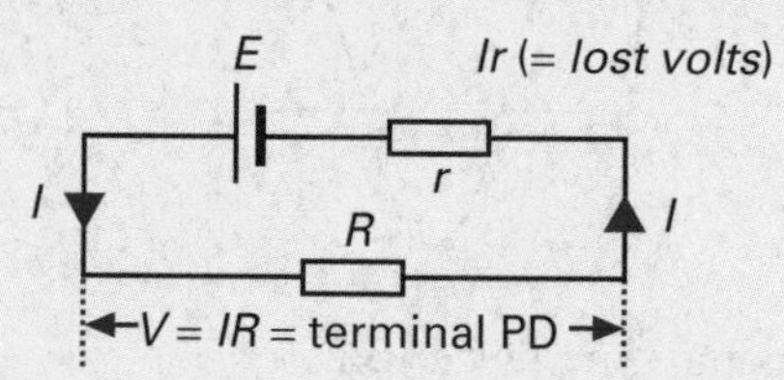

Connecting a voltmeter across a cell tells us about the energy gain provided by the EMF E minus the energy loss due to the internal resistance r. There is no way to separate the internal resistance from the EMF, and so there is no way to measure the EMF directly. To measure E and r indirectly, we apply Kirchhoff's second law to the circuit above:

$E = IR + Ir$ or $E = V + Ir$

$V = E - Ir$ or $V = -Ir + E$

A modified version of the circuit shown above is used in reality.

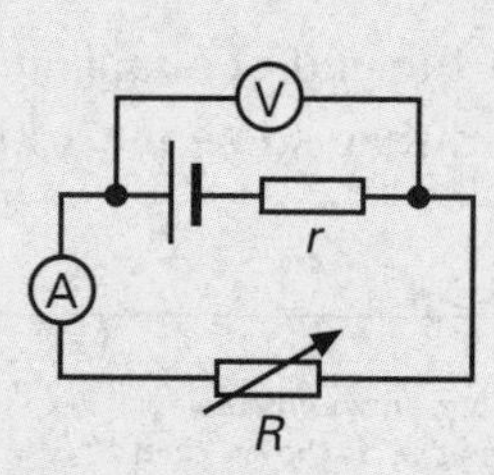

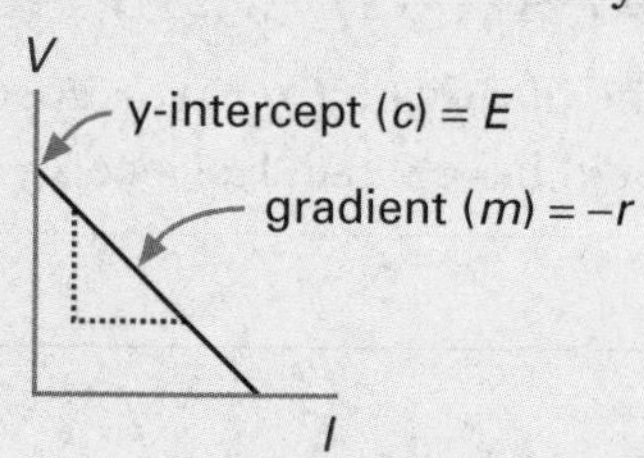

The resistance R is varied so that a series of values for terminal p.d. V and current I can be recorded using a voltmeter and ammeter as shown. If we compare our equation for this circuit ($V = -Ir + E$) with the equation for a straight line ($y = mx + c$), we can see that a graph of V against I will be a straight line, as shown above.

- Electrons lose a lot of KE to r if a big I flows through the cell.
- Lots of energy transferred to r means less will be available for the external circuit so terminal p.d. will fall (conservation of energy). For example, if a driver tries to start a car with the lights on, the starter motor will take a large current from the battery, the battery's terminal p.d. will fall and the car lights will dim.
- To transfer maximum energy to the external circuit, keep $r \ll R$.

Exam question answer: page 83

(a) Describe the energy transfers that occur in a series circuit of your choice.

(b) Why is the term *voltage* not particularly useful? Suggest improvements.

(c) Explain the condition for maximum energy transfer from a source of EMF to an external resistor. (20 min)

The jargon

The p.d. caused by the internal resistance is called the *lost volts*.

The jargon

The p.d. dropped between one terminal of the cell and the other is called the *terminal PD*.

Checkpoint 3

Why is there no way to separate the internal resistance from the EMF of a cell?

Examiner's secrets

Check your syllabus to find out what are the mathematical requirements for your course.

Checkpoint 4

Why does the terminal p.d. of a cell depend on the circuit the cell is connected to?

"The scientist is free, and must be free to ask any question, to doubt any assertion, to seek for any evidence, to correct any errors."

J. Robert Oppenheimer

Checkpoint 5

If you measure the EMF of an old non-rechargeable 1.5 V cell, it may still read close to 1.5 V. Try to use it, however, and you may find that it does not work. How can you explain this? (*Hint* what actually happens to a cell when it runs down?)

Alternating currents

Check the net

To find out more about the fight for AC go to www.parascope.com/en/0996/tesla2.htm

Checkpoint 1

What would be the benefit of using a cathode-ray oscilloscope (CRO) rather than the conventional voltmeter shown in the circuit on the right?

Checkpoint 2

Use $T = 1/f$ to find the time period T of an alternating supply of frequency f = 50 Hz.

Links

For more on power and AC see page 67.

The jargon

The *root-mean-square* (RMS) value of an alternating current, I_{RMS}, is the direct current that delivers the same average power as the alternating current. The RMS value is also known as the *effective* value.

Action point

Why $\sqrt{2}$?
Peak power is IV. Power varies between IV and zero, so average power = $IV/2$. Can you see that $IV/2$ is the same as $\frac{I_{RMS}}{\sqrt{2}} \times \frac{V_{RMS}}{\sqrt{2}}$?

Checkpoint 3

Explain, in a sentence, the benefits that come from using AC and transformers in the distribution of electricity.

Thomas Edison, the inventor of the electric light bulb was a very wealthy man at the head of a DC-based empire when an employee, Nikola Tesla, suggested that AC would be more efficient. Edison invented the electric chair and electrocuted cats and dogs to show how dangerous AC was.

Alternating current (AC) and voltage

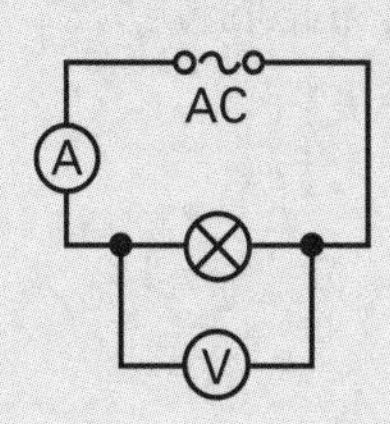

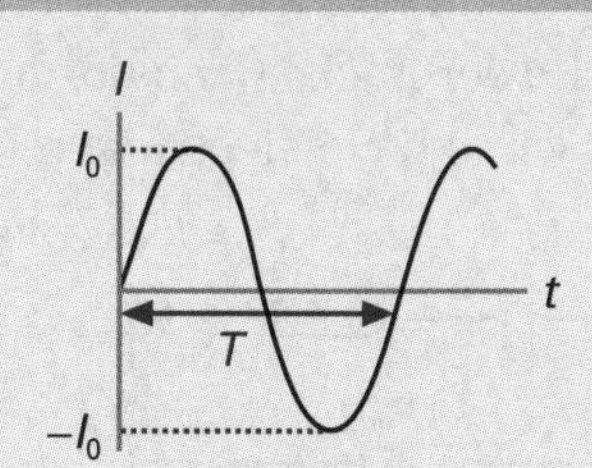

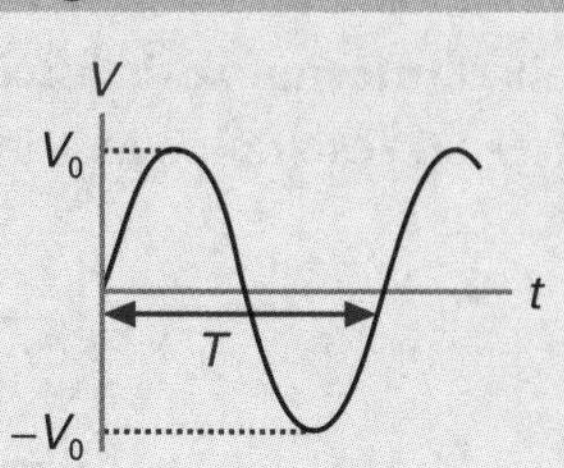

- AC is constantly changing direction.
- Frequency of mains AC is 50 Hz (changes direction 50 times/s).
- AC varies sinusoidally, $I = I_0 \sin 2\pi ft$, where f = frequency.
- In simple circuits, I and V vary in step or in phase.
- From the second graph, we can see that energy delivered varies.
- You cannot see a mains-operated bulb flickering as the energy to it varies, because 50 Hz is quite fast. The filament does not have time to cool before the next delivery of energy arrives.

Power

DC has the disadvantage that a customer at the end of a long line of users would have much less energy delivered than someone at the start.

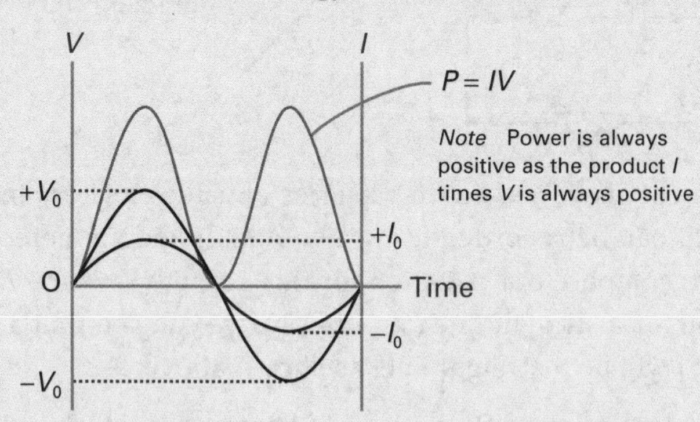

- AC power varies, but the average AC power = $I_{RMS} \times V_{RMS}$.
- The root-mean-square value of AC $I_{RMS} = I_0/\sqrt{2} \approx 0.707I_0$.
- $V_{RMS} = V_0/\sqrt{2} \approx 0.707V_0$.

Why does mains electricity use AC?

- Power loss from current-carrying wires = I^2R; i.e. $\propto I^2$.
- Power delivered = IV. To deliver sufficient power and minimize power loses, electricity is sent at high values of V and low values of I.
- Electricity is transmitted at up to 440 kV. This can be *stepped down* to 33 kV for heavy industry, 11 kV for light industry or 240 V for homes, using transformers that work on AC.
- Large transformers are approximately 99% efficient.

Rectification

Many electronic devices only work on DC. **Diodes** allow current to flow in one direction only. They can be used to convert AC into DC.

Half-wave rectification

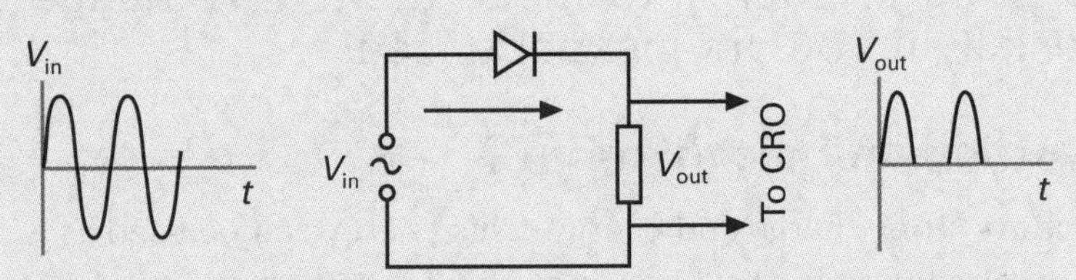

Diodes allow conventional current to flow through them only in the direction of the red arrow shown above. When the polarity of V_{in} changes, i.e. on the negative half of the V_{in} cycle, no current flows.

Full-wave rectification

Bridge rectifier circuits change AC to DC and let current flow all the time.

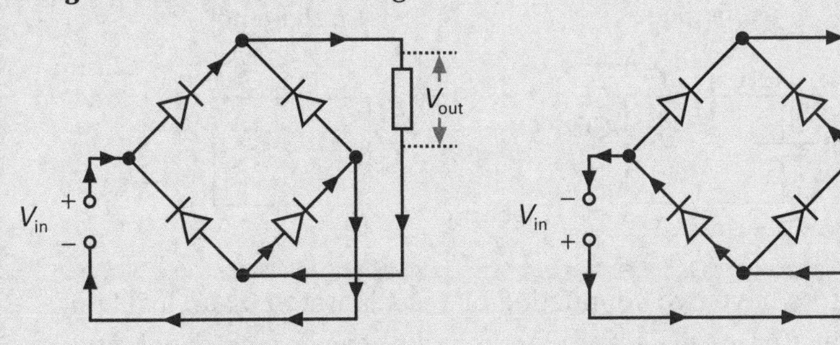

→ Current always flows through the resistor in the same direction.

→ AC is converted (rectified) into DC.

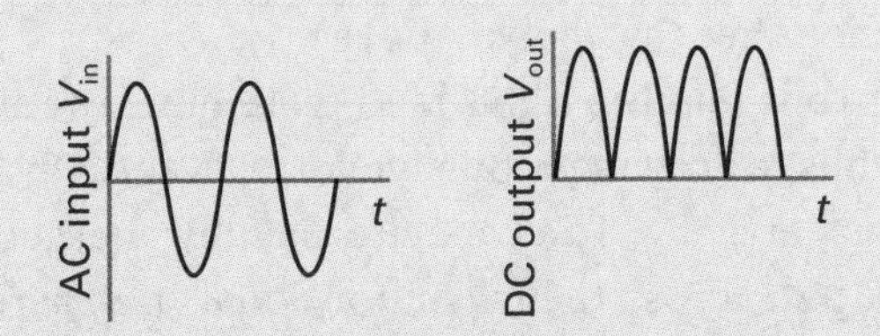

The output voltage has a ripple. This can be corrected (**smoothed**) by placing a large-value capacitor in parallel with the resistor. The rate at which capacitors discharge in a circuit like this is governed by the values of both the resistance R and the capacitance C. If the product RC is big, the capacitor discharges slowly so that the ripple disappears.

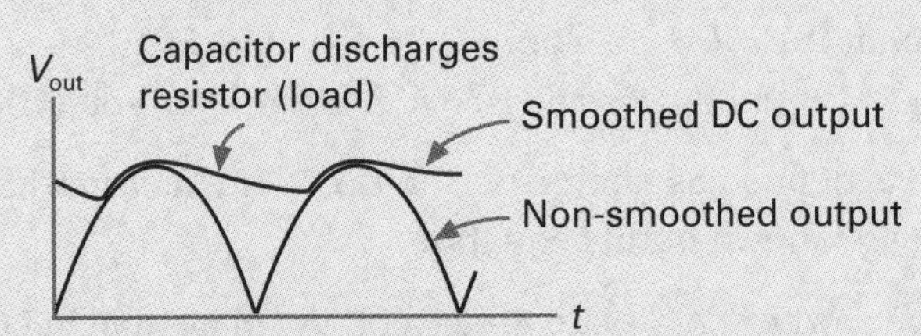

Exam question answer: page 84

(a) Why is it impractical to use DC electricity for large-scale electrical distribution?

(b) Draw a simple circuit to show how half-wave rectified AC can be smoothed. Explain how the smoothing effect takes place. (20 min)

Links

For more on electronics, see the *electronics options*, pages 164–9.

The jargon

Rectification means changing AC into DC. As with many words in physics, rectify comes from Latin. *Rectus* 'straight' and *facere* 'to make'; i.e. to make straight.

Checkpoint 4

What is the difference between AC that has undergone half-wave rectification as opposed to full-wave rectification?

Action point

Copy a diagram of a bridge rectifier circuit. On one circuit diagram, add arrows to show how conventional current flows during the positive half cycle of the input AC (shown on the left, opposite) and then use a different colour to show what happens during the negative half cycle.

Links

For more on *capacitors*, see pages 76–7.

Capacitors

Words of wisdom for any budding electrician includ 'Always keep one hand in your pocket' (to avoi letting current flow through your heart by touchin a live wire with both hands) and 'Never take th back off a TV'. Televisions contain large capacitors tha hold enough energy to make them very dangerous especially if you are inexperienced!

Charging and discharging

Capacitors store charge and energy. This energy can be used as a backup for computers, as a power supply for camera flash bulbs, etc. Anything that can store charge is a capacitor. When, in the 18th century, Stephen Gray hung a workhouse boy from silk threads and connected him to a static electric charge, the boy acted as a capacitor. (Gray wanted to show that the boy could lift pieces of paper from the ground beneath him using electrostatic attraction!)

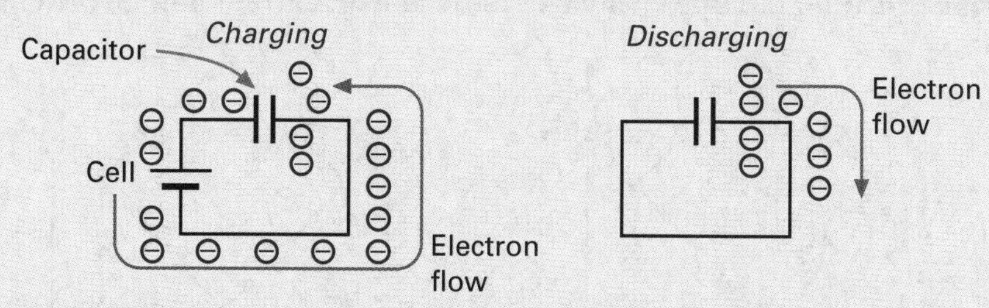

- Capacitors have two conducting plates separated by an insulator.
- The insulator, or dielectric, can be polythene, waxed paper, etc.
- To save space, capacitors are usually rolled up like a swiss roll.
- When charging, the cell pulls electrons off one plate and pushes them on to the other (as shown above).
- Initially, electrons join the right-hand plate quickly (high current).
- As the plate fills, a growing repulsive force eventually stops any more electrons arriving. The capacitor is fully charged. No current.
- At this point, p.d. across the capacitor equals p.d. across the cell. (The clockwise force equals the anticlockwise force.)

Capacitance

As the p.d. across a fully charged capacitor equals the cell p.d. V, the maximum charge that can be stored Q is proportional to V. So,

$Q = \text{constant} \times V = CV$ where C = capacitance

$C = Q/V$ (in units, 1 farad (F) = 1 coulomb per volt ($\mathrm{C\,V^{-1}}$))

Capacitance is defined as charge per unit p.d. The factors that affect capacitance are shown in this equation:

$$C = \frac{\varepsilon_0 \varepsilon_R A}{d}$$

where A = plate area ($\mathrm{m^2}$), d = plate spacing (m), ε_0 = a constant, the permittivity of free space ($\mathrm{F\,m^{-1}}$) ε_R = relative permittivity (no units)

Two or more capacitors are often connected to give the required capacitance. The following equations are used to calculate the combined capacitance C_T of three capacitors (C_1, C_2 and C_3):

In series $\frac{1}{C_T} = \frac{1}{C_1} + \frac{1}{C_2} + \frac{1}{C_3}$ *In parallel* $C_T = C_1 + C_2 + C_3$

Watch out!

Charge is a quantity of electricity, carried by electrons, *not* a force.

The jargon

A *dielectric* is an insulator, placed between the plates of a capacitor so that it can store more charge.

The jargon

Permittivity of free space, $\varepsilon_0 = 8.85 \times 10^{-12}\ \mathrm{F\,m^{-1}}$.

The jargon

The *relative permittivity* ε_R of a dielectric tells us how much more charge can be stored with the dielectric inside a capacitor, compared with empty space between its plates. ε_R of mica is 7, so a mica-filled capacitor can store seven times as much charge than a similar vacuum-filled capacitor. Its capacitance increases by a factor of 7.

Checkpoint 1

Calculate the capacitance of a pair of parallel plates, each measuring 0.2 m × 0.1 m, when they are separated by an air gap of 3 mm. (ε_R of air is 1.0005.)

Checkpoint 2

Calculate the combined capacitance of three 50 μF capacitors connected in series.

The energy stored in a capacitor

Capacitors store energy because work is done to push electrons on to their plates. Look at this graph of charge Q against p.d. V:

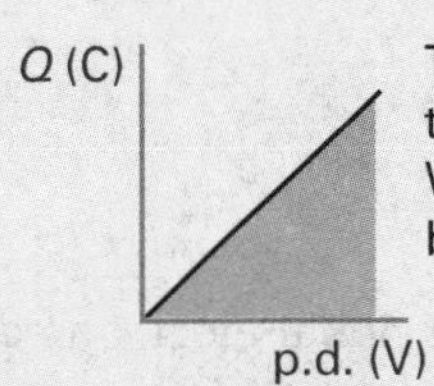

The energy stored by a capacitor equals the area under the graph. When it is charged to Q coulombs by a p.d. of V volts, then

energy = $\frac{1}{2}QV$

We can substitute $Q = CV$ in $E = \frac{1}{2}QV$ to give $E = \frac{1}{2}CV \times V = \frac{1}{2}CV^2$.
We can substitute $V = Q/C$ in $E = \frac{1}{2}QV$ so $E = \frac{1}{2}Q \times Q/C = \frac{1}{2}Q^2/C$.

> **The jargon**
>
> The *working voltage* of a capacitor is the maximum voltage that should be applied across it.

> **Checkpoint 3**
>
> Calculate the maximum energy that can be stored in a 10 000 μF capacitor when a PD of 20 V is applied across its plates.

The mathematics of charging and discharging

When a capacitor and a resistor are connected in series, the time for the capacitor to charge or discharge increases as the current in the circuit decreases. The circuit below is used to study charging and discharging.

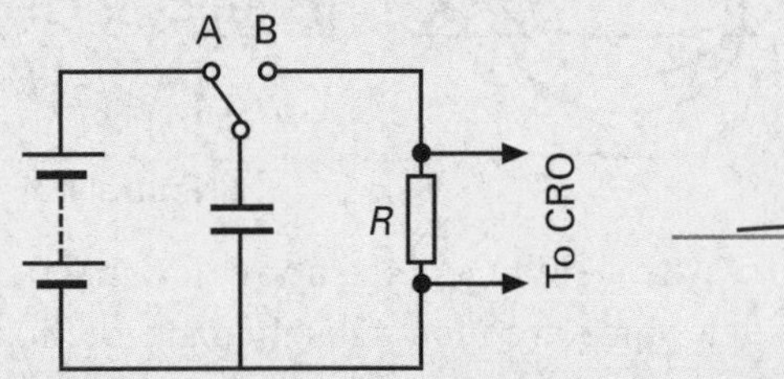

CRO trace

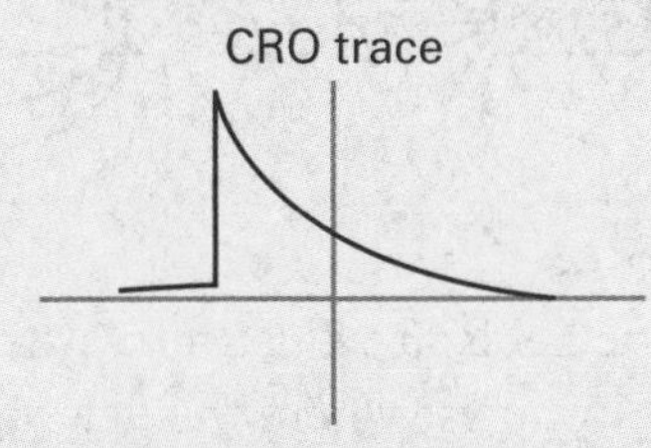

- → With the switch as shown, the battery charges the capacitor.
- → If the switch is flicked to B, the capacitor discharges through R.
- → A CRO trace is used to show how the p.d. across R changes.
- → As $V \propto I$ (remember $V = IR$), the trace also shows how I changes.
- → The shape of the curve is exponential.
- → The time taken for the current to halve (e.g. to go from maximum current I_0 to $I_0/2$ or from $I_0/2$ to $I_0/4$) is always the same.
- → The time constant RC (resistance × capacitance) is equal to the time taken for the charge to fall to 1/e of its initial value.
- → As $Q = It$, $I = Q/t = Q/(RC)$.
- → The charge Q on a capacitor at time $t = Q_0 e^{-t/(RC)}$, where Q_0 is the initial charge on the capacitor.

> **Checkpoint 4**
>
> What advantage does a CRO have, in comparison with a normal voltmeter, when used to study the charging characteristics of a capacitor?

> **Checkpoint 5**
>
> Why does the time for a capacitor to charge increase as the current in a circuit decreases?

And for the non-mathematically minded . . .

The rate of charge leaving from, or arriving on, a capacitor depends on how much charge is already there. Pushing more electrons on to a partially charged capacitor is harder than putting them on an uncharged capacitor. When a capacitor is discharging, the strong repulsive force provided by lots of electrons means that the exodus of electrons (i.e. the current) is quickest when the capacitor is fully charged. Charging and discharging are exponential changes.

> **Examiner's secrets**
>
> Capacitors deliver charge much the same as batteries and cells, but with one important difference: they cannot provide a constant current. You should learn this difference.

> **Exam question** answer: page 84
>
> (a) Why is the discharging of a capacitor an exponential change?
>
> (b) Construct a table showing the mathematical similarities between discharging a capacitor and radioactive decay.
>
> (c) Define the term *relative permittivity*. (15 min)

> **Examiner's secrets**
>
> Synoptic assessment is a feature of all A2 courses. You could be asked to compare two parts of your course, perhaps springs and capacitors.

Electromagnetism

"I have little to say on M. Oersted's discovery, for I must confess I do not quite understand it."

Michael Faraday

The jargon

A *magnetic field* is the area around a magnet in which it can exert forces on magnetic materials.

The jargon

Soft magnetic materials are easy to magnetize and easy to demagnetize.

Checkpoint 1

Explain why a named *hard* magnetic material would not be used as the core of an electromagnet.

Checkpoint 2

Describe and explain the magnetic field surrounding a long, straight, current-carrying wire.

Examiner's secrets

You will lose marks if you draw magnetic field lines crossing or touching. (Think what would happen to a compass needle if they did.)

The jargon

The force produced by a permanent magnet interacting with an electromagnet, as in the last diagram, is called the *motor effect*.

It has long been known that certain rocks (lodestone, an iron-rich ore) can attract or repel each other. In 1821, Oersted discovered that current in a wire exerts a magnetic force.

Magnetic fields

Magnetic fields are produced in two ways.

- *Permanent magnets* The movement of individual electrons in their atoms causes weak magnetic fields. In ferrous materials, e.g. iron, all the weak fields combine to generate a strong field.
- *Electromagnets* This type of magnetism is temporary because the field is only produced when a current flows. *Soft* iron cores are used to increase the electromagnet's field strength.

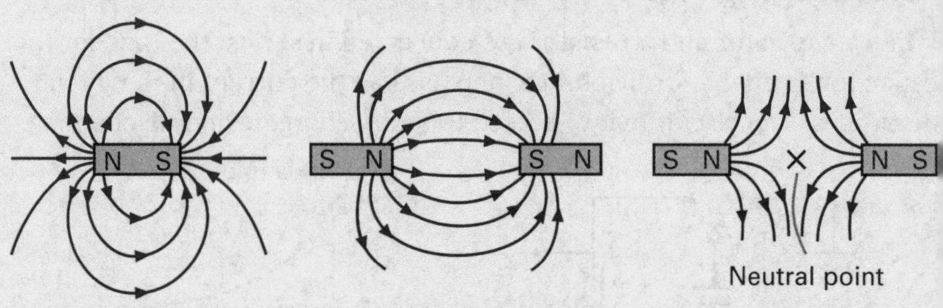

Magnetic fields are illustrated with field lines (see above). A magnetic field is strongest where the field lines are most closely packed. Most people know that a magnet is strongest at its poles, but the poles are actually a little way inside the magnet's surface.

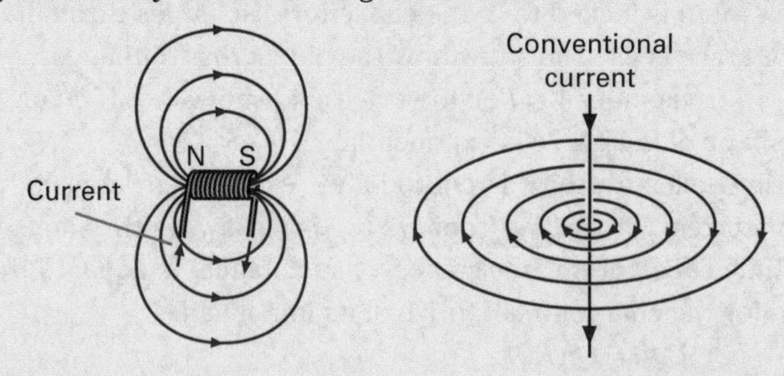

These diagrams show the magnetic fields around a current-carrying coil (a solenoid) and a current-carrying, straight wire. The field around the straight wire is weaker and changes direction if the current is reversed.

Magnetic force

Oersted discovered that when he switched on an electric current, a nearby compass needle moved. The temporary magnetic field around the current was interacting with the permanent magnet just as any two magnets would.

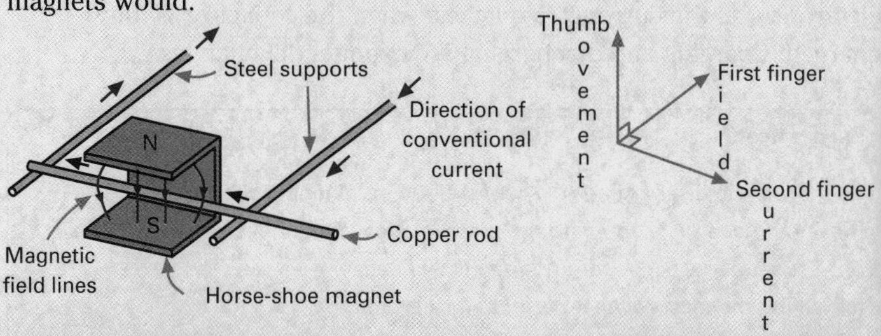

leming's left-hand rule

'he direction of the force (or movement of the copper rod) can be redicted by keeping the thumb and first two fingers of your left hand t 90° to one another, see the previous diagram. Point your first finger the direction of the permanent magnet's field (from north to south) nd your second finger in the direction of conventional current; your humb now shows the direction of the movement, or force, produced.

Magnetic flux ϕ and magnetic flux density B 

agnetic flux is an imaginary fluid that flows from the north pole of a nagnet to the south. It flows along the field or **flux** lines. B tells us how ightly packed the flux, or how strong the magnetic field, is.

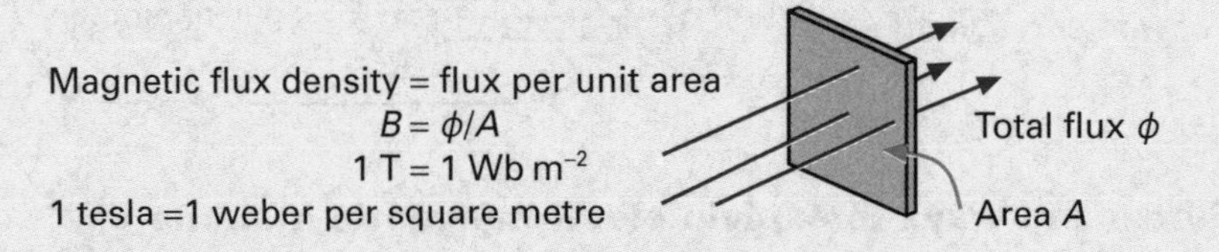

agnetic force

- $F = BIl$ — force on a wire of length l, carrying a current I.
- $F = BIl \sin \theta$ — current-carrying wire at angle θ to field.
- $F = BQv$ — force on a charge Q moving with speed v.
- $F = BQv \sin \theta$ — charge Q moving at angle θ to field.

efinition of magnetic flux density B

$F = BIl$ so $B = F/Il$

magnetic field has a strength of 1 tesla (1 T) if it exerts a 1 N force on a onductor 1 m long, carrying a current of 1 A at right angles to the field.

leasuring and calculating magnetic flux density B

lall probes can measure B. They contain a semiconductor, across which potential difference is established when the probe is placed in a nagnetic field. A stronger magnetic field will produce a larger potential lifference. Pre-calibrated probes convert this potential difference into reading of B. Just hold the probe so that the magnetic field lines are assing at 90° through its flat face. Equations for calculating B include:

$B = \mu_0 nI$ at the centre of a long solenoid
n = number of turns per metre on the solenoid

$B = \mu_0 I/2\pi r$ at a small distance r from a long straight wire

Exam question answer: page 84

(a) Explain how the direction of magnetic flux lines can be determined.

(b) Use a diagram to explain what a *neutral point* between two magnets is.

(c) Describe an experiment to study force on a current-carrying conductor in a magnetic field.

(d) How can the direction of the magnetic force on a charge moving through a magnetic field be predicted? Why does the charge follow a circular path? (35 min)

Checkpoint 3

Use Fleming's left-hand rule to predict the direction in which the copper rod in the diagram at the bottom of page 78 will roll.

Examiner's secrets

One of the strangest, and most frequent, sights in physics exams are row upon row of candidates grappling with Fleming's left-hand rule. Learn this rule! It crops up a lot!

The jargon

Flux means a flow or discharge. Think of an imaginary fluid that flows from north to south poles.

Speed learning

Try using flow diagrams to summarize your notes, e.g. the motor effect:

[magnetism] + [electricity] → [movement]

Checkpoint 4

Rearrange any of the equations for magnetic force to make B the subject of the equation. Is B a vector or a scalar quantity? Give a reason for your answer.

Check the net

Take a virtual tour of the Royal Institution, where Michael Faraday developed his ideas, at www.ri.ac.uk/

The jargon

μ_0 is a constant, the *permeability of free space*. $\mu_0 = 4\pi \times 10^{-7}\ \text{H m}^{-1}$.

Watch out!

Magnetic flux density and magnetic field strength are different names for the same quantity (B). Don't be fooled by this!

Electromagnetic induction

When Michael Faraday discovered electromagnetic induction he paved the way for electric transformers and generators, and for the whole electrical industry.

Take note

Michael Faraday was appointed at the Royal Institution only after a fight in the main lecture theatre led to the dismissal of his predecessor!

Check the net

To find out more about Michael Faraday, go to www.ri.ac.uk/History/M.Faraday/

The jargon

Flux linkage is defined as $N\phi$, where N is the number of turns on a coil ($N = 1$ for a straight wire) and ϕ is the amount of magnetic flux. Think of flux linkage as the amount of overlap there is between the conductor and the flux lines. If this amount changes, the flux lines must be being cut and an EMF will be induced.

Checkpoint 1

Rephrase Faraday's law of electromagnetic induction using the phrase *flux linkage*.

Checkpoint 2

Read the formal statement of Lenz's law again and then state what change would be observed if the magnet was pushed into and then pulled out of the coil, as shown above (right).

Generating electricity

If the following equipment is set up, we can show that:

- if either the wire or magnet moves vertically, a current is induced
- the current direction reverses if the wire moves up not down
- doubling the speed of movement doubles the current induced

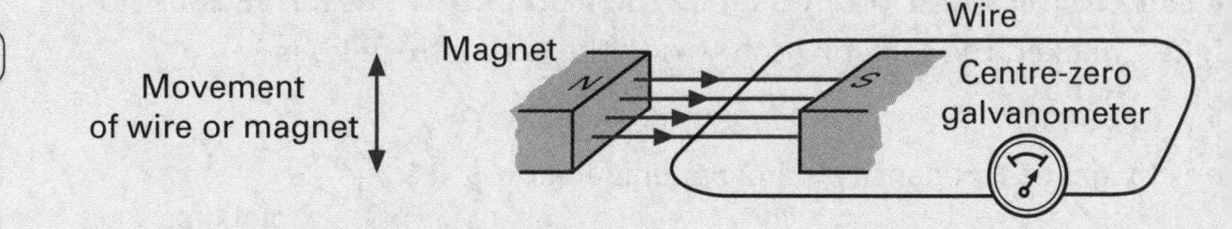

Short-hand ways to explain electromagnetic induction

- An EMF is induced when a conductor cuts magnetic flux lines.
- An EMF is induced when the amount of flux linkage changes.
- When the conductor forms part of a complete circuit, the induced EMF can cause an induced current to flow around the circuit.
- magnetism + movement $\rightarrow$ electricity

Laws of electromagnetic induction

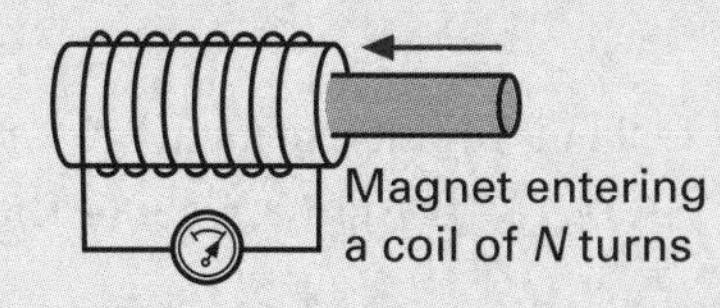

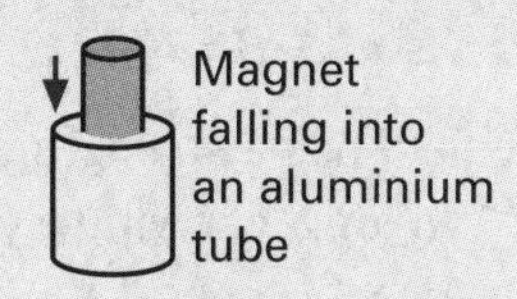

The experiment above should remind you of the first experiment on this page. An EMF is induced when there is relative movement between the coil and magnet. So, it does not matter whether the magnet or the coil moves. Faraday found that:

- the magnitude of the EMF (E) induced in a conductor is proportional to the rate at which magnetic flux is cut by the conductor, i.e.

$$E \propto d\phi/dt$$

The set-up shown above (right) is used to demonstrate Lenz's law. When the magnet is dropped it travels very slowly. Not as much of its potential energy is converted into kinetic energy as you might expect. Some is changed into electrical energy as an EMF is set up in the tube. Stated formally, **Lenz's law** is:

- the direction of the induced EMF is such that it opposes the change that caused it (in this example, the EMF is directed up)

Both laws can be combined in one equation:

$$E = -d\phi/dt$$

Or, for a coil consisting of N turns of wire:

$$E = -N\, d\phi/dt \qquad \text{(minus signs represents Lenz's law)}$$

Explaining electromagnetic induction

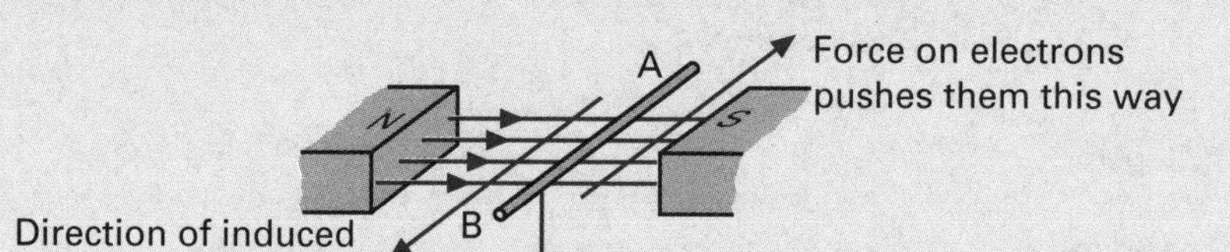

- A straight wire is shown falling through a magnetic field. Therefore the free electrons in the wire are moving down too.
- Free electrons move in the opposite direction to conventional current, and so the direction of conventional current must be up.
- We know the direction of the magnetic field and the conventional current, and so we can use Fleming's left-hand rule to predict the direction of the resulting magnetic force. It pushes the electrons within the wire from B to A.
- So, a conventional current has been induced (persuaded) to flow in the wire from A to B.

Transformers

Before the National Grid was established in the 1930s, each major town generated its own electricity. Nowadays, power stations generate electricity that is then stepped up to be distributed at hundreds of thousands of volts, as this wastes less energy. A network of electricity lines criss-cross the country rather like a spider's web. The voltages are then stepped down to different voltages to meet the needs of various users. Transformers step voltages up and down.

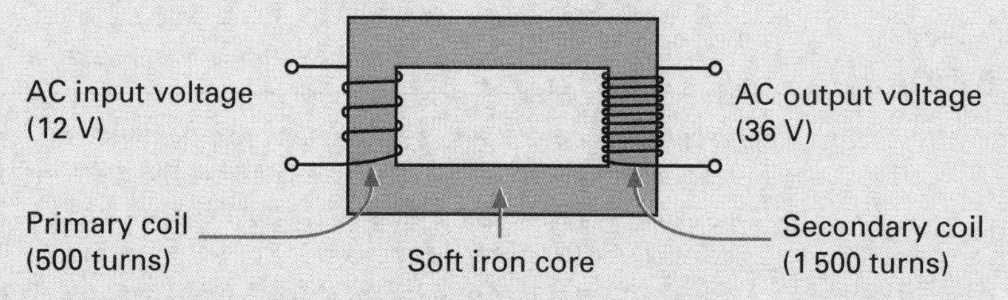

The alternating current in the primary coil produces a magnetic field that grows as the current increases and shrinks as the current value falls. The field lines associated with this changing magnetic field move forwards and backwards across the secondary coil. As we have relative movement between a magnetic field and a conductor, a current is induced in that conductor, i.e. the secondary coil. The soft iron core is easy to magnetize and demagnetize so it lets the magnetic field grow and shrink more easily, making the transformer more efficient.

Exam question answer: page 84

(a) Why is Lenz's law an example of conservation of energy?

(b) Thomas Edison wanted to distribute electricity as DC. He even electrocuted some cats and dogs to show that AC was too dangerous. Explain why AC is a good choice for electrical distribution.

(c) Explain how transformers work in terms of magnetic flux linkage. (25 min)

Checkpoint 3

When does the country's demand for electricity surge rapidly? How is this sudden demand satisfied?

Checkpoint 4

Would it make more sense to move the magnet or the conductor in a power station?

Examiner's secrets

Check whether your syllabus requires you to know about transformers.

Links

See pages 74–5 for more on alternating currents.

Action point

The following equation applies for ideal transformers: $V_1/V_2 = N_1/N_2$ where V_1 and V_2 are input and output voltages, and N_1 and N_2 are the number of turns on the primary and secondary coils. Use this equation to check that the figures quoted in the diagram on the left are correct.

Action point

You should know why transformers do not work with D.C.

The jargon

A *step up* transformer increases the voltage between the primary and the secondary coils. A *step down* transformer does the opposite.

Answers
Electricity and electromagnetism

Current as a flow of charge

Checkpoints

1. Electrons are moved, then remain *stat*ionary in *static* electricity. *Electr*ons flow like a river in current *electr*icity.
2. $Q = It = 12 \times 3 \times 60 \times 60 = 129\,600$ C
3. Resistance of an ideal ammeter is 0 Ω to avoid changing the current in question.

Exam question

(a) $t = Q/I = 4.68 \div 26 \times 10^{-3} = 180$ s

(b) Normally a conductor would contain many more free electrons than a semiconductor. If energy is transferred to the semiconductor (e.g. by heating a thermistor) many more free electrons can be generated.

(c) $v = I/nAe = 7 \div (10^{29} \times \pi r^2 \times 1.6 \times 10^{-19})$
$= 7 \div (10^{29} \times \pi r^2 \times 1.6 \times 10^{-19}) = 5.6 \times 10^{-4}$ m s^{-1}

Current, p.d. and resistance

Checkpoints

1. (i) The wires have no resistance so the graph is flat in sections that relate to the wires as no energy is being transferred.
   (ii) The electrons do not experience any resistance when flowing through the chemicals in the cell. The EMF of the cell is quoted as 6 V but the electrons would lose some of the energy they pick up just because of the cell's internal resistance.
   (iii) Current can flow, even when the circuit has been disconnected.
   (iv) Electrons give up all their energy before returning to the cell, otherwise the bulbs would get increasingly bright.
2. An ideal voltmeter has infinite resistance.
3. Insulators have very high resistances.
4. If one component in a parallel circuit blows only the arm containing that component will be affected.

Exam question

(a) In the water-pump model the cell is represented by a water pump, wires = water pipes, electrons = water, ammeter = water meter, voltmeter = pressure gauge, resistor = pipe narrows. One disadvantage of this model is that it suggests that only cross-sectional area affects the resistance of a wire, ignoring the effects of resistivity (i.e. the material), length, temperature and impurities.

(b) If you tried to count the number of passengers entering the Underground, slowing them down in any way could cause a queue that might put other travellers off thereby reducing the number of passengers using the service. An ideal ammeter would have zero resistance to achieve a 'God's-eye view'. Current could be checked without changing it, just as God is believed to be able to watch over the Earth without intervening in the actions of humans.

Resistors and resistivity

Checkpoints

1. $\rho = RA/l = \Omega$ m^2/m $= \Omega$ m
2. Ω^{-1} m^{-1}. Conductivities could be given in a table as conductivity is a property of materials not individual samples.
3. If the resistors are connected in series, electrons pass through all three, so the total resistance is 6 Ω. If the resistors were joined in parallel, each electron would only pass through one resistor so the total resistance would be less.

Exam question

(a) The graphs are on page 64. A metallic conductor, at constant temperature, is characterized by a straight line through the origin so current is proportional to p.d. This is a statement of Ohm's law. A metallic conductor, at constant temperature, is known as an ohmic conductor. The filament lamp graph is non-linear; a lamp is not an ohmic conductor. As current rises, the filament gets hotter, its resistance increases and the graph gets flatter. A thermistor gets hotter when more current flows. This causes it to release more free electrons; its resistance decreases. Diodes only allow a significant current to flow in one, 'forward' direction. Diodes have high resistances in the 'reverse' direction.

(b) Resistivity $\rho = RA/l$, so we must measure each of these values. The cross-sectional area A of a wire can be deduced after measuring its diameter with a micrometer screw gauge. NB $A = \pi r^2$ and r = diameter/2. The length of the wire can be measured with a metre rule and a multimeter, on its ohms range, can measure resistance.

(c) Superconductors are used to make superconducting magnets that produce exceptionally stable magnetic fields used in MRI scanners. Frictionless bearings are made possible by lowering a superconductor onto a magnet. The magnet induces a current in the superconductor, large enough (because the superconductor has zero resistance) to produce a magnetic field that is repelled from the original magnet so levitation (floating) is achieved. Superconducting power cables could transfer energy without power losses as there would be no heating effects in the cables.

Electrical energy and power

Checkpoints

1. The more powerful hairdryer can change electrical energy into other forms more quickly, and so your hair would dry faster.
2. $P = VI = 440 \times 10^3 \times 100 = 44$ MW. We are given the potential produced by the power station, not the p.d. across the section of wire involved. So we must use $P = I^2R$ not $P = VI$. $P = I^2R = 100^2 \times (20 \times 0.2) = 40$ kW.

3 $I = P/V = 4.2$ A, so use a 5 A fuse.
4 Electrical energy cost = kW h × cost of 1 kW h
= 5 × 2.5 × 10 p = £1.25

Exam question

(a) 0.1 kW × 24 h × 10 p = 24 p
(b) $E = I^2Rt = 25 \times 20 \times 60 = 30$ kJ
(c) $E = V^2/R \times t = 2 \times (9/15) \times 10 \times 60 = 720$ J

Kirchhoff's laws

Checkpoints

1 15 A
2 Because the EMF of the 20 V cell is greater than the EMF of the 5 V cell.
3 As a walker goes around a hill returning to the starting point, the walker's height above sea level might change. However, his/her final height above sea level (and so his/her PE) remains unchanged. As an electron goes round a circuit its potential also changes. As with the walker, when the electron returns to its starting point in the cell, its potential will once again be the same.

Exam question

(a) Σ EMF = ΣIR so 12 = 0.5R + (10 × 0.5) and R = 14 Ω
(b) Label the circuit with A,B,C,D,E and F as in the example on page 69. Applying Kirchhoff's first law at junction C gives $I_1 = 1 + I_2$ (eqn 1) so $I_3 = I_1$. Applying Kirchhoff's second law to loop ABEF gives $6 = 6I_2$ so $I_2 = 1$ A. Substituting this value in eqn 1 gives $I_1 = 1 + 1 = 2$ A. Applying the second law to ABCD gives 6 = 4 + (? × 1) so ? = 2 Ω.

Potential dividers and their uses

Checkpoints

1 $V_{out} = V_{in} \times (R_1/R_1 + R_2) = 8$ V
2 Thermistor is similar to thermometer; they are both associated with temperature.
3 A sensor (e.g. a thermistor), automatic switch (e.g. a transistor) and an actuator (e.g. a heater) are required.
4 The circuit is practical and cost effective as it senses and alerts automatically. As the current in the bell does not flow through the sensitive transistor, a more powerful bell, perhaps forming part of a completely separate circuit, can be used.

Exam questions

1

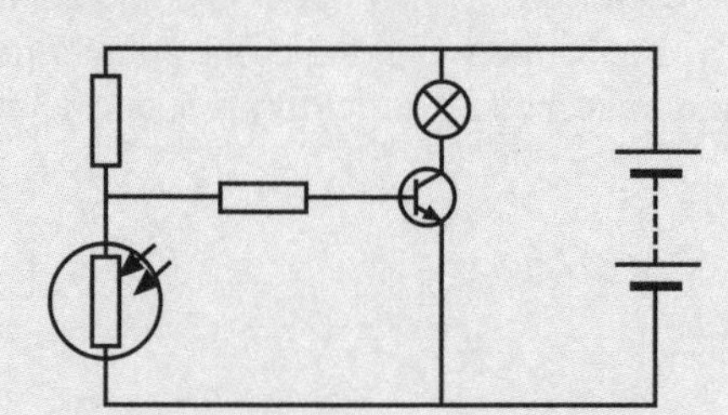

In daylight, the LDR has a low resistance and therefore takes a low share of the battery's 'voltage'. The transistor and bulb are off. As it gets dark, the LDR's resistance and share of the battery's 'voltage' increase. The transistor is switched on so that current can flow through the bulb. The base resistor stops dangerously large currents flowing into the transistor.
2 Current = p.d./total resistance
= $10 \times 10^{-3}/20 \times 10^3$
= 0.5×10^{-6} A
p.d. = $I \times R$
= $0.5 \times 10^{-6} \times 10 \times 10^3$
= 5 mV

EMF and internal resistance

Checkpoints

1 4 V across the 4 Ω resistor and 2 V across the 2 Ω resistor.
2 60 J
3 An EMF is produced as electrons travel through chemicals in a cell but the chemicals simultaneously provide resistance to the motion of the electrons.
4 The resistance of an external circuit affects the speed of electrons. This in turn affects the internal resistance of the cell and therefore the terminal p.d. of the cell. (See page 73 for more on this question.)
5 As a dry cell 'runs down' the internal resistance increases. Measuring the open circuit terminal p.d. will not indicate the extent of this increase as no current flows. As soon as a current in drawn from the cell, these will be 'lost volts' across the internal resistance which reduces the output p.d.

Exam question

(a) A series circuit consists of a cell and a bulb. Chemical energy is transferred into electrical energy as electrons pass through the cell. As they go around the circuit some of their energy is transferred to the wires, heating them. A lot of their energy is transferred to the bulb, heating it so much that it radiates light energy. Not all the electrons' energy is given to the external circuit as the cell has internal resistance. The cell can get warmer as a result of this energy transfer.
(b) Voltage can refer to electrons gaining or losing energy. EMF and p.d. are better as EMF is used when electrons gain energy and p.d. is used when electrons lose energy.
(c) To transfer maximum energy to the external circuit, internal resistance must be much less than external resistance so more volts drop across the external circuit.

Alternating currents

Checkpoints

1 A CRO provides a visual, analogue display that allows easy analysis of AC.
2 0.02 s

3 Transformers and AC supply electricity to a variety of users efficiently.
4 Half-wave rectification ensures that AC can only flow in one direction, but the current only flows 50% of the time. Full-wave rectification ensures that the current will flow in one direction only, all the time.

Exam question

(a) Power ($P = VI$) can be increased by increasing V or I. As electricity flows along the distribution network it loses power ($P = I^2R$), heating the cables. To reduce these losses and supply sufficient energy, electricity is distributed at high V, low I. V must be reduced to safe levels before reaching the consumers; this can only be done with AC.

(b) The capacitor charges up to a high p.d., then slowly discharges as the rectified p.d. drops. As soon as this p.d. rises again the capacitor charges once more. This means that the changes in p.d. across the load are reduced.

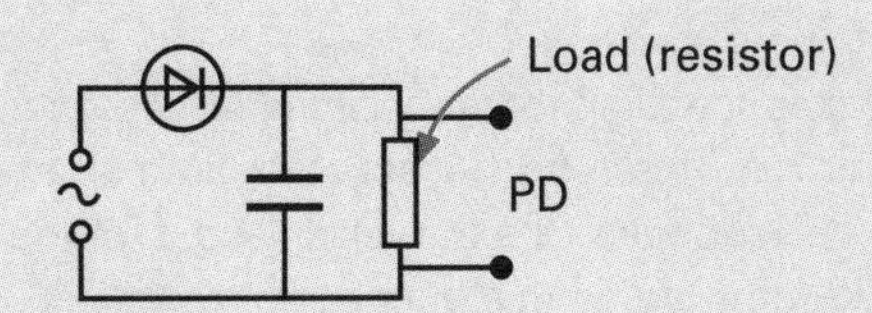

Capacitors

Checkpoints

1 5.90×10^{-14} F
2 16.7 μF
3 2 J
4 A CRO provides a visual record that makes comparison of alternating currents easy. It shows how changes occur against time.
5 Capacitors charge as more electrons reach their plates. Electrons arrive, and therefore charging occurs, slowly at low currents.

Exam question

(a) The rate of capacitor discharge at any particular time depends on how much charge remains at that time, and so it is an exponential change.

(b) Capacitors $\ln 2 = t_{1/2}/RC$ $\quad Q = Q_0e^{-t/RC}$ $\quad I = Q/RC$
Radioactivity $\ln 2 = \lambda t_{1/2}$ $\quad N = N_0e^{-\lambda t}$ $\quad A = \lambda N$

(c) Relative permittivity = (capacitance of an ideal parallel plate capacitor containing the material) ÷ (capacitance of the same capacitor with a vacuum between its plates).

Electromagnetism

Checkpoints

1 Hard magnetic materials, e.g. steel, are not used as they are hard to magnetize and demagnetize.
2 The magnetic field lines are in concentric circles. The spacing between circles increases further from the wire indicating that the magnetic field is getting weaker.
3 The rod rolls forward out of the magnet.
4 e.g. $B = F/qv$. F and v are vectors, so B is too.

Exam question

(a) Magnetic flux lines flow from north to south; i.e. the way a small, imaginary north pole would move.

(b) See page 78.

(c) A magnet is placed on top of a top-pan balance. Copper wire, connected to a low-voltage power pack, is lowered into the magnet's field. When no current flows, the balance shows only the mass of the magnet. When current flows, the force between the wire and the magnet increases or decreases the meter reading, depending on current direction.

(d) Use Fleming's left-hand rule (page 79). As the force is always perpendicular to the velocity of the charge, the particle will follow a circular path.

Electromagnetic induction

Checkpoints

1 The magnitude of an induced EMF is proportional to the rate of change of flux linkage.
2 An induced current flows in the opposite direction if a magnet is pulled out of, rather than pushed into, a coil.
3 When huge TV audiences watch events, millions of people can switch on their kettles simultaneously, e.g. for a half-time cup of tea during a World Cup final. Hydroelectricity satisfies this surge.
4 Most AC generators have stationary coils surrounding a rotating electromagnet.

Exam question

(a) When a magnet falls into an aluminium tube, some PE is changed into KE, some into electricity. As energy is conserved there is less KE than had the magnet fallen to Earth without the involvement of the aluminium tube.

(b) See page 81 and checkpoint 3 on page 74.

(c) Current is induced in a conductor when the amount of flux linkage in it changes. When AC flows through the primary coil of a transformer it has an associated magnetic field that grows and shrinks with the changing current. The flux linkage in the secondary coil changes, inducing a current.

Kinetic theory

The kinetic theory of matter should be very familiar to you. This theory pictures solids, liquids and gases as being made from particles that are constantly moving. Previously you will have used this theory to explain properties such as changes of state, gas pressure and diffusion. The kinetic theory model is used in both physics and chemistry. This section concentrates largely on gases. A more quantitative approach, as required at A-level, is developed.

Exam themes

- → *Evidence* in support of the kinetic theory, such as Brownian motion.
- → *Macroscopic properties* Using kinetic theory to explain properties like gas pressure.
- → *Quantitative expressions* Applying the theory to get expressions for pressure and temperature.
- → *Developing links* between kinetic theory and energy.

Topic checklist

○ AS ● A2	AQA/A	AQA/B	CCEA	EDEXCEL	OCR/A	OCR/B	WJEC
Temperature	○	●	●	○	●	●	●
Behaviour of gases	○	●	●	○	●	●	●
Ideal gas equations	○	●	●	○	●	●	●
Internal energy	●	●	●	○	●	●	●
Specific and latent heat capacities	○	●	●	○	●		●

Temperature

"You can't pass heat from a cooler to a hotter,
Try if you like, you far better notter,
'cause the cold in the cooler will get hotter as a ruler,
'cause the hotter body's heat will pass to the cooler."

Michael Flanders and Donald Swann

Speed learning

Breaking down words helps revision! Think of an old name for the Celsius scale, *cent*igrade. There are 100 *cents* in a dollar and 100 years in a *cent*ury. There are 100 equal divisions between the melting and boiling points of water on the *cent*igrade scale.

Action point

Redraw this chart as a scale diagram. Include the melting and boiling points of water on your chart.

Watch out!

Notice that we say 273 K, not 273 *degrees* Kelvin. A subtle difference, but physics can be subtle sometimes!

Test yourself

Learn this list, then close the book and see whether you remember every point. See if you can still remember it in a week's time.

Why is it that some objects in a room feel warmer than others? For example, assuming both objects have been at the same room temperature for some time, a wooden table feels warmer than a metal clamp stand. It is because metals are better conductors of thermal energy. They allow this energy to pass away from you more easily.

Celsius scale

The **Celsius** scale of temperature is flawed by certain problems that make it unsuitable for many scientific purposes: it is based on the melting point of pure ice and the boiling point of pure water. Unfortunately, both of these 'fixed' points change if pressure varies or if impurities are present.

The kelvin scale

Particles have energy associated with their movement. Hotter particles move faster. But what would happen if we removed all their energy? They would certainly be very cold! In fact, it has been calculated that the temperature would be –273.15 °C: **absolute zero**. This is as cold as it gets! No one has been able to achieve this temperature yet.

–273 °C (as it is normally rounded to) is quite an awkward number. We normally use the **kelvin** absolute or thermodynamic scale instead. On this scale, –273 °C equals zero degrees kelvin (0 K) and every 1 °C rise in temperature equals a rise of 1 K. A conversion chart for the kelvin and Celsius scales is shown below.

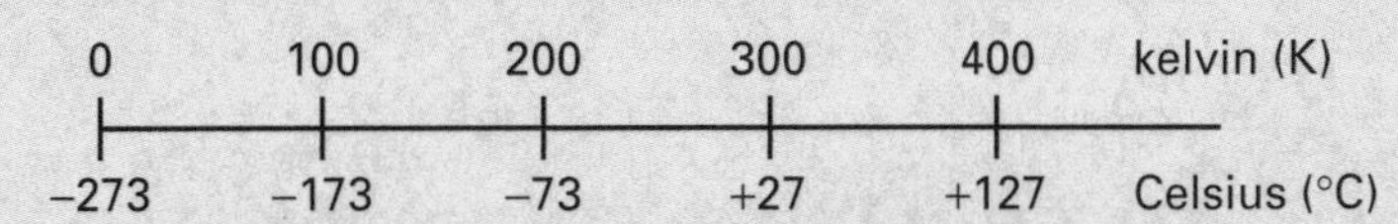

To change between temperatures in Celsius and kelvin, use these rules:

- temperature (°C) = temperature (K) – 273.15
- temperature (K) = temperature (°C) + 273.15

Finally, remember that the kelvin scale is better because one of its fixed points, absolute zero, has such great significance – it is not possible to have a lower temperature.

Properties that vary with temperature

Each of the following properties vary with temperature and provide the basis for a thermometer:

- *pressure* or *volume* of a gas, such as helium.
- *volume* of a liquid, e.g. mercury or ethanol.
- the *thermocouple* effect between metals, say copper and nickel.
- the *length* of a metal rod such as brass, e.g. an oven thermostat.
- the *resistance* of a semiconductor, e.g. silicon in a thermistor, or the resistance of a metal wire, such as platinum.
- the *magnetic* properties of a crystal like chromium potassium alum.
- the *radiation* from a hot object, like a star.

Different thermometers for different jobs

Each of these properties is useful for a different temperature range. Each property has key features that make it valuable.

- → The change produced by the temperature variation must be large enough to allow it to be measured accurately.
- → The property must always change in an entirely predictable way so that the same temperature will always result in the same reading.
- → If a property varies uniformly with temperature, the thermometer on which it will be based will be easy to calibrate. When two fixed points are established, a linear scale can be added easily.

Checkpoint 1

You have been asked to measure the core temperature of a nuclear reactor with an ethanol-in-glass thermometer. Give two scientific reasons to use an alternative.

Calibration

If you were given a capillary tube containing a mystery liquid whose volume was known to vary in a linear fashion, how could you set about adding a temperature scale to the side of your thermometer?

1. Start by measuring the height of the mystery liquid column at two fixed points, such as the melting point of pure ice and the boiling point of pure water.
2. Then plot a graph of column height against temperature; initially plot the two fixed points that we know are reproducible.
3. Now join up these points with a straight line that can be used to find the column height corresponding to any temperature between our fixed points.
4. If, however, the property varies irregularly, you will require many more calibration points on which to base your graph, or you could use a calibration equation, such as the one below.

Watch out!

Don't lose marks with silly errors on easy questions. Do the basic things well when plotting graphs – sharp pencil, label axes, appropriate scale . . .

Calibration equations

Calibration equations that have been designed to agree with observations are called *empirical*. An example is given below. To define the Celsius scale use:

$$\frac{t}{100} = \frac{l_t - l_0}{l_{100} - l_0}$$

t = unknown temperature
l_t = length of mercury (Hg) at t
l_0 = length of Hg at freezing point of H_2O
l_{100} = length of Hg at boiling point of H_2O

Checkpoint 2

What temperature t causes a mercury thread to rise to a height of 12.5 cm? l_0 = 4 cm, l_{100} = 14 cm.

Exam question answer: page 96

(a) Calculate the Celsius temperature t on a resistance thermometer using the empirical equation $\frac{t}{100} = \frac{R_t - R_0}{R_{100} - R_0}$ if $R_0 = 8.225\ \Omega$, $R_{100} = 11.450$, $R_t = 9.1\ \Omega$.

(b) Why can a baking tray feel hotter than the air surrounding it in an oven?

(c) Describe how the resistance of a thermistor changes. Refer to free electrons in your answer. (15 min)

Behaviour of gases

> *"Perfect clarity would profit the intellect but damage the will."*
>
> Pascal

What would happen if, after bumping your head, you could 'see' air? What would you see? How could you convince others to believe in your vision? Perhaps you could make use of the *Invisible Man* story. If an invisible man walked through a crowded street, he might be invisible, but his existence could be proved. How?

Brownian motion

If your audience was receptive, you might convince them that air existed by measuring the mass of a plastic bag, then re-measuring it again after blowing air into it, the difference being due to the air. Robert Brown's audience, of fellow scientists, were not easily persuaded.

In 1827, Brown watched pollen grains moving randomly in apparently still water. He concluded that the pollen grains were being bombarded from all sides by tiny, invisible water molecules.

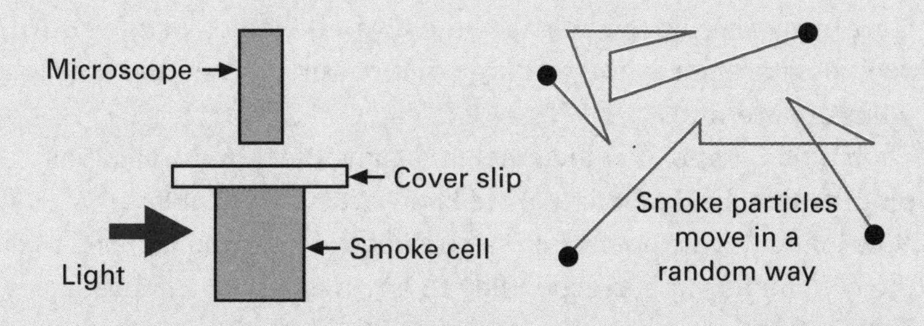

Checkpoint 1

Explain whether the observed movement of the smoke particles can be explained by:
(a) convection currents
(b) gravitational attraction between particles, or between particles and the Earth

In the updated version of Brown's work, shown above, smoke molecules (people in a busy street) are being bombarded by tiny, invisible air molecules (invisible man). The random nature of the smoke particles' motion, as shown above, provides indirect proof of the existence of air. (People falling over for no apparent reason betray the existence of the invisible man.)

It was not until 1906, when Einstein fully explained it, that many scientists finally believed in Brownian motion.

Checkpoint 2

What would happen to the motion of the smoke particles if the air was heated?

Kinetic model (as applied to gases)

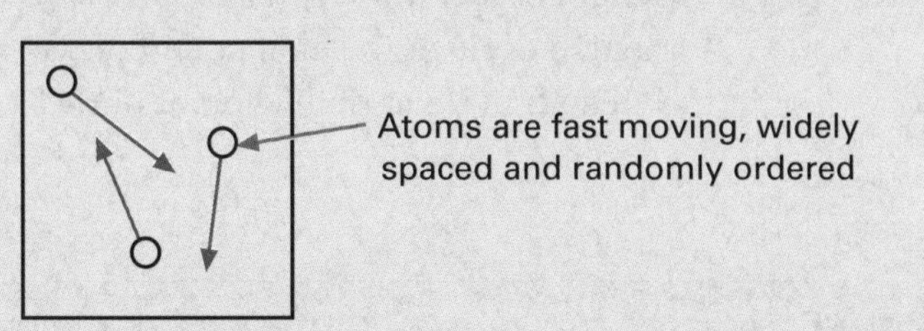

Watch out!

This model, as usual, isn't entirely accurate! It's only a 2-D representation of a 3-D effect. If the diagram had been drawn to scale, we would have been lucky to capture one atom in an entire A4 snapshot!

This theoretical model is based upon three assumptions:

- → matter is made of tiny particles
- → these particles attract each other
- → the particles are usually moving

Macroscopic and microscopic

Large-scale properties of materials are called **macroscopic** because they can be felt or measured by instruments. One of the great strengths of the kinetic theory is that it is a **microscopic** (small-scale) model that can be used to explain many macroscopic properties, like pressure and density.

Action point

Study the diagram at the bottom of page 88. It shows the kinetic theory of atoms in gases. Draw similar diagrams to illustrate how the kinetic model represents solids and liquids.

Gas pressure

We think of atoms in gases as fast moving, hard spheres. We know that they create gas pressure when they collide with each other, or with the walls of their container. Pressure is measured in pascals (Pa).

$$\text{pressure} = \frac{\text{force}}{\text{area}} \quad \text{or} \quad P = \frac{F}{A} \qquad (\textit{Note} \quad 1\ \text{Pa} = 1\ \text{N m}^{-2})$$

Action point

Rewrite these points in full sentences:

- average speed of air molecules is $400\ \text{m s}^{-1}$
- speed of sound is $330\ \text{m s}^{-1}$
- sound waves move when . . .
- and so the closeness of the two numbers is not a coincidence!

Density

Density tells us how tightly packed atoms in a material are found to be. The density of a substance is the mass of $1\ \text{m}^3$ of that substance.

$$\text{density} = \frac{\text{mass}}{\text{volume}} \quad \text{or} \quad \rho = \frac{m}{V} \qquad (\textit{Units} \quad \text{kg m}^{-3})$$

The mole and Avogadro's constant

A **mole** is just a pile of atoms or molecules. The number of particles (atoms or molecules) in the pile always remains constant: there are always 6.02×10^{23}. This number is called **Avogadro's constant**.

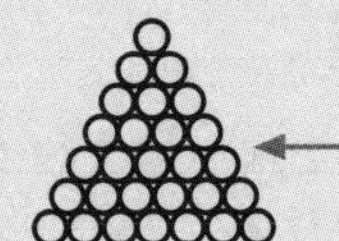

There are 6.02×10^{23} particles in this pile!

Check the net

To find out more about the extraordinary life of Pascal go to www-groups.dcs.stand.ac.uk/history/Mathematicians/Pascal.html

It can be useful to have a benchmark for comparison. Athletes often compare each performance against their personal best. Scientists compare the mass of atoms or molecules against the mass of 1 mole (that is 6.02×10^{23} atoms) of carbon-12. Carbon was chosen as the benchmark because it is quite common and easy to work with. One mole of carbon-12 atoms has a mass of 12 g or 0.012 kg.

Checkpoint 3

A sample of helium-4 has a mass of 4.0 kg. How many atoms are contained within this sample?

Exam question

answer: page 96

(a) What qualitative deduction can be made about the speed of air particles (mass = 1×10^{-25} kg) that collide with smoke particles (mass = 1×10^{-15} kg) in a Brownian motion demonstration? How did you arrive at this answer?

(b) What effect does an increase in temperature of a gas have on the pressure that the gas exerts? Why does this change take place?

(c) In general, solids are denser than liquids. Suggest an exception to this trend. Why does this example behave in such an unusual way? (20 min)

(d) A sample contains 12×10^{24} molecules. How many moles is this? (2 min)

Ideal gas equations

Have you seen a demonstration to show that atmo spheric pressure can support water in an upturne glass? Air pressure can hold up a 10 m tall colum of water. Quite a tall glass! Now, we will see kineti theory produce quantitative descriptions of gas prop erties like pressure.

Why look at gases?

Atoms in gases move freely, quickly and in straight lines. So we can think of them as small, hard spheres. This helps keep things simple.

The gas laws

Speed learning

Sometimes just reading the same information, presented in a different way, can make it stick!
Pour une masse donnée, à température constante, le produit p × V *est constant. C'est la loi de Boyle-Marriotte.*

Boyle's law

Imagine a gas held in a flexible box. What would happen to the pressure p exerted by the gas if we reduced the volume V of the box? More frequent collisions would result and the pressure would increase.

→ The volume of a fixed mass of gas is inversely proportional to its pressure, provided the temperature stays constant.

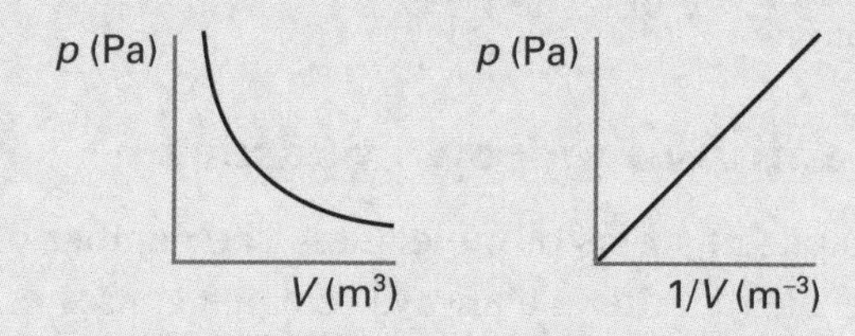

Checkpoint 1

How can Boyle's law be used to explain the movement of gases into, and out of, the lungs?

Boyle's law p is proportional to $1/V$ or pV = constant.

Charles' law

What would happen if gas held in a flexible container was heated? The particles would begin to move faster and push the walls out.

→ The volume of a fixed mass of gas is directly proportional to its absolute temperature (T), provided pressure remains constant.

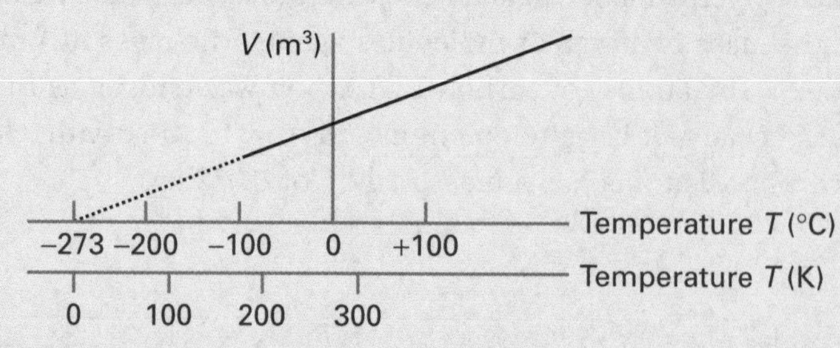

Examiner's secrets

Many candidates make careless mistakes in calculations. One of the most common is ignoring the fact that in this topic temperature is always measured in kelvin (K).

Charles' law V is proportional to T or V/T = constant.

The pressure law

What if we increased the temperature of a gas held in a box with fixed walls? The particles would move faster and pressure would increase.

→ The pressure of a fixed mass of gas is directly proportional to its absolute temperature, if volume stays constant.

→ All three gas laws can be combined as: pV/T = constant, or

$$\frac{p_1V_1}{T_1} = \frac{p_2V_2}{T_2}$$

where subscript 1 refers to an initial set of conditions and subscript 2 a later set

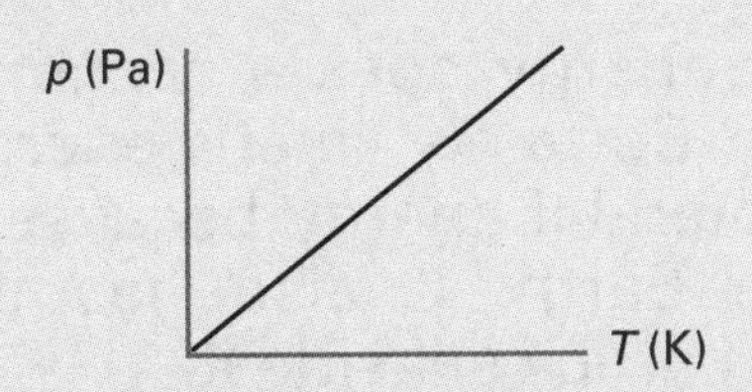

Pressure law p is proportional to T or p/T = constant.

Ideal gases

An **ideal gas** obeys all three gas laws. The laws work well for gases like air, helium and nitrogen when studied at normal atmospheric pressure and around room temperature. But *real* gases can be found at lower temperatures and higher pressures. Real gases do not obey these laws.

Equation of state (for an ideal gas)

This equation combines all three gas laws and takes account of the amount of gas, n moles, being examined: $pV = nRT$, where R is a constant, the universal gas constant. $R = 8.31\ \text{J mol}^{-1}\,\text{K}^{-1}$.

Kinetic theory

This theory can be used to apply the laws of mechanics to get expressions for pressure and temperature in terms of particle speed, mass and number. When we consider our ideal gas we have to assume that:

- the volume of the molecules is small compared to the volume of the gas (this avoids having to consider different-sized molecules!)
- there are no intermolecular attractions (which would decrease the pressure produced on the walls of the gas container)
- all collisions are elastic (otherwise pressure would fall)
- the impact time is much less than the time between collisions (this allows us to consider force as rate of change of momentum)

The pressure exerted by the molecules of an ideal gas is given by:

$$p = \tfrac{1}{3}Nm\langle c^2\rangle/V$$

Where N molecules, each of mass m, are moving with velocity c in volume V. Or:

$$pV = \tfrac{1}{3}Nm\langle c^2\rangle$$

(*Note* $\langle c^2\rangle$ = average value of c^2 for all the molecules in the gas.)

Let us compare these two expressions for pV:

$$pV = \tfrac{1}{3}Nm\langle c^2\rangle = nRT \qquad \text{so} \quad \tfrac{2}{3}N \times \tfrac{1}{2}m\langle c^2\rangle = nRT$$

meaning that the **average translational kinetic energy** is proportional to the absolute temperature (as $\tfrac{2}{3}$, N, n and R are all constants).

Exam question answer: page 96

(a) $5 \times 10^{-2}\ \text{m}^3$ of a gas are held at a temperature of 300 K and a pressure of 1.5×10^6 Pa. If the pressure is increased to 2.0×10^6 Pa, while the temperature remains constant, calculate the new volume occupied by the gas.

(b) A cylinder of volume $3 \times 10^{-3}\ \text{m}^3$ holds gas at 1×10^6 Pa and 327 K. Calculate the number of moles and then the number of molecules present.

(c) Explain the difference between a macroscopic and microscopic property.

(15 min)

Watch out!

Even though ideal gases play an important role in physics, none exists! You might wonder why we bother with them. Well, at high temperatures many real gases behave like they are ideal.

Checkpoint 2

A consequence of one of these assumptions is that an ideal gas would never be allowed to change into a liquid. Which assumption causes this consequence?

Examiner's secrets

Doubling the temperature (in K) increases the average speed of the molecules ($\langle c^2\rangle$) by about 1.4. This is a common *show that* question. ($\sqrt{2} = 1.414$)

The jargon

Translational kinetic energy refers to the energy a molecule has because it is moving along. (Molecules may also spin or tumble – rotational kinetic energy.)

Internal energy

> **The jargon**
> *Heat* is a process, not a form of energy! Heating or cooling an object changes its internal energy.

What forms of energy does a cannonball fired from a cannon have? Certainly kinetic energy because it is moving and potential energy because it is above the surface of the Earth. It also has internal energy because of the particles inside it.

What is internal energy?

At any temperature above 0 K, the particles within a substance vibrate randomly. They have kinetic energy. They also have potential energy because their motion keeps them apart and opposes the bonds holding them together. This diagram shows that the particles in a solid are tightly packed.

> **The jargon**
> *Internal energy* is the sum of the kinetic energy of all the particles and the potential energy stored in all the bonds.

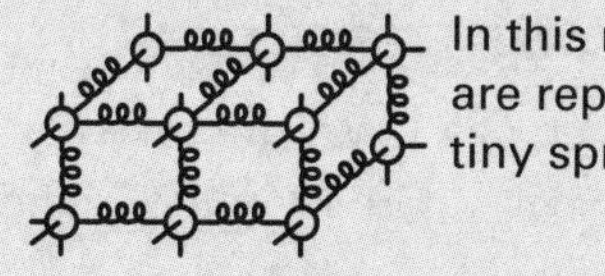

In this model bonds are represented by tiny springs

Here are some other important points about internal energy:

> **Action point**
> Rephrase the points in this list into an essay-style paragraph.

- solids, liquids and gases all have internal energy
- the energy within a substance is continually changing between potential and kinetic; e.g. if the particles move further apart, their bonds oppose the motion, and so kinetic energy decreases and potential energy increases
- the total internal energy of a substance is constant at a particular temperature
- at absolute zero (0 K) all objects have minimum internal energy (internal energy is the sum of all potential and kinetic energies)

> **Don't forget**
> The spontaneous flow of energy from hot to cold is called a *thermal transfer of energy*.

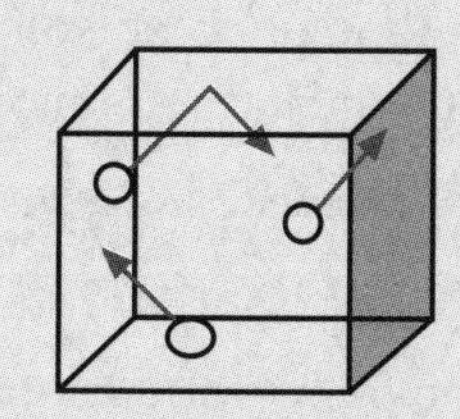

All the gas particles in this box have potential and kinetic energies

Two ways to increase the internal energy of a gas

> **Watch out!**
> Internal energy is sometimes referred to as *thermal energy*. This can be misleading, so don't do it!

- *Heat it* the walls of the container get hot, making the particles inside move more quickly so their kinetic energy increases.
- *Compress it* the walls of the container move in so that the gas particles bounce off more quickly, increasing their kinetic energy.

The first law of thermodynamics

Can you remember the conservation of energy law? (Energy can never be made or destroyed, but only changed from one form to another.) The first law of thermodynamics just restates the conservation of energy!

gain in internal energy of a system	=	energy transferred to system by heating	+	energy transferred to system by compressing (doing work)
ΔU	=	ΔQ	+	ΔW

> **Jargon**
> *Thermodynamics* is the study of energy changes that involve heating or cooling.

Two ways of expressing the first law of thermodynamics

There are two expressions for this law. One is used when work is done *on* the system, the other is used when work is done *by* the system.

Work done on the system

$$\Delta U \text{ (increase in internal energy of system)} = \Delta Q \text{ (energy applied by heating)} + \Delta W \text{ (work is done } on \text{ the system)}$$

Work done by the system

$$\Delta U \text{ (increase in internal energy of system)} = \Delta Q \text{ (energy applied by heating)} - \Delta W \text{ (work is done } by \text{ the system)}$$

Work done by, or on, a gas

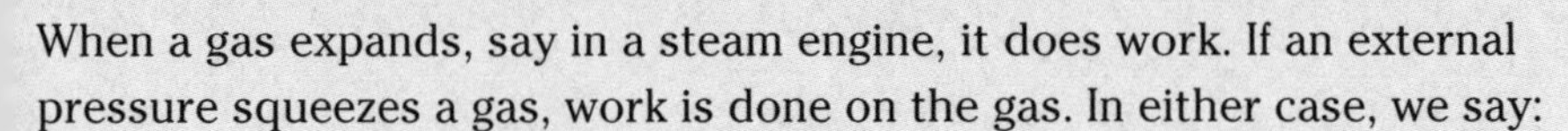

When a gas expands, say in a steam engine, it does work. If an external pressure squeezes a gas, work is done on the gas. In either case, we say:

work done = pressure × change in volume

$$W = p\Delta V$$

The work done by a gas is equal to the area under a p–V graph.

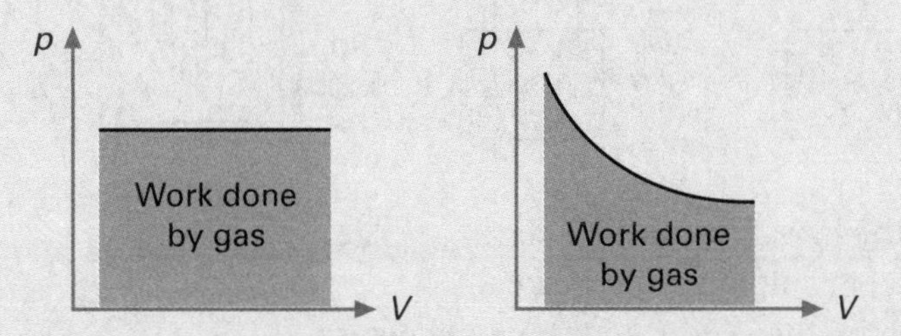

What is temperature?

When we heat an object, the particles within it move quicker. They gain kinetic energy and so there is an overall increase in internal energy. You can think of temperature as a measure of the average kinetic energy of the particles. We have already seen that for ideal gases, the average translational kinetic energy is proportional to the absolute temperature.

It can be shown that:

$$\tfrac{1}{2}m\langle c^2\rangle = \tfrac{3}{2}kT = 3RT/N$$

Where k is the Boltzmann constant ($1.38 \times 10^{-23}\,\text{J K}^{-1}$), R is the universal gas constant and N is the number of molecules.

Exam question answer: page 96

(a) In 1993, copper was cooled to just seven-millionths of a degree above 0 K. What is internal energy? What would copper's internal energy be at 0 K?

(b) A gas is held in a circular cylinder of radius 5×10^{-3} m at a pressure of 3.03×10^5 Pa. What work is done by the gas if the cylinder's piston moves out 1 mm? How would your answer be different if the piston had moved in 1 mm?

(c) Why is the first law of thermodynamics a restatement of the principle of conservation of energy? (20 min)

Action point

Place a rubber band between your lips. Do work on the rubber band by stretching it. You will notice that the band gets warm.

Checkpoint 1

Predict what will happen to the temperature of the rubber band, previously stretched, if it is subsequently released.

Examiner's secrets

A graph of *p against V* is also known as an *indicator* diagram.

Action point

Draw a $p - V$ graph and show on it the following processes:
(a) constant pressure
(b) constant volume
(c) constant temperature

Checkpoint 2

It has been rumoured that James Joule wanted to measure the water temperature at the top and bottom of a waterfall while on honeymoon in Switzerland. What scientific link was he trying to prove?

Examiner's secrets

Many physical processes can be described on a *microscopic* scale. It is often useful to relate these to a *macroscopic* scale, that is on a scale that we can measure.

Specific and latent heat capacities

How could you speed up making a cup of tea? If the kettle had to supply less energy that would help! You could heat less water or not wait for the water to reach 100 °C. Or you could use alcohol rather than water. (Alcohol gets hot quicker – strange tea though!)

The jargon

Specific refers to 1 kg of a substance and means per unit mass.

Watch out!

Specific heat capacity is now often known as *specific thermal capacity* to get away from the old idea that *heat* is the energy inside matter.

Examiner's secrets

Marks are available for good spelling, punctuation and grammar. For example, when a metre is a unit of measurement it ends in *–re*. When a meter is a measuring instrument it ends in *–er*.

Action point

Rearrange $c = \Delta Q/(m\Delta\theta)$ to give an expression for the amount of energy ΔQ required to raise the temperature of a substance.

Watch out!

Several solids, such as ice, contract on melting. The important point is that the particles in a solid stay in fixed relative positions.

Specific heat capacity *c*

If you supply the same amount of energy to 1 kg of alcohol, it will get hotter than 1 kg of water treated the same way.

→ The **specific heat capacity** of a substance is defined as the amount of energy required to raise the temperature of 1 kg of the substance by 1 °C.

Determining a solid's specific heat capacity *c*

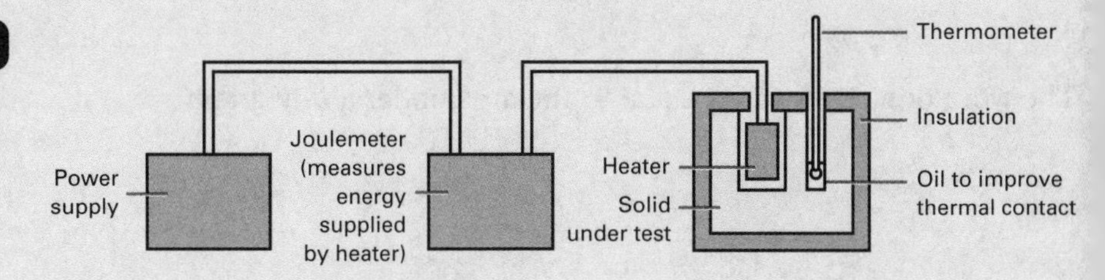

→ Find the mass m of the solid.
→ Record its initial temperature.
→ Switch on the power and the joulemeter.
→ Record the change in temperature $\Delta\theta$ in a given time.
→ Record how much energy ΔQ has been supplied in this time.
→ Calculate specific heat capacity c using $c = \Delta Q/m\Delta\theta$.

Changes of state

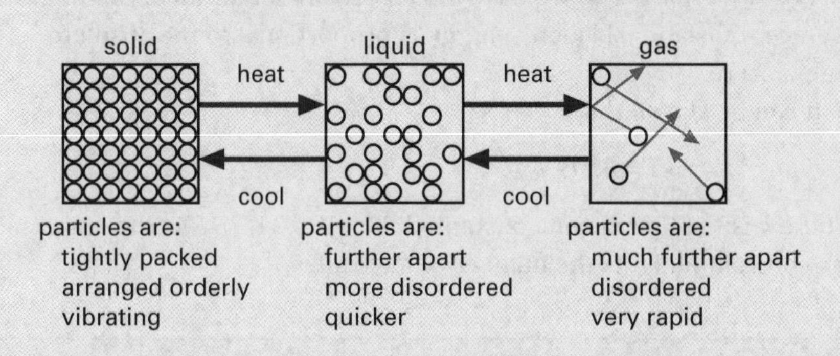

As a solid is heated, the network of bonds that holds its particles together breaks down. In liquids, the particles can wander around. This allows liquids to flow into any shape. Gas particles have the greatest kinetic energy, because they move so quickly. They also have the greatest potential energy because they are the furthest spread apart.

If we could supply energy to ice at a steady rate we would get idealized results and a graph like this:

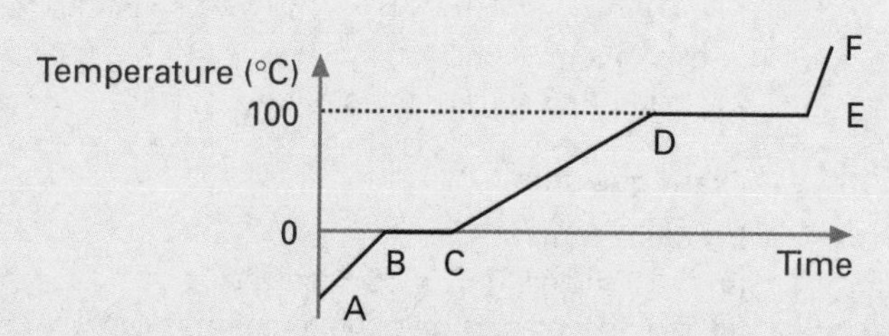

You can see from the graph that there is no change in temperature during changes of state (between B–C and D–E). The energy transferred to the water, during these periods, pulled particles apart by breaking the bonds between neighbours. It increased the potential energy of the particles in the process.

Look at the other sections of the graph (A–B, C–D and E–F). You can see that between changes of state, the energy supplied to the water increased its temperature. This allowed the particles to move faster so that their kinetic energy increased. All the time that energy was transferred into the system, its internal energy was increasing.

Checkpoint 1

How can you tell from this graph that boiling water requires more energy than melting ice?

Specific latent heat *L*

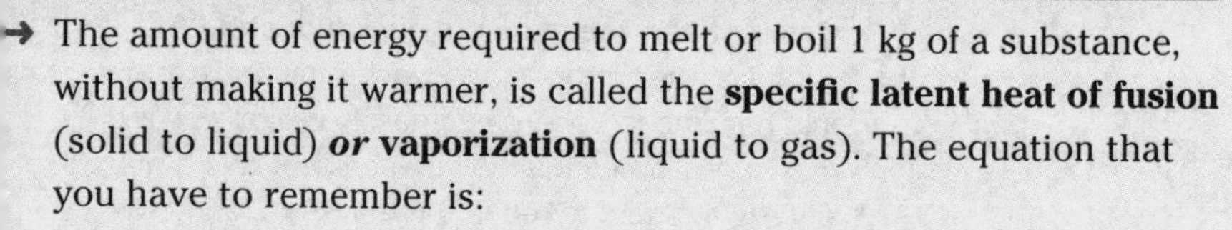

→ The amount of energy required to melt or boil 1 kg of a substance, without making it warmer, is called the **specific latent heat of fusion** (solid to liquid) ***or*** **vaporization** (liquid to gas). The equation that you have to remember is:

$L = \Delta Q/m$ or $\Delta Q = mL$

The jargon

Latent means hidden. When you melt or boil a substance, its temperature doesn't rise – the energy supplied seems to have disappeared!

Evaporation

Evaporation is not the same as boiling! It is a surface effect during which the liquid does not have to be heated to its boiling point. The gas formed by *evaporation* is a *vapour* – a gas below its boiling point. When you perspire, the sweat evaporates from the surface of your skin making use of evaporation's cooling effect. There are two ways to explain this effect. You can say that the sweat takes latent heat from your body so that it can start evaporating, thus lowering your temperature. Or, you can explain that it is the most energetic particles that leave the surface of the liquid, reducing the average energy of those left behind so that the liquid is now colder.

Checkpoint 2

Why does an aerosol can cool down when you use it?

Exam question answer: page 96

(a) A 20 kg hot water tank, made of copper (c = 390 J kg^{-1} °C^{-1}), holds 70 kg of water. Calculate how much energy is required to raise the temperature of the tank by 1 °C, first when it is empty and then again when it is full of water (c = 4 200 J kg^{-1} °C^{-1}).

(b) Why does melting ice require much less energy than boiling the same mass of water?

(c) Draw diagrams to show the differences between evaporation and boiling. (20 min)

Examiner's secrets

You should always be prepared to answer questions based on your practical work. There are a number of common class practicals involving heat capacity and latent heat. Look back over your lab book!

Answers
Kinetic theory

Temperature

Checkpoints

1 Ethanol-in-glass thermometers cannot be read remotely and only read up to 323 K.
2 $t = 85$ °C

Exam question

(a) 27.13 °C
(b) The baking tray is a better heat conductor. Free electrons transfer energy quickly as they move from hotter to colder parts of it (electron diffusion). It also transfers energy by lattice vibrations. Air relies solely on vibration of its atoms. It feels colder as it transfers energy to you more slowly.
(c) Thermistors are made from semiconductors. As temperature increases, more free electrons are released. The current in the thermistor increases; we can say that its resistance has fallen.

Examiner's secrets

Diagrams could help your answer to parts (b) and (c).

Behaviour of gases

Checkpoints

1 Convection and gravity would both cause the smoke particles to move along predictable routes; neither can explain the haphazard motion that is observed.
2 More violent collisions would force the smoke to be even more agitated.
3 6.02×10^{26}

Exam question

(a) The air molecules are much lighter than the smoke particles, yet they still push them around! The air molecules must be moving a lot faster than the smoke particles.
(b) If a gas is heated its particles move more quickly. They collide with each other, and the walls of their container, more often and with greater force. Gas pressure would increase.
(c) Water is denser as a liquid (1 000 kg m^{-3}) than as a solid (920 kg m^{-3}). We imagine water molecules having a non-spherical shape. When they pack together to form ice, there is more empty space between particles so ice is less dense.
(d) $n = \dfrac{12 \times 10^{24}}{6.02 \times 10^{23}} = 19.93$ (about 20 moles).

Examiner's secrets

Part (a) could be extended quantitatively, referring to conservation of momentum.

Ideal gas equations

Checkpoints

1 During inhalation, the diaphragm (a wall of muscle below the lungs) moves down, increasing the lungs' volume so their internal pressure falls (Boyle's law). As external pressure is now greater than the pressure inside the lungs, air is forced in. The reverse is true for exhalation.
2 No intermolecular forces of attraction.

Exam question

(a) 0.037 5 m^3
(b) 1.10 moles, 6.62×10^{23} molecules
(c) Macroscopic (large-scale) properties can be measured by instruments, e.g. pressure. Macroscopic properties are often explained by microscopic (small-scale) properties involving atoms, molecules and ions. Gas particles colliding to cause pressure is a microscopic property.

Internal energy

Checkpoints

1 It becomes cooler.
2 A link between energy and temperature.

Exam question

(a) Internal energy is the sum of the kinetic energy of all the particles and the potential energy stored in all the bonds in a system. Internal energy is a minimum at 0 K.
(b) 0.024 J. If the piston had moved in, rather than out, work would have been done *on*, rather than *by*, the gas.
(c) When a gas is heated and compressed, two amounts of energy are transferred into the system, some by heating ΔQ and some by doing work ΔW. Internal energy, $\Delta U = \Delta Q + \Delta W$ (first law of thermodynamics). This is a restatement of the conservation of energy as no energy is lost or gained, only changed into another form.

Specific and latent heat capacities

Checkpoints

1 Section DE is longer than BC. So it takes longer to boil the water than melt the ice. As energy is supplied at a constant rate, more is supplied to boil the water.
2 In evaporation, energetic particles escape first, forming a gas and reducing the average energy of the liquid particles. The liquid, and its container, both cool.

Exam question

(a) Empty = 7 800 J, full = 301 800 J.
(b) When a solid melts, only one or two bonds are broken per molecule. Each molecule is still bonded to most of its neighbours. When a liquid boils, each molecule has to break free from all its neighbours; eight or nine bonds may have to be broken so more energy is required to boil rather than melt.
(c)

Evaporation is a surface effect

Boiling is a bulk effect, it happens throughout the liquid

Heat

Waves and oscillations

We live in a wavy world! If you look around, listen to a radio or speak on a telephone your life involves waves. Whether you use a microwave oven, dip your toe in a bath or play a musical instrument, it is hard to avoid waves. Waves are very important in physics (and science generally). Health physics, astrophysics, particle physics, electricity, light, sound . . . the list goes on and on. What do they have in common? Waves!

Exam themes

- → *Mathematical description* of waves, e.g. the wave–speed equation.
- → *Wave properties* Description and understanding of wave properties, e.g. diffraction, interference.
- → *Natural world* Examination of aspects that involve waves, such as light and sound.
- → *Applications in modern physics* of wave ideas, e.g. wave–particle duality.

Topic checklist

○ AS ● A2	AQA/A	AQA/B	CCEA	EDEXCEL	OCR/A	OCR/B	WJEC
Types of waves and their properties	●	○	○	●	○	○	○
Electromagnetic spectrum	○	○	○●	●	○	○	○
Reflection and refraction	○	○	○	●	○	○	○
Total internal reflection and fibre optics	○	○	○	●	○	○	○
Diffraction and resolution	●	○	○	●	○	○	○
Superposition	●	○	○	●	○	○	○
Interference	●	○	○	●	○	○	○
Standing waves	●	○	○	●	○	○	○
Planck's constant	○	●	○	●	○	○	○
Photoelectric effect	○	●	○	●	○	○	○
Quantum behaviour	●	○			○	○	○
Atomic line spectra	○	●	○	●	○	○	○
De Broglie's equation and atomic models	○	●	○	●	○●		○
Particle diffraction and probing the atom	○	●	○	●	○●	○●	○●

Types of waves and their properties

> *"Almost everything that distinguishes the modern world from earlier centuries is attributable to science."*
>
> Bertrand Russell

It has been said that if the population of China jumped up and down in harmony, they would set up a tidal wave that could sweep the Earth. Waves carry energy from place to place.

Transverse waves

Transverse waves are caused by vibrations moving at right angles to the direction of the wave motion. All electromagnetic waves, such as light waves, are transverse. When a stone is dropped into water, the waves that spread out are transverse.

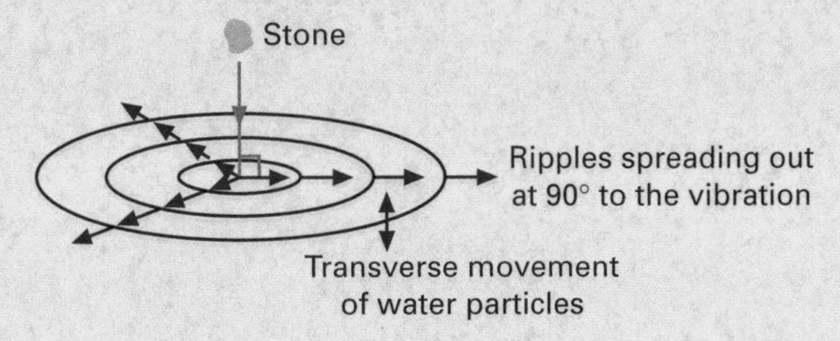

You can also use a slinky to demonstrate transverse waves.

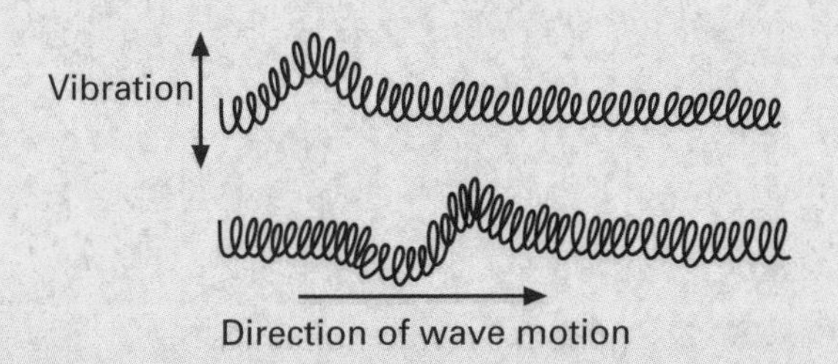

Jargon

A *slinky* is a coiled wire.

Checkpoint 1

Earthquakes also produce transverse waves. What would happen to both types of seismic waves if they reached a region of denser rock within the Earth?

Checkpoint 2

When waves are produced, do they transfer energy or matter?

Longitudinal waves

Vibrations moving in the same direction as the wave motion cause **longitudinal waves**. When your vocal cords vibrate back and forth, the resulting sound waves are longitudinal. The fastest (primary) waves produced by earthquakes are longitudinal. A slinky can also be used to demonstrate longitudinal waves:

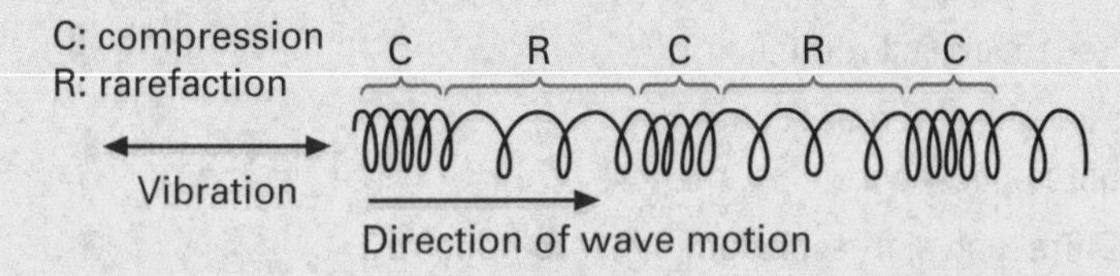

Describing waves

You need to learn the following to fully describe waves:

- **Displacement** x the distance between a point on the wave and the line of zero disturbance. It is measured in metres.
- **Amplitude** a maximum displacement. It is also in metres.
- **Period** T the time taken for one complete vibration to be made. It is measured in seconds.
- **Frequency** f the number of vibrations made per second. It is measured in hertz (Hz), 1 Hz = 1 vibration per second. ($f = 1/T$)
- **Wavelength** λ the distance between two similar points on a wave (e.g. crest to crest or trough to trough). It is usually in metres.
- **Speed** c for a wave, speed equals distance/time.

Examiner's secrets

The examiner expects you to include an appropriate unit when you define a physical quantity, e.g. speed in m s^{-1}.

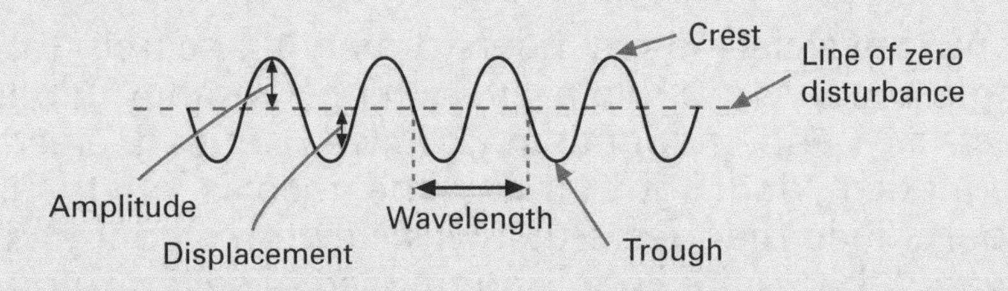

Classifying waves

We have already met one way to distinguish between waves: they are transverse or longitudinal. Waves can also be described as:

- **Mechanical** requiring a material to travel through, e.g. sound.
- **Electromagnetic** (EM) not requiring a material to travel through, e.g. light waves.

Waves can also be thought of as:

- **Progressive** spread out energy from the source vibration into the surrounding space, e.g. when a ripple is made on the surface of a puddle. As the energy spreads out, so its intensity decreases.
- **Stationary** or **standing** waves. The positions of their peaks and troughs do not move. Think of a guitar string. Some parts of the string vibrate while other parts (e.g. the ends) do not.

Wave properties

All waves can be **reflected** (bounced back), **refracted** (bent) and **diffracted** (spread out). As waves can spread out, they can also **interfere** with one another. All these properties are dealt with later in this section. Transverse waves, but not longitudinal, can be **polarized**. This means that they can be forced to oscillate in one fixed direction only. In this example, the transverse wave has been polarized in the vertical plane:

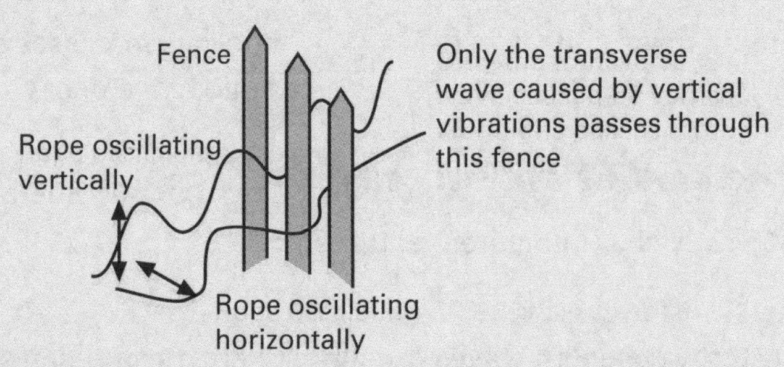

The wave equation

The speed, frequency and wavelength of any wave are related by:

speed = frequency × wavelength or $c = f\lambda$

Exam question answer: page 126

(a) Describe how transverse and longitudinal waves differ.

(b) Explain why it is impossible to polarize one of these wave types.

(c) Give an example of both transverse and longitudinal waves.

(d) Give one example of a standing transverse wave, and then name a standing longitudinal wave. (20 min)

Test yourself

Study the terms and diagram used to describe waves. Then cover this page with a blank sheet of paper and see how much you can remember.

Action point

Divide a page into four squares. Label the rows *transverse* and *longitudinal*. Label the columns *electromagnetic* and *mechanical*. Now place all the waves you know in the appropriate squares, but remember, the waves must match the labels. Where would you place a *Mexican* wave?!

Links

For more on EM waves see pages 100–1.

Watch out!

Don't get confused by the term *stationary wave*. It does not mean that nothing is moving.

Checkpoint 3

What are the units of each of the physical quantities in the wave equation?

Electromagnetic spectrum

"First we had the luminiferous ether. Then we had the electromagnetic ether. And now we haven't e(i)ther"

The Strange Story of the Quantum

Jargon

Superunified theories aim to simplify observations of the universe by unifying the fundamental forces of nature, i.e. electrical, magnetic, nuclear and gravitational.

Test yourself

Try to memorize the wavelength range for all the principal radiations shown here.

Links

For more on X-rays see the *medical and health physics option 2*, pages 160–1. For gamma rays see *properties of ionizing radiation*, pages 46–7.

Don't forget

$10^0 = 1$ and you must key 10^7 (for example) into your calculator as 1 EXP 7.

Checkpoint 1

After studying this double-page spread, try to establish what the only real difference is between an X-ray and a gamma ray of the same wavelength.

Examiner's secrets

The usual way to represent the electromagnetic spectrum is on a *logarithmic* scale. The numbers do not go up in even steps, but in equal ratios. A linear scale for this would stretch as far as Pluto! You should be able to explain what a *logarithmic* scale is and why it is useful.

Many physicists have been driven to search for a single theory to explain everything! Faraday unified electricity and magnetism. Maxwell unified electromagnetism and light. Today the race is on to find superunified theories. But it was Newton who first discovered that white light is composed of seven colours, the visible spectrum. Now we know that this forms part of the electromagnetic spectrum.

Electromagnetic (EM) waves

EM waves are produced when charged particles (e.g. electrons) vibrate or lose some of their energy. You need to know about the main radiations listed below.

Radiation	*Wavelength (m)*	*Produced by*
Radio	$>10^6$ to 10^{-1}	electrons oscillating in a transmitting aerial
Microwave	10^{-1} to 10^{-3}	Klystron (electron tube) oscillators
Infrared (IR)	10^{-3} to 7×10^{-7}	hot solids emit IR; in fact all objects emit some IR because of the motion of their particles
Visible	7×10^{-7} (red) to 4×10^{-7} (violet)	the Sun and vibrating atoms in other light sources, e.g. bulbs
Ultraviolet (UV)	4×10^{-7} to 10^{-8}	high-temperature matter emits some of its energy as UV beyond visible violet, hence ultraviolet
X-rays	10^{-8} to 10^{-13}	bombarding metal targets with fast moving electrons
Gamma rays	10^{-10} to 10^{-16}	nuclear processes such as radioactive decay

Properties of all EM waves

All EM waves share common features:

- they transfer energy
- shorter wavelength waves are more energetic and dangerous
- they can be reflected, refracted or diffracted
- they are transverse waves
- they can all travel through a vacuum at 3×10^8 m s^{-1}
- wave speed = frequency × wavelength

Features and uses of EM waves

The features, and therefore uses, of EM waves gradually change as their wavelengths, frequencies and energies change. You are expected to know about the main features of each region of the EM spectrum.

Radio waves

Stars produce radio waves that can cause hissing noises in televisions or radios that have not been tuned to a particular station.

- → *Ultra high frequency* (UHF) waves transmit TV signals.
- → *Very high frequency* (VHF) waves are used by local radio, police and ambulance communications.
- → *Medium and long wave* radio waves send messages over longer distances. As they have long wavelengths, they can diffract around the curve of the Earth and any hills that might be in their path.

Microwaves

Microwaves are sometimes considered to be a sub-group of radio waves. They are often associated with cooking but do not forget about:

- → *TV and communications satellites* Microwaves are the most energetic radio waves and are therefore ideal for this application.
- → *Radar* originally used in World War Two, radar is now used for air-traffic control around the world.

Infrared (IR) radiation

IR cameras are used by fire fighters to detect IR radiation coming from warm objects, like humans! IR radiation is also used in:

- → *Optical fibres* IR radiation can be used in telephone networks to code and send information, such as speech, along these fibres.

Light

This is the only form of radiation that is visible to the human eye. Therefore it is very important in communicating images or ideas.

Ultraviolet (UV) radiation

UV rays can cause tanning, skin cancer, eye damage and fluorescence.

X-rays and gamma rays

X-rays can cause cancer, but in concentrated beams they can be used to kill cancer cells. Gamma rays have similar effects and uses.

The nature of electromagnetic radiation

An EM wave is a disturbance in electric and magnetic fields in space. A change in an electric field can cause the associated magnetic field to change too. The resulting continuous cycle can keep the EM wave going. But the wave aspects of EM radiation are only half the story; you can start to find out about the particle aspects on page 114.

Exam question answer: page 126

The development of an understanding of our physical environment has taken place over many centuries. With reference to one area of physics, describe how the earlier work of scientists led the way for those who followed them. (25 min)

Action point

Use $c = f\lambda$ to check that as velocity is constant for all EM waves, a shorter wavelength means a higher frequency.

Action point

Use $E = hf$, where E is energy, h is a constant and f is frequency, to check that higher frequencies are more energetic.

Watch out!

Photon energy is also given by $E = hc/\lambda$, which you can get by combining $E = hf$ and $c = f\lambda$.

Watch out!

Short wavelength IR is often called IR *light* even though it isn't visible!

Examiner's secrets

Try to be specific when giving information. Don't say that UV radiation causes cancer; make sure you mention skin cancer.

Reflection and refraction

"God created everything by number, weight and measure."

Isaac Newton

Checkpoint 1

Why can we only normally see the Moon by night?

Action point

This diagram shows *regular* reflection, with light bouncing off a smooth surface. Draw a similar diagram to show *diffuse* reflection from a crumpled surface, e.g. cooking foil.

The jargon

If three lines all lie in the same plane, they are said to be *co-planar*. They can all be drawn on one flat sheet of paper.

The jargon

A *slanting angle* is taken to be an angle of incidence other than 0°.

Checkpoint 2

What would happen to the car, or ray, if the angle of incidence was 0°?

Examiner's secrets

By understanding how and why phenomena like refraction happen, you will find it easier to remember the equations associated with them.

Physics is about explaining the world around us. Why is it that when an explosion damages property in a narrow street, we often find that only every other window, on each side of the street, has been broken? How are rainbows formed after heavy rain? The phenomena used to explain these examples are reflection and refraction.

The laws of reflection

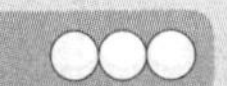

Light travels in straight lines called **rays**, with many rays forming a **beam**. This can be seen when we examine laser light or see sunlight streaming through trees. Luminous objects, which are normally hot (e.g. the Sun), give out their own light. But most objects are non-luminous (e.g. the Moon); we can only see them if light bounces off them (reflects) and enters our eyes.

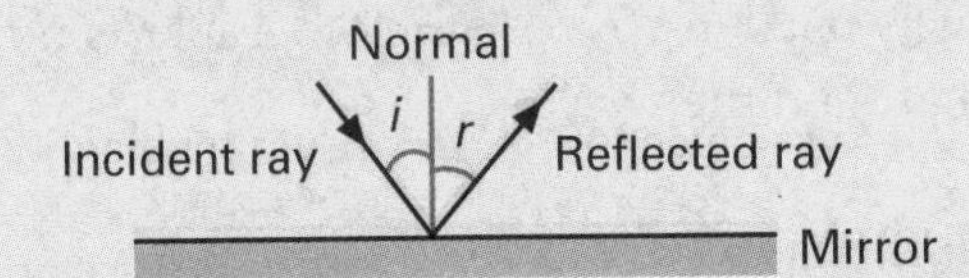

The **laws of reflection** are that:

- The angle of incidence i is equal to the angle of reflection r.
- The normal, incident and reflected rays all lie in the same plane.

Refraction

Are bears more intelligent than is generally recognized? They certainly seem to have grasped that light is bent (**refracted**) as it leaves the river having **reflected** off their lunch! Why does light bend when it leaves one *medium* (e.g. water) to enter another (e.g. air) at a *slanting angle*?

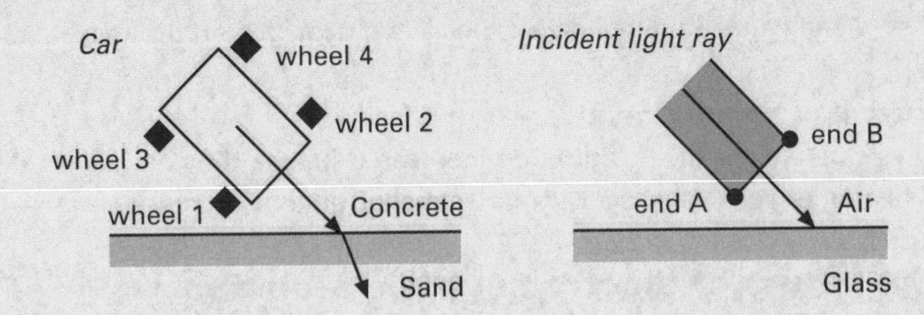

Imagine a car leaving a road to enter a beach, as shown above. As each wheel enters the sand, it will slow down. The diagram above shows a bird's-eye view of the scene. We can see that wheel 1 will reach the sand first and we can imagine that the car will change direction as shown. Consider a light ray entering glass:

- Which end of the ray enters the glass first (A or B)?
- What will happen to the speed of that end, on entry to the glass?
- What will happen to the path of the light ray?

Think again of a car driving into sand; you should see that if the speed of the wheels changed more dramatically on going from one medium to another, then the car would swerve even more. The ratio of the two speeds, called the **refractive index** (n), tells us about the degree of bending.

The laws of refraction

Try to remember that when a ray enters an optically denser medium, e.g. travelling from air into glass, it bends towards the normal.

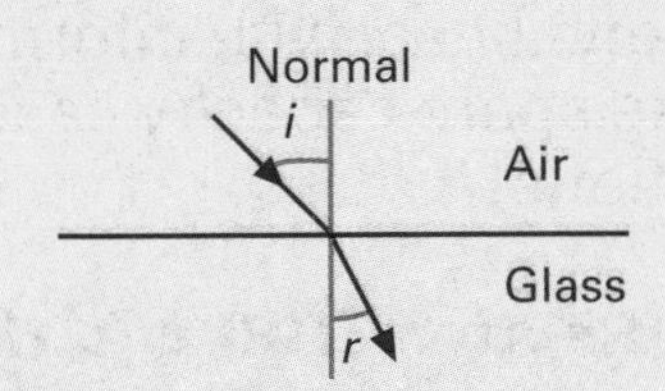

The **laws of refraction** are:

- $n_1 = \dfrac{\text{speed of light in a vacuum } c}{\text{speed of light in medium 1 } c_1}$

  Where n_1 is called the **absolute refractive index**.

- refractive index ${}_1n_2 = \dfrac{\text{speed in one medium } c_1}{\text{speed in second medium } c_2}$

  Where ${}_1n_2$ is the refractive index for light, or the car, travelling from medium 1 into medium 2.

- ${}_1n_2 = \dfrac{\sin i}{\sin r}$ (Snell's law) and ${}_1n_2 = \dfrac{\text{real depth}}{\text{apparent depth}}$

  (i.e. actual depth of the bear's fish/the depth it appeared to be at)

- ${}_1n_2 = \dfrac{n_2}{n_1}$ and ${}_2n_1 = \dfrac{1}{{}_1n_2}$

Example

An ultrasound wave travelling through muscle approaches bone at an angle of 15°. As ${}_{\text{muscle}}n_{\text{bone}} = 0.377$, calculate the angle of refraction.

$${}_1n_2 = \sin i \div \sin r$$
$${}_{\text{muscle}}n_{\text{bone}} = \sin 15° \div \sin r = 0.377$$
$$\sin r = \sin 15° \div 0.377 \quad \text{so} \quad r = 43.4°$$

Waves: reflecting and refracting

Ripple tanks can be used to show reflection and refraction of waves. The thin lines, in the diagrams below, represent wave crests. Notice that when the waves refract in the second diagram, there is a change in wavelength due to a change in speed in the shallow water.

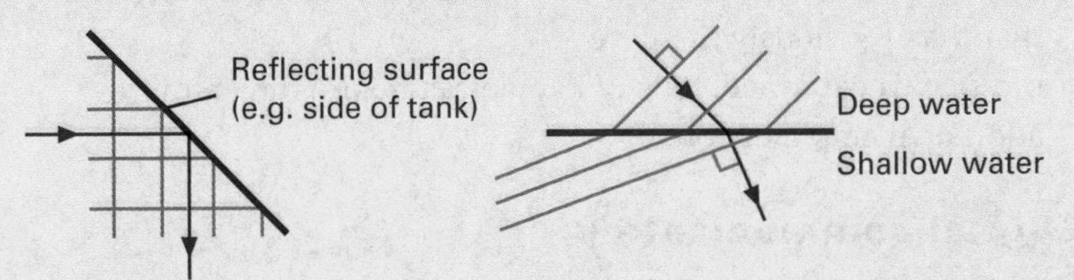

Exam question answer: page 126

(a) The refractive index of glycerol is 1.47. Calculate the angle of refraction if light enters glycerol at an angle of incidence of (i) 5°, (ii) 15°.

(b) Violet light has a wavelength of 400 nm in a vacuum. When it enters an unknown material, its speed decreases to 2.5×10^8 m s^{-1}. Calculate the frequency of violet in both media and its wavelength in the unknown material. (15 min)

Checkpoint 3

In what direction would a ray bend if it entered an optically less dense medium at a slanting angle?

Watch out!

Don't get confused! If you look up the value of a refractive index in a book, you will find the absolute refractive index quoted.

"What is it that breathes fire into the equations and makes a universe for them to describe?"

Stephen Hawking

Examiner's secrets

The work of many students can be let down by howlers. This candidate was describing a refraction experiment: 'We *filled* the ripple tank with *shallow* water.'

Examiner's secrets

Every year examiners report that many students fail to gain marks because of the poor quality of their diagrams. Make sure you are not one of these students by practising your drawings until they are clear and accurate.

Total internal reflection and fibre optics

"I think there's a world market for about five computers."

Thomas Watson, Founder of IBM

Telecommunications and personal computers are becoming ever more popular. Internet shopping and banking, home-working and virtual classrooms are now a growing reality for many. Humans and computers now share telephone lines, with information from computers filling in the spare capacity caused by pauses in human conversations!

Total internal reflection (TIR) and the critical angle

When light travels from one medium (e.g. glass or plastic) into a less dense medium (e.g. air), it bends away from the normal. If the angle of incidence i is greater than a critical angle C, the light will be totally internally reflected at the boundary of the two media.

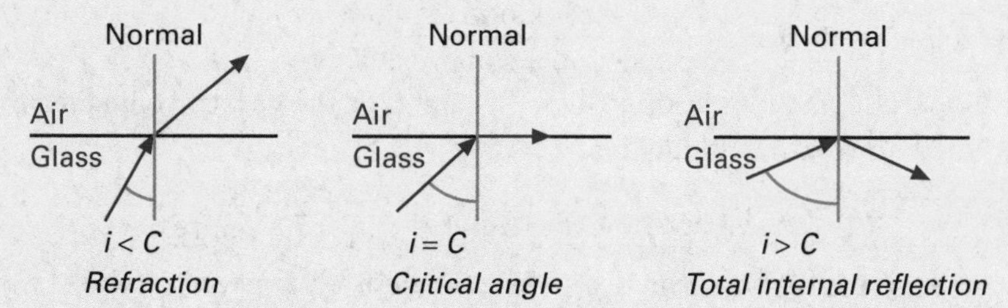

Checkpoint 1

Calculate the critical angle for the sample of glass shown in the central figure opposite, where $_{air}n_{glass} = 1.50$.

Consider the diagram above. When the angle of incidence equals the critical angle, the ray is refracted along the boundary of the two media:

$$_{glass}n_{air} = \sin i \div \sin r = \sin C \div \sin 90^\circ = \sin C$$

Examiner's secrets

It is easy to get the ratios the wrong way up. If you get a refractive index less than 1, you must have made a mistake. Check again if this happens to you!

We already know that $_1n_2 = 1/_2n_1$ so:

$$_{air}n_{glass} = 1/\sin C$$

Fibre optics

One of the most important uses of total internal reflection is in the field of telecommunications. Information is sent, as pulses of light, along optical fibres. These optical fibres have some important advantages:

- each one can potentially carry more than 30 000 calls at a time
- they are lighter, smaller and cheaper than copper cables
- the signals carried are more secure, reliable and of a better quality than was previously the case
- the signals can travel tens of kilometres before they require additional amplification

Action point

Produce a concept map to show the links between each of the words printed here in **bold** and *italic*.

Digital communication

But how can a human telephone conversation be converted into pulses of light and for what purpose? The microphone in your telephone converts sound waves into electrical signals: a sequence of *high* or *low voltages*. This code can easily be transformed into a sequence of *light pulses* – very similar to *Morse code*, which uses just *dots* and *dashes* to send information. When there are just two states (high or low, light or dark, dot or dash) we have **digital communication**. Digital communication is fast, accurate and great for computers as they use a digital (*binary*) code (0s and 1s) anyway!

Transatlantic telephone calls

When you make an international call, your message could be travelling through every one of these media: through the air as sound waves; along wires as electricity; along coaxial cables as radio waves; through air and space as microwaves, and down fibre optics as light pulses. Nowadays, transatlantic calls go via underwater fibre optic cables.

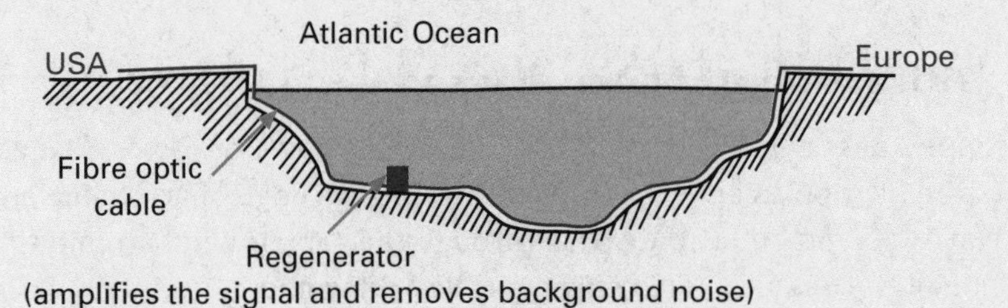

Other advantages of using fibre optics in this way are that:

- → they do not have to be laid on a flat surface
- → they are relatively cheap to put in place, allowing cheaper calls

Fibre optics – up close

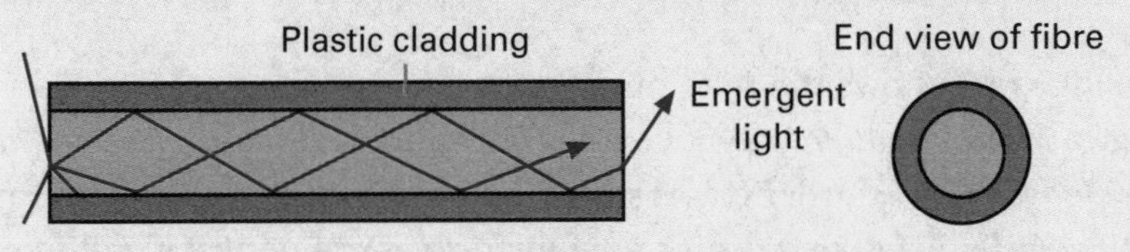

Optical fibres are often grouped together in bundles. Each fibre is protected by a polyurethane cover, as shown above. Light, whose angle of incidence is greater than the critical angle, is 'trapped' and sent down each optical fibre.

Optical fibres must let light pass through as easily as possible. Very pure sodium borosilicate glass is used, as a 20 km length is as transparent as a window pane. However, fibre optics do have practical problems.

- → They must use monochromatic (single-wavelength) light to stop some pulses travelling faster than others. (Lasers are used to send monochromatic infrared pulses. They can be switched on and off at a rate of up to 2×10^8 Hz, meaning that millions of pulses can be sent every second.)
- → The fibres must be very thin (no thicker than a human hair) to reduce the number of possible paths the light pulses can take. (This also helps to prevent some pulses arriving before others.)

Exam question answer: page 126

(a) Write down six points which could form the basis of an essay on 'The social, economic and technological changes which have arisen as a result of the development of fibre-optic transmission of information'.

(b) Name four other uses of fibre optics.

(c) Draw diagrams to illustrate total internal reflection and the critical angle. (20 min)

The jargon

A *coaxial cable* is a cylindrical wave guide used to carry radio waves.

Checkpoint 2

A material used to produce optical fibres absorbs 1% of the available light for each metre of its length. Calculate what percentage of the original light would reach a regenerator placed 50 m from the start of the fibre.

Checkpoint 3

State three changes that light undergoes when it travels from air into an optical fibre, at a slanting angle.

Checkpoint 4

If the refractive index of a glass fibre is 1.5, the speed of light in it will be 2×10^8 m s^{-1}. By how much does the wavelength in the fibre change?

Diffraction and resolution

Why can we hear around doorways and yet we canno see around them? When Christiaan Huygens suggested his wave theory of light, this question was a big issue. Others argued that if light was composed o waves, it should be able to diffract, or spread out, like water waves. We now know that if we look closely diffraction of light can be seen.

Diffraction of water waves

Ripple tanks can be used to show diffraction of water waves. Plane (straight) ripples are sent towards a gap or an edge of an obstacle. Waves spread out as they pass through the gap, or bend around the side of the obstacle. This effect is called **diffraction**.

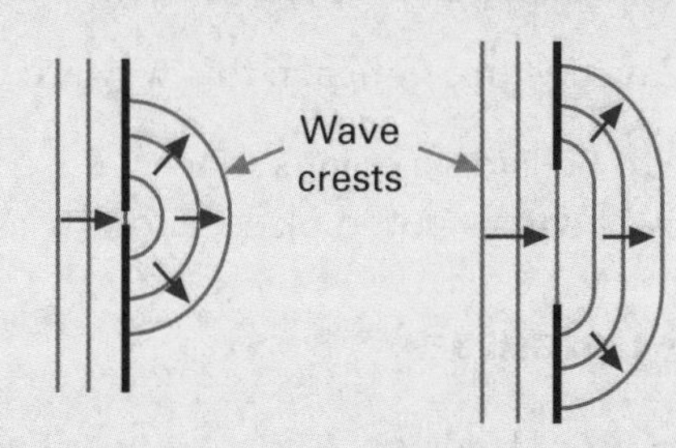

This diagram shows the relationship between gap width and wavelength. *Maximum diffraction* occurs when the *gap width is equal to the wavelength*. The wavelength of sound waves is approximately the same as the width of a doorway and so sound waves are noticeably diffracted allowing you to hear around the opening!

When a plane wave reaches a narrow gap, it makes the water in the gap go up and down. As far as the water in the gap is concerned, this is no different to a point source (like a stone dropping in the water), and so a circular wave is sent out.

Explaining diffraction – Huygens' construction

Huygens explained diffraction by splitting the wavefront within the gap into an arbitrary number of secondary sources, like lots of stones falling into the water along the length of the wavefront. Each secondary source sends out its own little wavelets, which all combine to form the next wavefront.

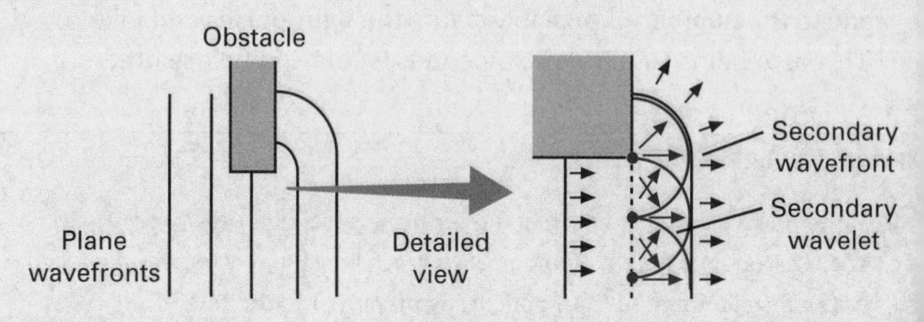

Diffraction is a feature of all waves. Designers can put a knowledge of diffraction to good use. For example, microwave ovens have metal grids in their doors to keep the microwaves in and let light waves out. You are kept safe but you can still keep an eye on your food!

Watch out!

When using a ripple tank, don't worry if your results are slightly different from those shown in the diagram opposite. Idealized results are shown here.

Checkpoint 1

If you were trying to eavesdrop on two people chatting around a corner from you, would their sound waves reflecting off walls help or hinder you?

Examiner's secrets

You should be able to describe what happens to waves passing through a gap as the gap is progressively narrowed.

Jargon

A *wavefront* is a surface at 90° to the wave's motion and a *wavelet* is a small wave.

Diffraction of light waves

he wavelength of visible light lies between 4×10^{-7} m and 7×10^{-7} m.

- A very narrow gap is required to diffract light successfully.

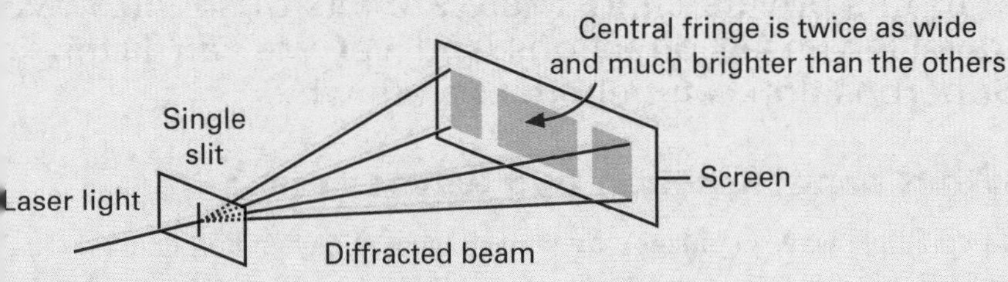

y adjusting the slit width and wavelength of light (by putting colour lters in front of the pin hole), in the experiment shown below, we find nat diffraction increases:

- as slit width decreases (at constant wavelength)
- as wavelength increases (if slit width remains constant)

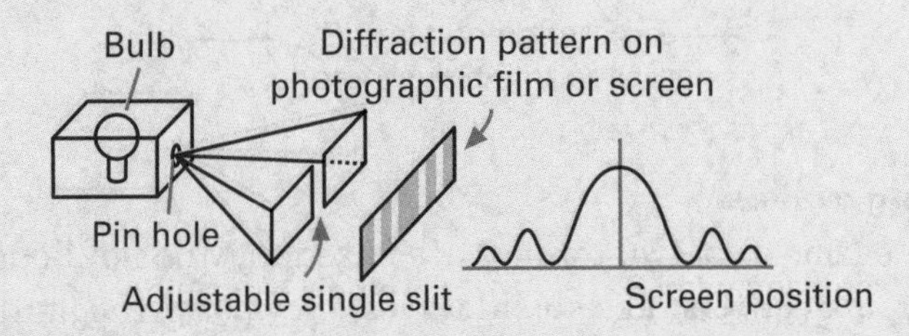

Resolution

'he diffraction pattern produced by a circular hole, such as the pupil of our eye, is a bright central spot surrounded by bright and dark rings. 'he angle of the first dark ring is given by:

$$\sin \theta = \frac{1.22\lambda}{D}$$

Vhere D = diameter of hole, λ = wavelength and θ = angle subtended by ing radius to hole.

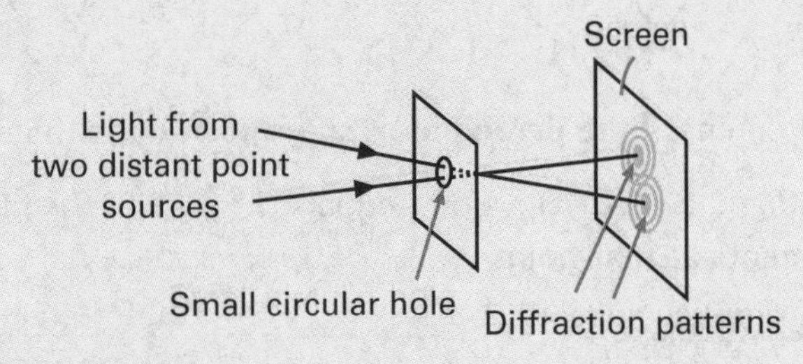

'he **Rayleigh criterion** for resolving two points (i.e. being able to tell hem apart) is that their angular separation θ must be greater than the 'alue given by $\sin \theta = 1.22\lambda/D$.

Exam question answer: page 127

(a) Why do harbours (natural and artificial) normally have circular shorelines? (Use a diagram in your answer.)

(b) A microwave oven uses microwaves whose wavelength is 0.12 m. The door of the oven contains a metal grid with grid gaps of 0.005 m. How does this design protect us and let us see inside the oven? (15 min)

Links

To understand how diffraction produces bright and dark fringes see pages 108–11.

Action point

To observe diffraction of light waves, look at a distant street lamp through a pinhole in a piece of cooking foil. You should see rings of light around the pinhole.

Examiner's secrets

Drawing fringes in diffraction patterns can be confusing. You must label the light and dark bands so the examiner is clear which is which. Otherwise the examiner must assume a shaded band is a dark fringe. This may be the exact opposite to what you intended!

Checkpoint 2

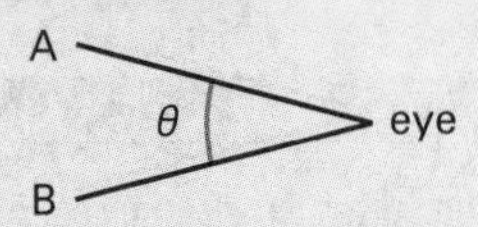

Calculate the minimum angle θ at which two objects, A and B, can be identified as separate. Use $\lambda = 5 \times 10^{-7}$ m, and the diameter of the observer's pupil $D = 2 \times 10^{-3}$ m.

Action point

Learn the definitions of *diffraction*, *refraction* and *reflection*. They sound similar, but have different meanings.

Superposition

Links

Superposition applies to all waves. It is used to explain diffraction (pages 106–7), interference (pages 110–11) and standing waves (pages 112–13).

Have you ever been in a crowded room when your attention has suddenly been seized by a conversation taking place at the other side of the room? Almost amazingly we can shut out all other noises and listen in to this private chat! There are lots of sound waves travelling in all directions and yet we can listen in. Superposition can help to explain why.

What happens when two waves cross?

If two single waves (**pulses**) are sent down a slinky, simultaneously and from opposite ends of the spring, the *pulses pass through each other* and then continue to carry on their way as if the crossover had never happened! In just the same way, sound waves can cross through each other allowing a private chat to pass through a sea of other sound waves before arriving at your ear!

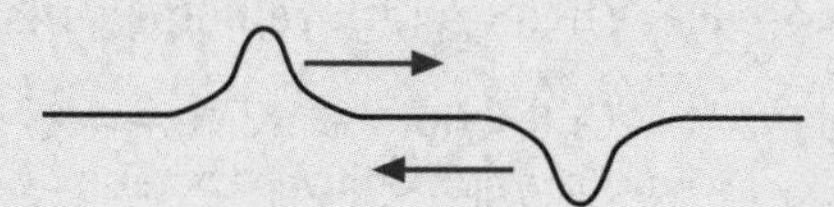

Links

To remind yourself of what moves when waves travel, see pages 98–9.

Combining waves

If the light beams from two torches are mixed they do not bounce off each other. Instead, just like sound waves, they pass straight through one another. Remember that energy moves when waves travel, not matter, and so it is not like two cars crashing into one another!

The principle of superposition

Checkpoint 1

If two up pulses (crests) are sent from opposite ends of a slinky, what would happen at the instant that they cross?

At the instant that the two wave pulses passed, in the diagram above, their displacements combined. For a fleeting moment a crest met an equal-sized trough and it was as if there were no waves travelling down the slinky – and then the two waves carried on. Consider an alternative

When two troughs meet they produce . . .

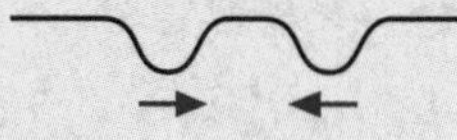

. . . a 'supertrough'

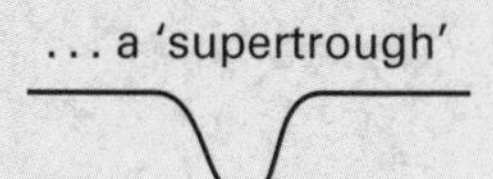

Watch out!

Displacement is a vector quantity so remember to consider whether each displacement is positive or negative when using the principle of superposition.

A formal statement of the **principle of superposition** is that:

→ the total displacement at a point equals the sum of the individual displacements at that point

In less formal language:

→ crest + crest = supercrest
→ trough + trough = supertrough
→ crest + trough = zero

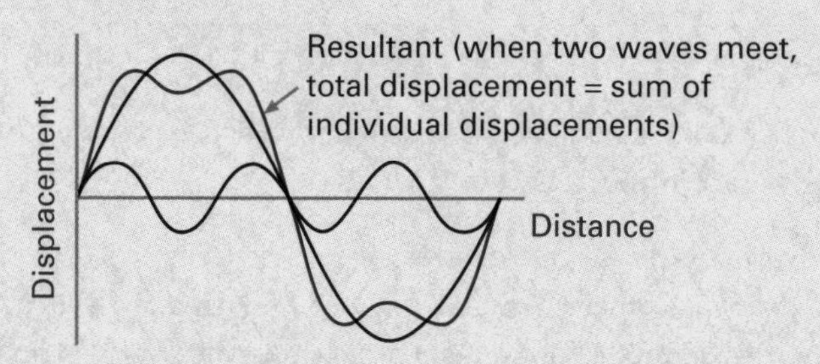

Explaining single-slit diffraction

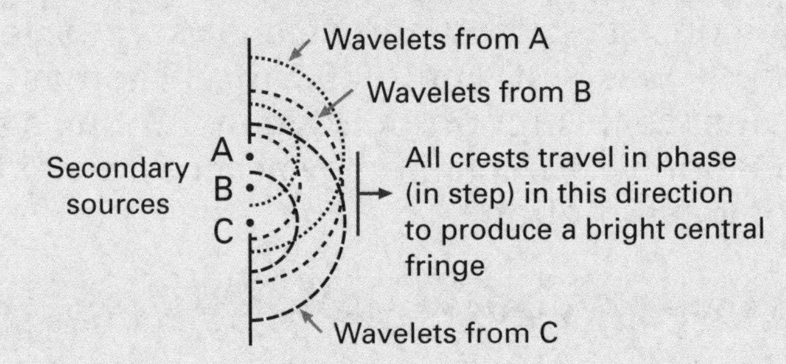

this diagram, three secondary sources are sending out wavelets. We an use the principle of superposition to find the combined effect of all ıese wavelets in any direction. In some directions they add together to roduce **bright fringes** (where crests arrive with other crests to form upercrests) and in other directions they cancel out to form dark inges.

In the straight-ahead direction all the crests are travelling **in phase**. ach crest leaves A at exactly the same time as crests leave B, C and very other secondary source along the gap width. They all travel at ıe same speed and they all have the same distance to cover to reach ıe screen (**path difference** equals zero). Therefore they all reach the creen together to produce a supercrest, resulting in a bright fringe.

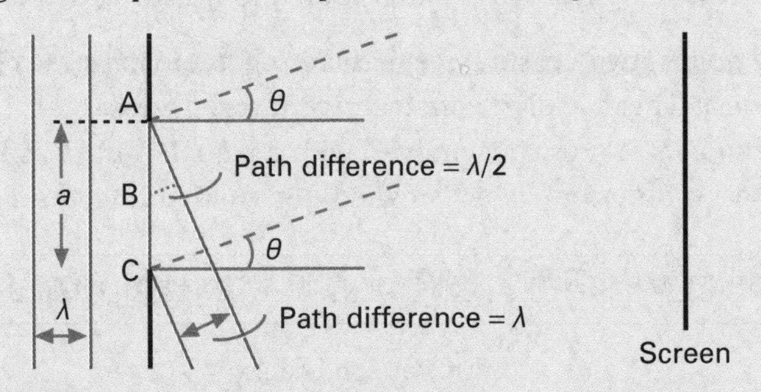

his diagram shows that at angle θ, wavelets from C have to travel ırther than wavelets from A in order to reach the screen. They have to ravel an extra wavelength λ. Point B is halfway between A and C. The vavelets from A and B are in **antiphase**. So when a crest arrives at the creen from A, it is cancelled by a trough arriving from B. For every econdary source between A and B, there is another between B and C roducing wavelets in antiphase causing a dark fringe. **Dark fringes** are roduced at all angles θ when $\sin\theta = n\lambda/a$, where $n = 1, 2, 3 \ldots$. At oints between these angles not all the light is cancelled and so light ands occur, the intensity of which decrease as θ increases.

The jargon

Travelling in phase means travelling in step.

The jargon

Path difference means how much further one wave has to travel compared with a second wave.

Examiner's secrets

If the path difference is a *whole number of whole wavelengths* ($n\lambda$), there will be a bright fringe. Dark fringes result from a path difference of an *odd number of half-wavelengths* $(2n + 1)\lambda/2$. (Take $n = 0, 1, 2 \ldots$ etc.)

Checkpoint 2

What is the path difference between wavelets sent from A and C, and from A and B?

Jargon

In antiphase means exactly out of step.

Exam question

answer: page 127

(a) Superposition of waves is used to combat noise pollution by making antisound. Describe how a sound wave could be cancelled in this way.

(b) Describe how hand movements can explain superposition of wave pulses. (15 min)

Examiner's secrets

Complete cancellation requires two waves to be in *antiphase*. Sometimes waves that superpose do not have exactly the same amplitude and so complete cancellation does not occur.

Interference

"The yeoman work in any science is done by the experimentalist, who must keep the theoreticians honest."

Michio Kaku

Checkpoint 1

Both loudspeakers are connected to the same signal generator operating at 1 000 Hz. The speed of sound in air is 330 m s^{-1}. Calculate the wavelength of the sound waves coming from both loudspeakers.

The jargon

Phase difference means how much one wave leads, or lags behind, another. If the *phase difference* equals zero, neither wave leads. They are said to be *in phase*.

Examiner's secrets

Of course, lasers hadn't been invented in Young's day! If you don't have a *coherent* source in your set-up, you must have a single slit before the double slits.

Examiner's secrets

This experiment frequently appears in examinations because of its importance in support of the wave theory of light.

As we have seen, Huygens' wave theory of light wa not readily accepted by others. Newton's proposa that light consists of tiny particles, or corpuscles, wa preferred. It was not until 1801 that Thomas Youn found evidence in support of Huygens. Young's exper ment could only be explained by accepting that light i a form of wave motion.

Interference of sound waves

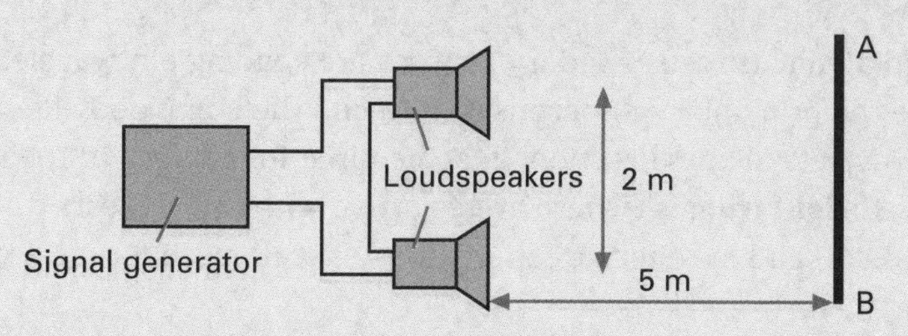

To demonstrate interference of sound waves:

- connect the equipment as shown, to produce waves in phase
- set the signal generator to 1 000 Hz
- ask observers to walk along the line AB

The results and conclusions are that loud and quiet regions exist.

- At *loud points*, two crests have arrived together (in phase) forming a supercrest. This is called **constructive interference**.
- At *quiet points*, a crest has arrived with a trough (antiphase). They cancel each other out. This is called **destructive interference**.

A modern version of Young's experiment

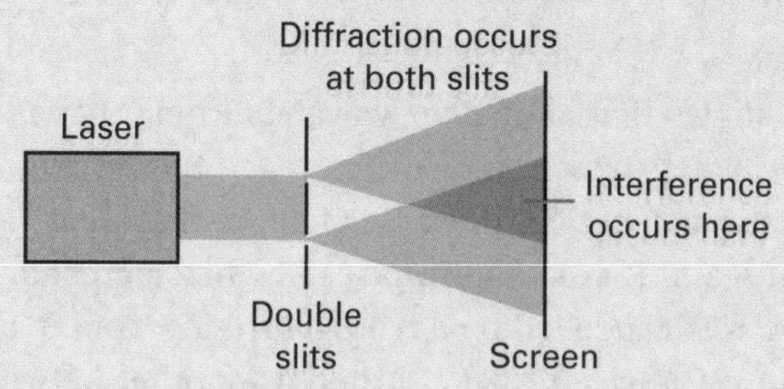

In this modern version of Young's experiment, laser light spreads out a each of the two slits. The light waves from each slit overlap and interfer with one another. The principle of superposition can be used to predic or explain the resulting interference pattern.

Using a laser allows us to get an interference pattern because:

- As only one laser is used, the waves from both slits are **coherent** (meaning that the phase difference between them remains constant)
- Lasers produce **monochromatic** (single wavelength) light. This ensures that both sets of waves have the same wavelength.

Explaining the experiment

The interference pattern is caused by waves from either slit travelling different distances to reach each point on the screen. The following diagram can be used to help explain double-slit interference.

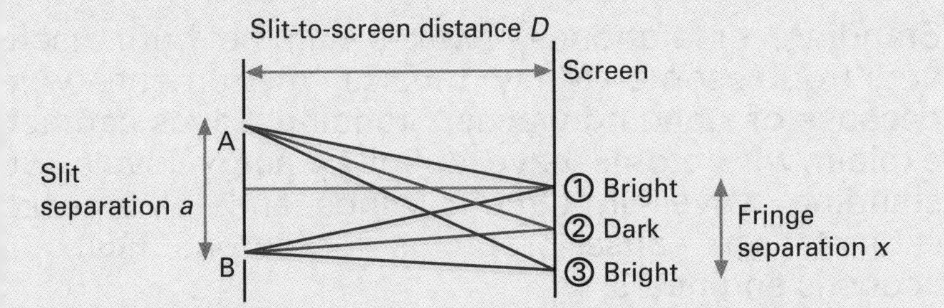

- Point ① is equidistant from both slits. Both sets of waves travel the same distance to reach point ① (path difference = 0). As both waves left the slits in step (in phase), they reach point ① in step. They will interfere constructively to produce a bright fringe here.
- Point ② is the centre of the first dark fringe. Light from slit A has had to travel an extra half wavelength to reach point ② (path difference = $\lambda/2$). As this light is $\lambda/2$ behind, it interferes destructively with light from slit B, causing a dark fringe here.
- Point ③ is at the centre of the next bright fringe. Light from A has had to travel a complete extra wavelength to reach this point. So, the two sets of waves arrive in step here to produce a bright fringe.

Other important points

- The double-slit experiment can be used to find the wavelength of light using the formula $\lambda = ax/D$ (see the diagram above).
- This experiment supports the wave theory of light because it can only be explained using wave properties: diffraction and interference.
- Similar experiments use a ripple tank or microwaves.

Diffraction gratings

Passing monochromatic light through more than two slits can also cause interference. Diffraction gratings have up to 10 000 slits per cm.

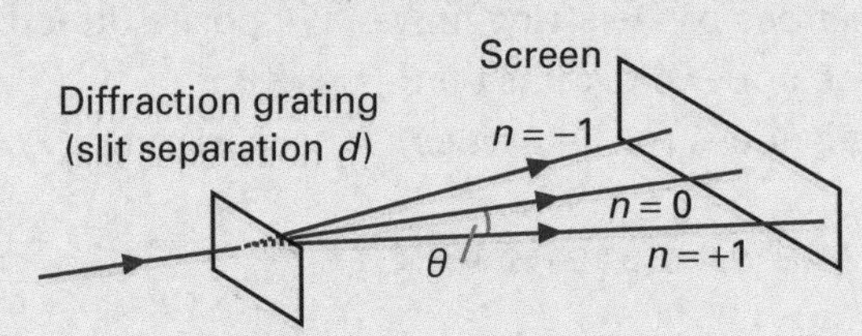

For diffraction gratings, wavelength λ and θ are related by:

$d \sin \theta = n\lambda$ $\quad$ d = slit separation, n = order of the maximum

Exam question answer: page 127

(a) By explaining the term *coherence*, explain why we do not observe many more clear interference effects in our everyday lives.

(b) Give one example of both coherent and incoherent light sources. (15 min)

(c) The slit separation (d) of a diffraction grating is 0.4 mm. The wavelength of light is 600 nm. Calculate the angle made by the second order fringe. (5 min)

Checkpoint 2

The slits used in an experiment similar to that shown opposite were 1 mm apart. The distance between slits and screen was 3 m and the width of 10 fringes was 3 m. Calculate the wavelength of light used.

Test yourself

When you have finished working through this section, close the book and see how much you can remember. In a week's time, refresh your memory and then test yourself again!

Watch out!

If asked how many maxima will be produced by a certain diffraction grating, remember to include the zeroth-order maximum and all those for which n is negative.

Examiner's secrets

Each syllabus is slightly different. Does your syllabus require a derivation of $d \sin \theta = n\lambda$?

Action point

If a grating has a slit spacing of 0.25 mm, how many lines (slits) per m does it have? (*Hint* $N = 1/d$)

Standing waves

"There is no higher or lower knowledge, but only one, flowing out of experimentation."

Leonardo da Vinci

Links

The terms standing or stationary waves and transverse waves were first introduced on pages 98–9.

The jargon

The simplest way a guitar string can vibrate is with its *fundamental* frequency, sometimes called the *first harmonic*. The next frequency to produce a standing-wave pattern is called either the *first overtone* or the *second harmonic*.

Checkpoint 1

Sketch a diagram showing a string vibrating at its second harmonic.

Examiner's secrets

To get full marks when describing how to produce standing waves you must have three basic ingredients:

(i) Something to start the initial vibration.
(ii) A method of changing the frequency of vibration.
(iii) Knowledge that there could be different wave patterns.

Make sure you learn all of these.

Standing, or stationary, waves can be both usefu and troublesome. Many musical instruments wor because of standing waves. Standing waves can hel explain why atoms have definite energy levels. Bu standing waves in aircraft wings and loudspeake cabinets can cause problems for aeronautical an acoustic engineers.

Standing-wave patterns

If you tie a slinky to a table leg, it is easy to see standing waves. Flick the other end of the slinky from side to side to send transverse waves along. They reflect off the fixed end and by varying the frequency you can get different patterns, as shown below.

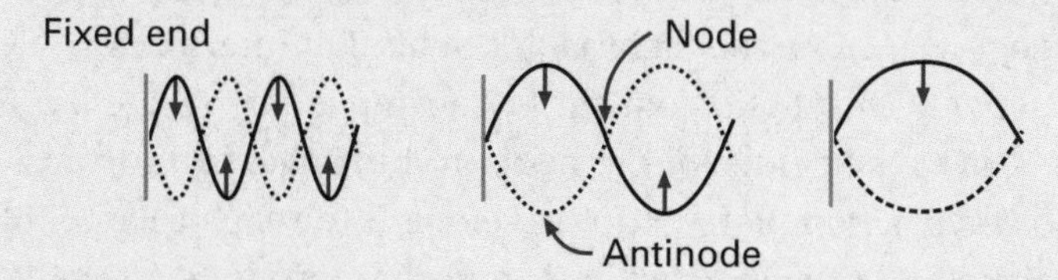

Nodes and antinodes

The simplest way a guitar string can vibrate is shown below. As with al standing waves, a node is a point where the amplitude is zero. A point where it is a maximum is always called an antinode.

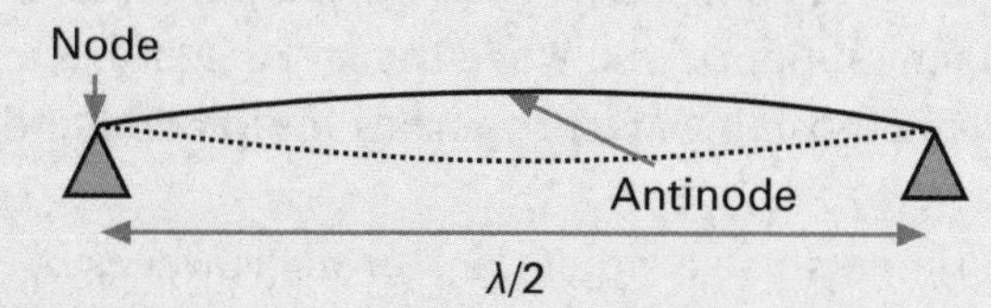

Other important points about nodes and antinodes include:

- the string appears as a series of loops separated by nodes (see the upper diagrams on this page too)
- adjacent sections of the string move in opposite directions (see arrows in the upper diagrams on this page)
- neighbouring nodes (or antinodes) are separated by $\lambda/2$

Standing-wave experiments

Stretched strings

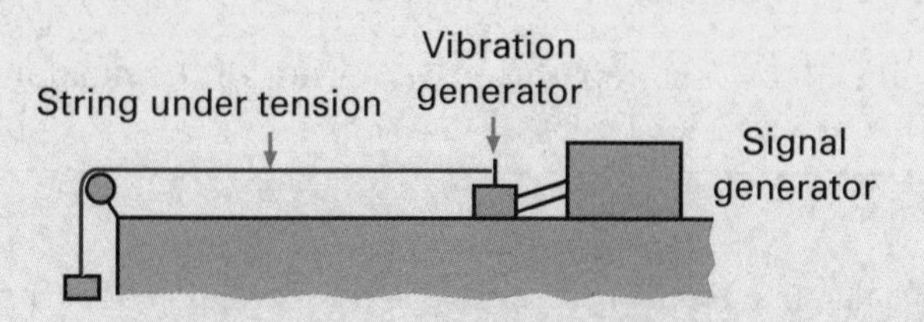

This diagram shows Melde's experiment. Weights attached to one side of the string, via a pulley, maintain the tension. As the frequency of the vibration generator is increased, all the standing-wave patterns discusse above (and more) can be seen. The wavelength of the waves can be calculated by measuring the distance between nodes. String length, tension and other factors can be studied.

Microwaves

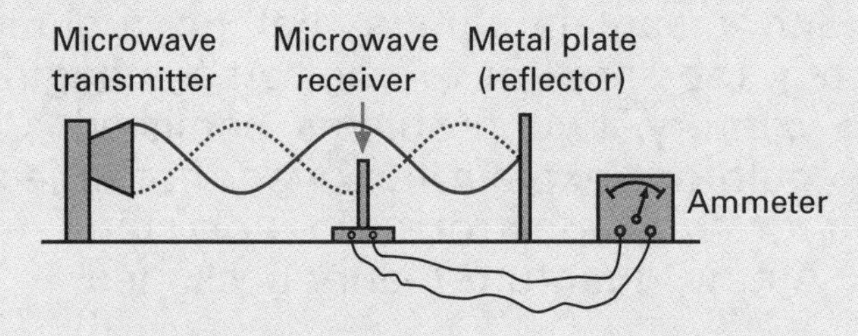

- This diagram shows how you could investigate standing waves produced by microwaves. Position the transmitter 50 cm from the reflector.
- Move the receiver around in the region between transmitter and receiver to observe alternating points of high and low intensity.

Sound waves

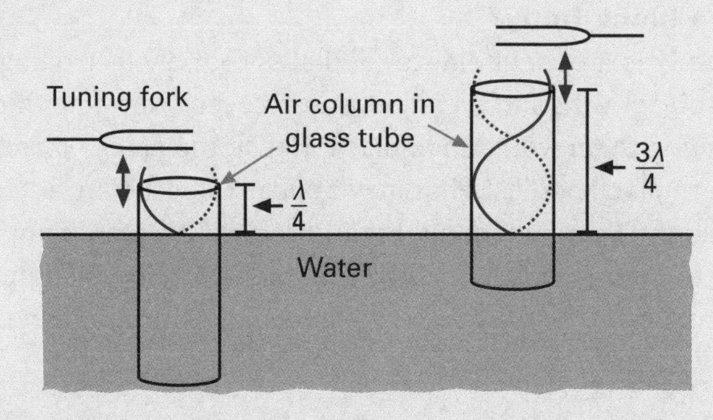

This apparatus can be used to deduce the wavelength of sound waves:

- hold a vibrating tuning fork over the glass tube
- move the tube up and down to vary the air column length
- the sound made by the tuning fork will get louder and quieter
- the shortest air column that produces a loud sound = $\lambda/4$

Explaining standing waves

Standing waves get their name because they do not *appear* to be travelling in either direction (forwards or backwards). In fact they are caused by two identical waves travelling in opposite directions: the original wave and a reflected wave. In the example above, the tuning fork sent out the original wave, which was reflected at the water surface.

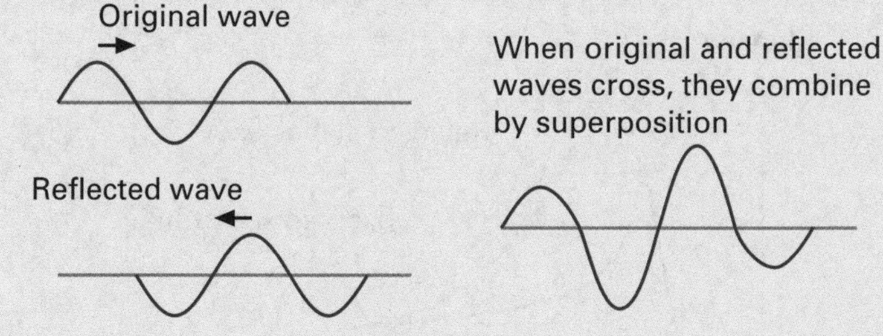

Nodes (and antinodes) are caused by destructive (and constructive) interference.

Exam question answer: page 128

Write a series of revision notes showing how the wavelength and speed of sound can be found using standing waves. Mention the errors that are inherent in this experiment and describe an alternative experiment. (20 min)

Checkpoint 2

In an experiment like the one opposite, the shortest air column to produce a loud (resonant) sound was 16 cm long. What was the wavelength of the note produced? The tuning fork was labelled 512 Hz; what was the velocity of the sound? By examining the diagram carefully, explain why your calculations are really only estimates.

Action point

Kundt's tube can be used to demonstrate standing waves in pipes. Sound waves are sent down a horizontal tube, closed at one end. A layer of fine powder on the bottom of the tube moves into piles. Would the piles be at the nodes or the antinodes? Explain your answer.

Examiner's secrets

The equipment for practical exams normally has to be easily available to schools. Experiments that require standard equipment, like the one above, are often used.

Examiner's secrets

Make sure you can convince your fellow students that you know the difference between *standing* waves and *progressive* waves.

Planck's constant

Classical physics can describe 90% of the universe But it cannot explain things that go too fast, tha are too big, too small or too hot. At the beginning o the 20th century, two problems persisted. Classica physics could not explain how very hot objects emi radiation or why electrons do not spiral into the nucleus. A new, quantum theory beckoned.

The ultraviolet catastrophe 

Careful experimentation led to a series of laws regarding the emission of radiation from hot objects. One such law was the **Rayleigh–Jeans law** that sought to relate the intensity I of radiation at a particular wavelength λ to the absolute temperature T of a black body.

What is a black body?

A **black body** is a perfect emitter and absorber of electromagnetic radiation. An object with surfaces that absorb every wavelength of radiatio incident upon them would look black. Hence the name – black body.

Stars are black bodies! That may appear to be a contradiction so let us think about it for a moment. Light directed at a star is not reflected because the star is a perfect absorber – a black body. But stars also produce and emit their own energy. In fact, they are perfect emitters – black bodies!

What is the simplest type of black body?

The simplest black body is a hole in a box that has been painted black on the inside. Radiation enters and bounces off the internal walls so many times, losing energy each time, that it is eventually absorbed.

Radiation enters the black body (box)

Radiation is trapped within the black body

And the ultraviolet catastrophe?

Unfortunately the Rayleigh–Jeans law only worked at long wavelengths! The catastrophe was that classical physics could not explain why.

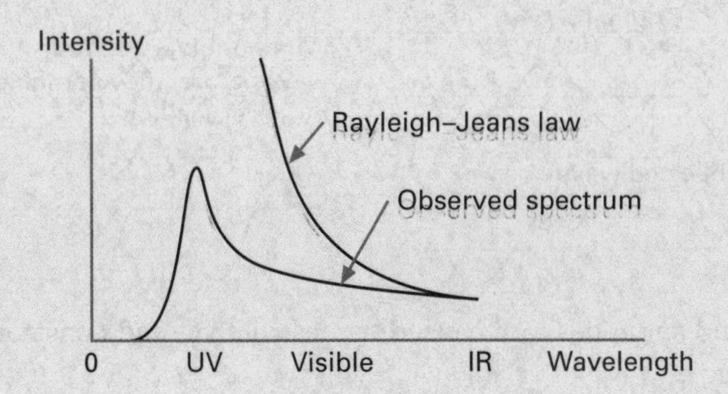

"God does not play dice with the universe."

Albert Einstein

Checkpoint 1

The Rayleigh–Jeans law states that the intensity of electromagnetic radiation (I) is inversely proportional to its wavelength to the power 4, i.e. $I \propto \frac{1}{\lambda^4}$. According to this law, what would happen to intensity as wavelength gets very small?

Check the net

To get a simple introduction to modern physics, including quantum theory, go to www.kapili.com/physics4kids/modern/index.html

Action point

The *intensity* of radiation is actually defined as *power per unit area*. What units does it have?

Links

For more on the *electromagnetic spectrum*, see pages 100–1.

Max Planck's solution

In 1900, Max Planck explained how black bodies emit electromagnetic radiation. Everyone knew that electromagnetic waves were produced when electrons and other charged particles vibrate. Unfortunately, they all incorrectly assumed that they could oscillate at any frequency. Planck's idea was that the oscillators would be more like the strings on a guitar, only able to oscillate with certain frequencies.

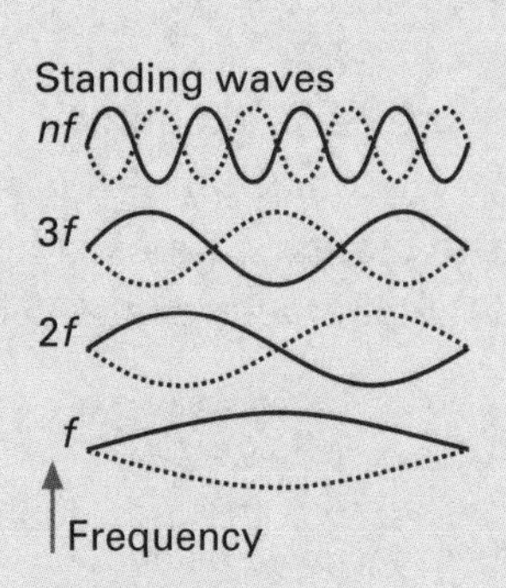

Planck used this idea to show that there is an upper limit to the highest frequency (shortest wavelength) radiation. He demonstrated that the energy of the oscillator E was proportional to its frequency f.

$$E = hf \qquad \text{where } h = 6.63 \times 10^{-34}\ \text{J s}$$

The constant h is called **Planck's constant**.

Vibrating strings have a certain number of set frequencies at which they can produce standing-wave patterns and now Planck was suggesting that oscillators have discrete frequencies too. If there are only certain allowable values for f, $E = hf$ means that there are only certain allowable values for E, namely energy levels or energy states. As f changes in small, discrete steps so too must E. These steps are called **quanta**.

The steps for frequency are $f, 2f, 3f, \ldots, nf$ (where n is a whole number – the **quantum number**). Therefore the steps for energy are hf, $2hf, 3hf, \ldots, nhf$ (see the diagram above).

A footnote

By his own admission, Planck did not fully understand his own work at first. He did not like the idea of energy coming in small packets (quanta) and he kept trying to find an explanation which would allow any value of energy E. Planck received the Nobel Prize for his work in 1919. You can get some insight into Planck's personality by reading his own description of his work.

Exam question answer: page 128

Give a brief definition of each of the following words or phrases, all of which can be associated with the work of Max Planck:

Classical physics; modern physics; black bodies; discrete; continuous; quanta; quantum number; quantum theory; standing waves; energy states. (25 min)

Links

To remind yourself about how *electromagnetic waves* are produced, see pages 100–1.

Links

For more on *standing waves*, see pages 112–13.

Checkpoint 2

Replace f in $E = hf$ by rearranging $c = f\lambda$. Remember that c = velocity of light.

Check the net

To find out more about Max Planck go to www-groups.dcs.st-andrews.ac.uk/%7Ehistory/Mathematicians/Planck.html

"... the whole procedure was an act of despair because a theoretical interpretation had to be found at any price, no matter how high that might be."

Max Planck

Photoelectric effect

> *"Some say that the only thing quantum theory has going for it is that it is unquestionably correct."*
>
> Michio Kaku

Check the net

To find out more about Einstein go to physics.gla.ac.uk/introPhy/Famous/einstein/einstein.html

Examiner's secrets

If you are asked to describe the experiment shown opposite, make sure you state that the gold-leaf electroscope is first charged. By the way, shining light onto it does not charge it!

Checkpoint 1

After reading about Einstein's explanation of the photoelectric effect, explain why the minimum frequency of light required depended upon the metal used.

Action point

Construct a table with two columns. Make a note of each of these puzzling observations in one column and Einstein's explanations in the other.

It has been said that new theories are never accepted, it is just that old men die! Newton believed that light was composed of tiny particles. Young's experiment could only be explained using Huygens' wave theory. But when the photoelectric effect was discovered, the wave theory failed. An unknown, Albert Einstein, had the answer.

Einstein's big year!

Before 1905, Albert Einstein was a humble civil servant. But in one year he published three papers that were to revolutionize physics and immortalize himself. He explained Brownian motion, proposed his theory of special relativity and explained the **photoelectric effect**.

The photoelectric effect

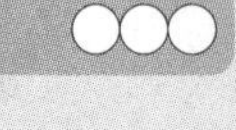

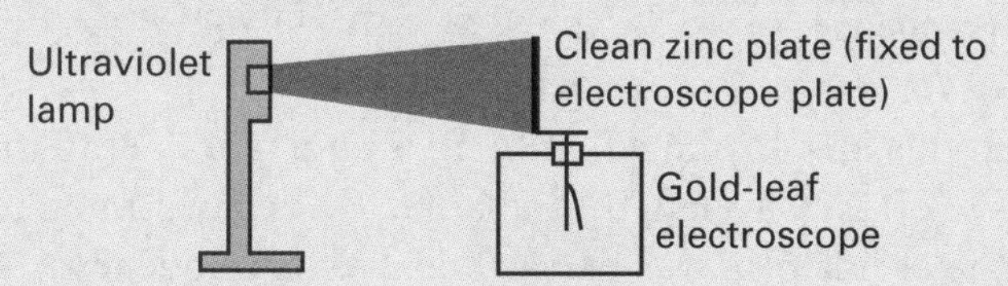

To demonstrate the photoelectric effect:

- use a high-voltage supply to give the electroscope a negative charge, the leaf will deflect
- shine an ultraviolet lamp on the zinc and the leaf will slowly fall
- mysteriously, the light helps the electrons supplied to escape, and so the leaf loses its charge and falls back down – the escaping electrons are called **photoelectrons**

Observations that the wave theory could not explain

- Electrons were not flicked out unless the frequency of light used was above a minimum value (which depended on the metal used). The wave theory suggested that if enough weak light waves were used they should be able to release an electron eventually.
- It did not seem to matter how bright the light was. Some electrons left the metal with more kinetic energy (KE) than others. The maximum kinetic energy (KE_{max}) was dependent on the frequency of the light used, not its intensity.
- Even a very weak beam of high-frequency light released electrons almost immediately. How could such a weak wave, spread evenly over a large area, pick out just a few electrons for release?

Setting the scene

Planck had suggested that oscillating charges could only vibrate with certain set frequencies so that energy too would come in small packets. Einstein then said that radiation is quantized in small packets as well.

Einstein's explanation

Einstein proposed that the electromagnetic radiation (ultraviolet in our demonstration) reached the metal in packets of energy – now called **photons**. He said that the energy carried by the radiation E was proportional to the frequency of the light f, since $E = hf$. His theory was that all a photon's energy is given to just one electron.

It is quite significant that the photoelectric effect takes place from metals since they have so-called ***free* electrons** that are not bound to any one atom. If an electron is sufficiently loosely bound to its metal, some of an incoming photon's energy can be used to break it free. If there is any energy left over, the escaping electron can flee the surface of the metal as the remaining energy reappears as its KE. The minimum amount of energy required to let the electron reach the surface of the metal is called the **work function** W. A metal that holds onto its electrons more strongly would have a larger work function.

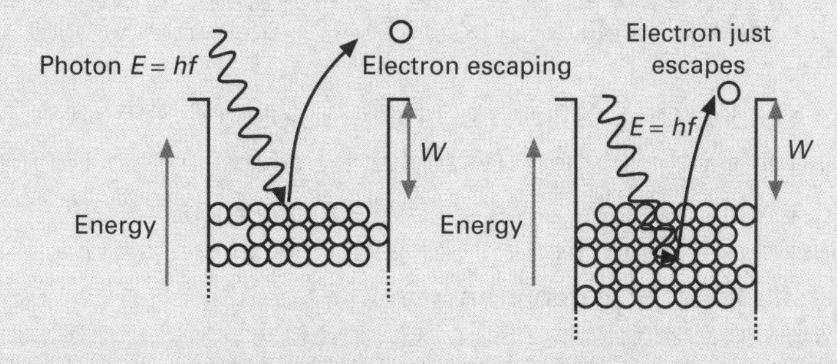

Electrons escape with a variety of kinetic energies

$$\begin{pmatrix}\text{energy of incoming}\\ \text{photon } E\end{pmatrix} = \begin{pmatrix}\text{work done to remove}\\ \text{electron from metal}\end{pmatrix} + \begin{pmatrix}\text{KE of escaping}\\ \text{electron}\end{pmatrix}$$

Electrons from the metal surface require least energy (the work function W) to break them free. Hence they have most kinetic energy, KE_{max}.

$$KE_{max} = E - W = hf - W \qquad \textbf{(Einstein's equation)}$$

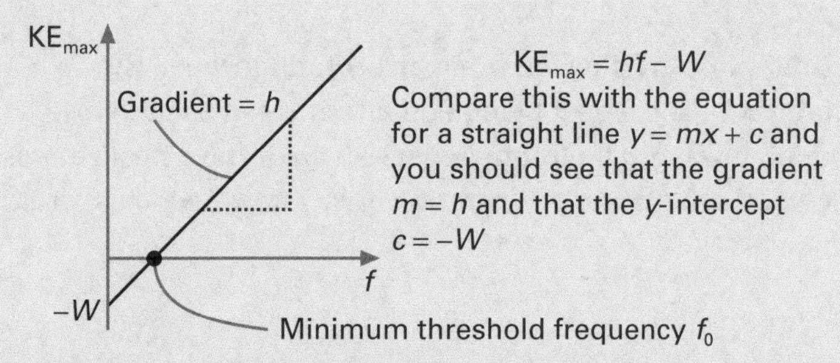

The graph above illustrates Einstein's photoelectric equation.

Exam question

answer: pages 128–9

(a) Use the term *work function* to explain why photons are sometimes unable to remove certain electrons from metals, can get some electrons to the metal's surface and can release others with a variety of kinetic energies.

(b) Why is there such a short delay between shining appropriate light on a metal and the escape of the first electrons?

(c) Why does intensity not affect KE_{max} of photoelectrons? (20 min)

"Science can purify religion from error and superstition. Religion can purify science from idolatry and false absolute."

Pope John Paul II

Checkpoint 2

What other scientific phrase can be used to replace the term *free electrons*?

Watch out!

Work function is often quoted in electron-volts (eV).

Examiner's secrets

The graph of KE against frequency for a different metal would have the same gradient but the intercept would be different.

Don't forget

To get the gradient of a graph you have to wise up! Gradient = y/x.

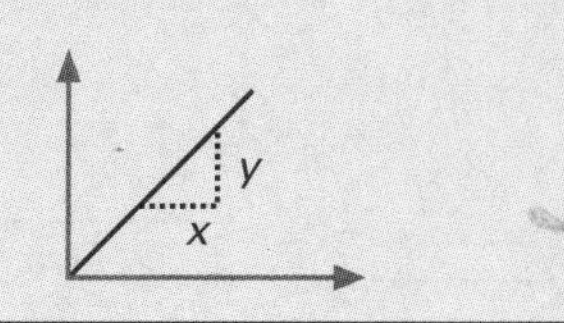

Quantum behaviour

*"The **most** we can know about is in terms of probabilities."*

Richard Feynman

We know that light travels in straight lines, right? We also know that light has a split personality – it can behave as a wave or as a particle. So what is this chapter about? Well, light and all particles of matter, including electrons and quarks, share the same quantum behaviour. This chapter looks at this behaviour in a new light!

Getting about

To understand quantum behaviour we need to challenge some of the ideas we already hold about light and other particles of matter. What we see and measure are the consequences of a pattern of behaviour that appears to obey few rules.

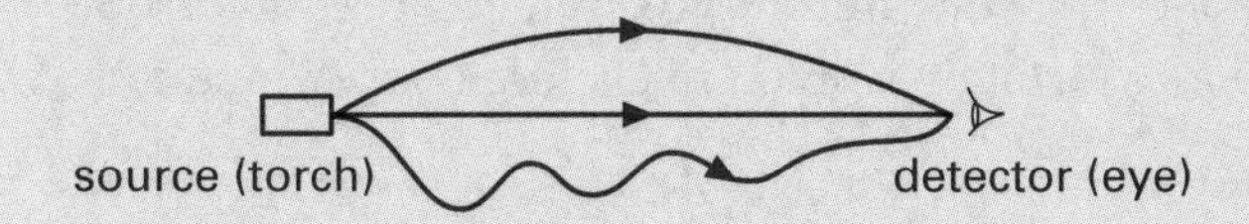

Links

Have a look at the spread about *Planck's constant* on pages 114–15.

We learn that light travels in straight lines as photons. We also know that, as a wave, light diffracts. Both of these effects are easy to produce. What we need to consider is that actually photons explore *all* pathways from source to detector. None of the pathways shown above is forbidden in the strange quantum world.

Rotating clock

Have you ever used a trundle wheel to measure distances? The wheel measures out a distance of 1 metre for each revolution using a rotating arrow.

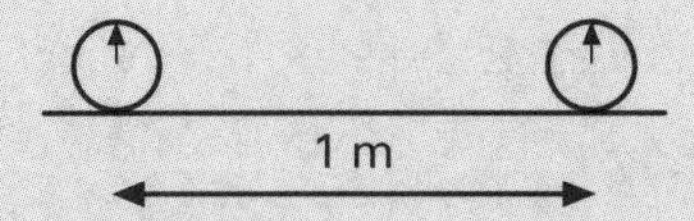

Imagine that photons have an on-board rotating wheel, the circumference of the wheel being equal to the wavelength of corresponding wave. The frequency with which the wheel rotates is the frequency of the light and the rotating arrow now becomes a rotating clock.

$$f = \frac{E}{h}$$

Links

See pages 108–9 for a more traditional treatment of superposition.

With this rotating clock idea we can consider the arrow as an indication of the phase of the wave. When two photons from the same source arrive at a detector in phase, the arrows of the clock point in the same direction. This gives rise to constructive superposition. Arrows in antiphase result in destructive superposition.

So, this is just a new way to look at an old idea. The rotating arrows are called *phasors*.

Checkpoint 1

What is the rate of rotation of the phasor with an energy of 3×10^{-19} J?

Going everywhere

If you accept that photons explore all possible pathways then you are bound to ask how this fits with reflection and refraction. After all, the evidence couldn't be clearer that light goes only one way!

To appreciate the 'explore all paths' rule you need to consider the contribution each path makes to the overall effect.

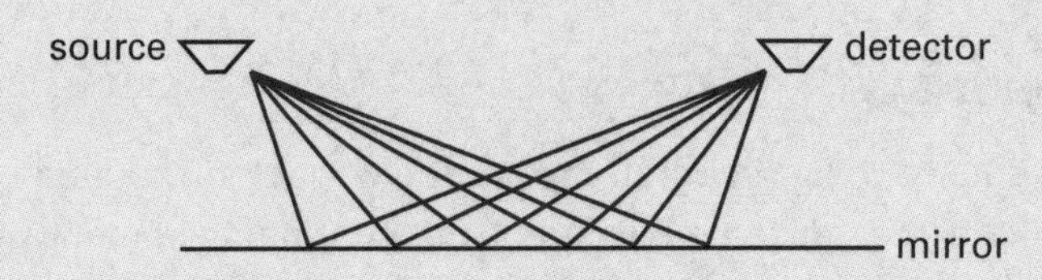

The diagram above shows some of the paths photons can take from source to detector, reflecting off a mirror. Imagine a trundle wheel or rotating clock moved along each path. On arrival at the detector the phasors can be added together just like vectors. The sum of all the phasors is related to the amplitude of the final wave.

It turns out that most of the final amplitude comes from paths near the middle of the mirror – so reflection works! Paths from the ends of the mirror have large differences in trip time and so the corresponding phasors almost cancel out.

Probability and amplitude

The length of the resultant phasor from exploring all paths is used to calculate the probability that a quantum of energy will arrive. If the probability is high, then the intensity is large. In fact, probability is proportional to the square of the resultant phasor amplitude. So, we don't have waves and we don't have particles – just probability, or quantum behaviour!

"The electron is just a smear of probability."

R H Coats

Links

See pages 4–5 for vector addition.

Checkpoint 2

In a double slit experiment, photons take one of two paths A and B. The resultant phasor amplitude for path A is four times that of path B. What is the probability that photons will take path A compared with path B?

Exam question answer: page 129

A beam of electrons has an energy of 16×10^{-17} J.

(a) What is the speed of these electrons?

(b) What is the frequency of the rotating phasor for each electron?

(c) How far will the electrons travel for one rotation of the phasor? (8 min)

Atomic line spectra

"An expert is a man who has made all the mistakes which can be made, in a very narrow field."

Niels Bohr

Checkpoint 1

This may not be the first time that you have come across a technique that uses colour to identify chemicals. What are *flame tests*?

Action point

Produce a flow diagram to show how theories about the nature of light changed over time. Improve this chart by adding really important extras – breakthrough experiments, puzzles that new theories could explain, etc.

Examiner's secrets

It is a common error for students to get absorption and emission the wrong way round. *Down and out* might help you remember it!

Action point

After reading this spread, draw a diagram, like the one opposite, to show how this imaginary element could absorb a more energetic photon.

We know that white light consists of seven colours, the visible spectrum. By the mid-1800s astronomers were using the fact that hot gases emit only certain characteristic colours as fingerprints to identify the elements present in stars. But it was not until 1913 that Niels Bohr was able to use quantum theory to explain these characteristic spectra.

Line spectra

A narrow source of light containing a hot gas, such as sodium or neon, can produce a recognizable line spectrum containing some wavelengths of light but not others. Spectrometers are used to view line spectra.

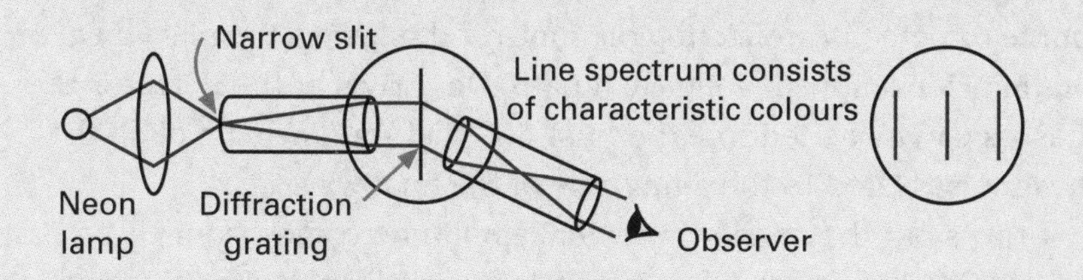

There are two types of line spectra.

- **Emission spectra** show the light emitted by hot gases, as previously described.
- **Absorption spectra** are obtained when white light passes through a cool gas. The gas absorbs certain wavelengths from within the white light. The absorbed wavelengths are characteristic of the gas and can therefore be used to identify it.

What causes line spectra?

It was obvious that atoms of a particular element could only emit or absorb certain wavelengths of light. The question was why?

Einstein had suggested that electromagnetic (including light) radiation is carried in small packets or **photons**. When a photon strikes an atom, the target atom's electrons absorb its energy. When electrons lose energy, they emit photons.

The fact that electrons in particular atoms can only absorb, or emit, very specific wavelengths of light means that they can only absorb, or emit, very specific amounts of energy carried in very specific photons.

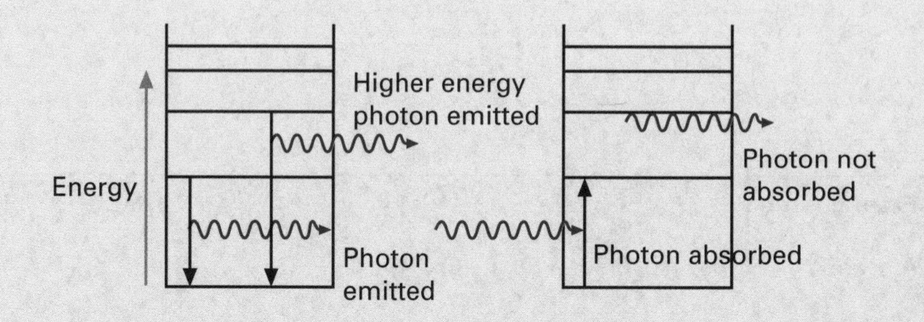

These **energy-level diagrams** (or *ladders*) show that electrons can only emit or absorb specific photons. Each element has its own unique energy-level diagram.

Try to remember that:

- → it is as if each element has its own ladder
- → every ladder has rungs in different places so an atom's electrons can only have certain amounts of energy
- → each element is characterized by the energy levels available to its electrons

When an atom emits light, one of its electrons falls from a higher to a lower energy level. A single photon is emitted whose energy equals the energy difference between the two levels. This means that a bigger fall for the electron will produce a more energetic photon.

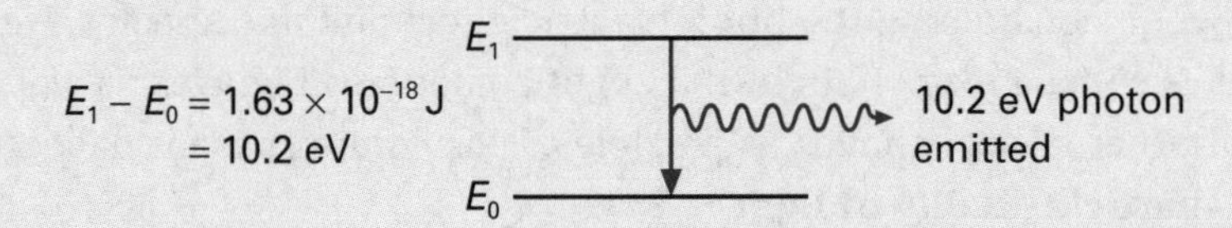

This diagram shows an electron falling from E_1 to E_0 in a hydrogen atom. To pump the electron back up to E_1 (its **first excited state**) requires a photon of exactly 10.2 eV to be absorbed. The atom would then be unstable and the electron would very quickly fall back to E_0 (its **ground state**), emitting a 10.2 eV photon.

This idea of elements having energy ladders with characteristically fixed rungs, allowing only certain jumps or falls, explains why hot gases emit *signature* photons. As these photons have specific energies, the wavelengths of light that can be observed are limited.

Similar ideas apply to absorption spectra:

- → if an incoming photon is to be absorbed, it must deliver exactly the right amount of energy to lift an electron from one energy level, or rung, to another
- → if it does not deliver this amount, it will not be absorbed
- → energy-level spacings are characteristic of elements, so each element can be identified by the photons it absorbs

Line spectra of solids and liquids

So far, we have only mentioned gases. The line spectra of solids and liquids are much more complicated because their atoms are more tightly packed. This allows electrons of neighbouring atoms to affect one another producing spectra with many, many wavelengths.

Ionization

If an atom absorbs a photon of sufficient energy, it can be possible for an outer electron to leave the atom altogether. This is called **ionization**.

The jargon

An *electron volt* (eV) is the amount of energy an electron gains when it is accelerated across a potential difference of 1 volt. 1 eV = 1.60×10^{-19} J

Checkpoint 2

Use $E = hf$ to calculate the frequency of the photon emitted in the diagram opposite. Now calculate its wavelength.

The jargon

The *ground state*, or *level*, is the lowest energy state available to an electron.

Examiner's secrets

Very often the highest energy level is not shown. This is because it has the value *zero*. An atom is ionized when an electron has enough energy to jump to this level from the ground state.

The jargon

The *first excited state* is the first energy level, above the ground state, that is available to an electron.

Examiner's secrets

You are expected to know specifically about the energy levels for the *hydrogen* atom.

Speed learning

Try to understand how line spectra are produced; the theory will then be easier to remember.

Exam question answer: page 129

(a) Stars produce both emission and absorption spectra. With reference to energy-level diagrams, explain how both types of spectra are produced.

(b) Why can spectral analysis identify individual elements? (20 min)

De Broglie's equation and atomic models

> *"If one does not feel dizzy when discussing the implications of Planck's constant* h *it means that one does not know what one is talking about."*
>
> Niels Bohr

In 1913, Bohr's work had been a major breakthrough. For the next ten years or so, others tried to build on his results, without success. Then Prince Louis de Broglie took a huge step forward. He suggested that all particles, including electrons, had split personalities. They could exist not only as particles but also as waves!

Wave–particle duality of light

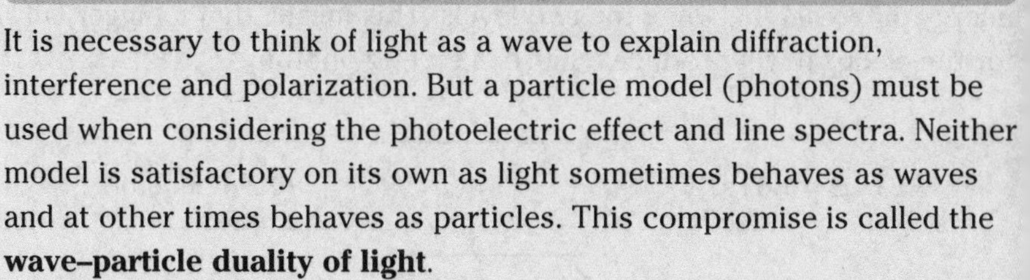

It is necessary to think of light as a wave to explain diffraction, interference and polarization. But a particle model (photons) must be used when considering the photoelectric effect and line spectra. Neither model is satisfactory on its own as light sometimes behaves as waves and at other times behaves as particles. This compromise is called the **wave–particle duality of light**.

You should use *two rules to decide how to think of light* (or any other form of electromagnetic radiation) in a given situation:

→ use the particle model when light interacts with matter, e.g. when it strikes the surface of a metal in the photoelectric effect
→ use the wave model when light goes through a gap of similar width to its own wavelength, e.g. single-slit diffraction

Checkpoint 1

Why can wave–particle duality be thought of as a complementary principle?

Wave–particle duality of all particles

De Broglie's great advance was to suggest that *all particles, not just light, might have a dual nature*. He related the wavelength λ of a particle to its momentum p in his equation:

$\lambda = h/p$ Where h is Planck's constant

This equation shows the dual nature of particles, as wavelength is a wave property whereas momentum is associated with particles. As

momentum = mass × velocity or $p = mv$
$\lambda = h/p$ becomes
$\lambda = h/mv$

This means that if velocity is constant, wavelength is inversely proportional to mass. The wavelength of electrons is approximately 1×10^{-10} m, roughly the spacing of atoms in solids that can therefore diffract them.

Action point

Work out your own de Broglie wavelength. Find the product of your mass and the fastest speed you can run. Divide this into Planck's constant. Hence explain why *you* can never diffract! (*Hint* diffraction is greatest when the gap is the same size as the wavelength.)

Check the net

To find out more about the work of Prince Louis de Broglie go to www.chembio.uoguelph.ca/educmat/chm386/rudiment/tourquan/broglie.htm

The Rutherford–Bohr model of the atom

The quantum theory, and Niels Bohr, had another trick to reveal! The nucleus contains positive charge, electrons are negative – why does electrical attraction between them not cause the atom to self-destruct?

Rutherford's original model of the atom could not explain this mystery. It could not show why electrons do not lose energy continuously and spiral into the nucleus. Bohr's use of quantum theory, with fixed electron orbits, solved the problem.

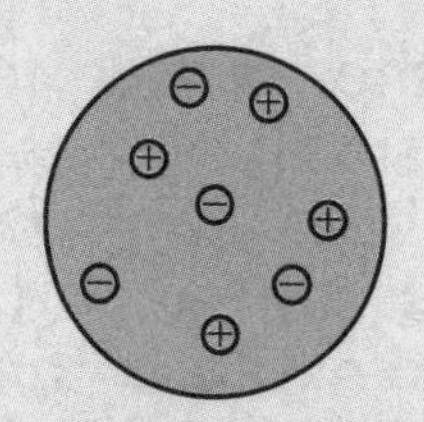
Thomson's 'plum-pudding' model of the atom (1898)

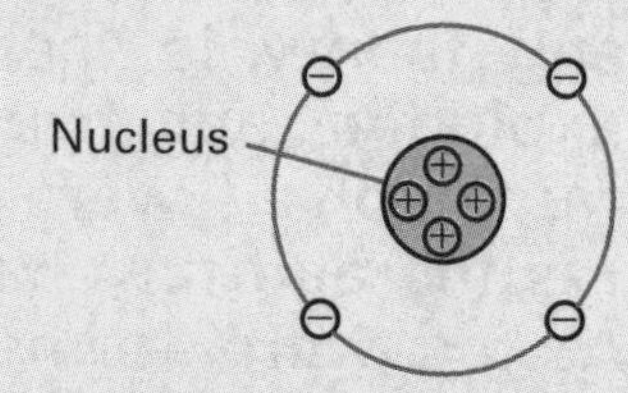

Rutherford's model of the atom (1911) included orbiting electrons

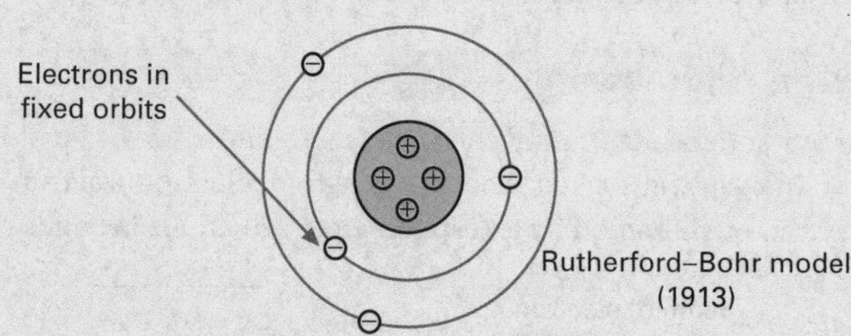

Rutherford–Bohr model (1913)

Models of the atom developed over a number of years. The main points about each model shown are as follows:

- *J. J. Thomson* viewed the atom as a ball of positive charge containing electrons dotted inside, like plums in a pudding!
- *Rutherford* knew that the atom was mostly empty space with positive charge at its centre. He thought that the much lighter electrons would orbit the nucleus like planets around the Sun.
- *Bohr* refined Rutherford's model using quantum theory. As energy could only be absorbed or emitted in well defined packets, he proposed that electrons in higher orbits would have more energy than those below. Energy coming in defined quanta meant that only certain energy jumps could be made, allowing a limited number of electron orbits. This became known as the Rutherford–Bohr model.

Other advances

- Rutherford discovered positive protons in 1919.
- Bohr's theory worked only for hydrogen. In 1925, Schrödinger developed a wave-mechanics model to solve this problem.
- In 1932, James Chadwick discovered neutrons.

Why electrons do not spiral into the nucleus

Electrons can be considered as waves. If an electron is to squeeze towards the nucleus, its length (or wavelength) must decrease causing its energy to increase. If small enough to fit inside the nucleus, an electron would be too energetic for the protons to hold it!

Exam question answer: pages 129–30

(a) Why would it be unlikely for one to observe matter behaving as waves in everyday life? Why do atoms not self-destruct? Draw the Rutherford–Bohr atomic model and a corresponding energy-level diagram. (20 min)

(b) Calculate the speed and de Broglie wavelength of electrons of energy 1.0 eV. (8 min)

Checkpoint 2

In what way do the Thomson and Rutherford models agree? In what way do they disagree?

Action point

Sketch the Rutherford–Bohr model of the atom as shown here. Update it by including appropriately labelled protons and neutrons.

Links

To find out more on Rutherford (and alpha-particle scattering), see *the nuclear atom*, pages 40–1.

Examiner's secrets

Find the learning method that suits you best! Some people find it easier to remember images. A football stadium could represent Rutherford's nuclear model of the atom. Most of the ground (atom) is empty, the nucleus is like a 10 pence coin at the centre of the pitch and the electrons are like flies buzzing around the perimeter.

Particle diffraction and probing the atom

"I recognize that many physicists are smarter than I am . . . I console myself with the thought that I have broader interests than they have."

Linus Pauling

Watch out!

Some books refer to the pattern of rings produced in this experiment as a diffraction pattern, others call it an interference pattern. The waves had to diffract before they could overlap and interfere, so both are correct!

Links

Use this result to confirm 'why electrons do not spiral into the nucleus' – see page 123.

Checkpoint 1

What is the range of wavelengths for X-rays? Why are X-rays suitable for studying the structure of crystals?

Test yourself

Try to remember the prefixes used to describe powers of ten. For example, nanometres (nm) = ?

The big-bang theory is that all the matter in ou universe exploded out from a small space. Th temperatures involved were huge! As time passec temperatures fell and tiny particles started to clum together. Atoms were formed. To explore the tin particles that formed atoms we have to mimic th big bang using particle accelerators.

Electron diffraction

For diffraction to be significant, the wavelength of the waves must be similar to the gap width used. The wavelength of electron waves is suc that they can be diffracted by the spaces between atoms in solids.

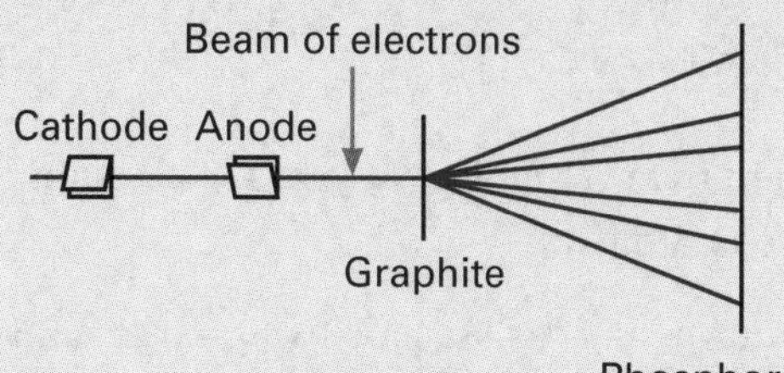

This diagram shows electron diffraction by graphite. The experiment provides evidence for the dual nature of electrons:

- the light and dark interference pattern is characteristic of waves
- the interference pattern is made up of lots of individual flashes on the screen, each caused by an individual electron particle
- a magnet can bend the electron beam, shifting the whole pattern an providing evidence of electrons as charged particles

There are other important results and conclusions from this experiment.

- Increasing the potential difference between anode and cathode gives the electrons more kinetic energy. The rings then shrink in size, becoming brighter. The electron's wavelength has decreased.
- Just one electron in the beam, at any one time, is enough to produce an interference pattern. One electron can pass through both slits as a wave, then recombine to cause a single flash on the screen.

As the planes of atoms act like a diffraction grating consisting of many slits, $d\sin\theta = n\lambda$ can be used to find the interatomic spacing d (θ is the angle through which the electrons have been diffracted, n is the order of the maximum and the wavelength can be calculated from the anode–cathode potential difference).

X-ray diffraction

X-rays are diffracted by the parallel planes of atoms found in crystals. The layers of atoms form the equivalent of a diffraction grating. This technique suggests that the layers are separated by approximately 0.1 nm. X-ray diffraction has been used to study complex molecules like DNA, providing fundamental information for the field of genetics.

Probing deeper into atoms

To probe the nucleus, short wavelengths are required. Gamma rays have shorter wavelengths than X-rays and they originate in the nucleus. But unfortunately it is difficult to produce a useful beam of gamma rays.

High-energy alpha particles

Early alpha-particle scattering experiments used natural radioactive sources. Higher energy alphas could get closer to the nucleus. At such close ranges they were deflected by large angles that electrical repulsion alone could not explain. The strong nuclear force was responsible.

High-energy electrons

Large linear accelerators can produce electrons of such high energies that their wavelengths are short enough to allow them inside the nucleus. As they are charged, electrons are attracted by protons and so give information on charge distribution within the nucleus. Electrons are leptons; therefore they do not interact with the strong nuclear force.

Protons and neutrons

Protons and neutrons can be used to investigate the nucleus. They both experience nuclear forces, and the proton also experiences electrical repulsion. Therefore they provide different information from electrons. Slow neutrons, from nuclear reactors, can be used to investigate the atomic structure of solids.

Particle accelerators

Accelerating particles close to the speed of light can provide the high energies required to probe inside nuclei. The particles are crashed into nuclei or other particle beams. New particles can be created as energy transfers into mass. These experiments are used to search for fundamental particles that are not made of other particles. Our current standard model suggests that electrons and quarks are fundamental particles. Protons and neutrons both contain two types of quarks.

- **Linear accelerators** speed up charged particles in straight lines. They can only accelerate the particles once, as they travel along the machine. Therefore they can be several kilometres long.
- **Circular accelerators** include cyclotrons and more modern synchrotons. Synchrotons use electromagnets to accelerate charged particles repeatedly as the particles go around a circular path. The synchrotons at CERN, in Geneva, use so much electricity that they are only used in the summer when electricity is cheaper!

Exam question answer: page 130

(a) How can electron diffraction identify an unknown metal?

(b) What is a fundamental particle?

(c) What fundamental particles make up protons, neutrons and electrons?
(15 min)

Links

To find out more about the conclusions from Rutherford's early alpha-particle scattering experiments, see pages 40–1.

Checkpoint 2

The following list of particle accelerators and their related discoveries has been muddled. Rearrange it properly:

cathode ray tube	bottom quark
electron microscope	evidence for quarks
cyclotron	structure of nuclei
linear accelerator	structure of viruses
synchroton	mass and charge of the electron

Check the net

The CERN homepage is a good starting point to find out more on this topic.
Go to www.cern.ch/

"I do not know with what weapons World War 3 will be fought, but World War 4 will be fought with sticks and stones."

Albert Einstein

Answers
Waves and oscillations

Types of waves and their properties

Checkpoints

1 Both would travel faster, causing them to bend or refract.
2 Waves transfer energy.
3 Speed in m s^{-1}, frequency in Hz, and wavelength in m.

Exam question

(a) Vibrations at 90° to the direction of the wave motion cause transverse waves. Vibrations moving in the same direction as the wave motion cause longitudinal waves.
(b) It is not possible to polarize longitudinal waves because they only vibrate in the direction of the wave's motion. If vibrations were blocked in this direction, there would be no wave.
(c) Any electromagnetic wave, e.g. light, is transverse. Sound waves are longitudinal.
(d) See page 113 for diagrams showing how transverse standing waves (microwaves) and longitudinal standing waves (sound waves) can be produced.

Examiner's secrets

Try to pick out the keywords used by the examiner to tell you what to do. In this question, the examiner is expecting you to spend longer on parts (a) and (b), as parts (c) and (d) only require examples to be given.

Electromagnetic spectrum

Checkpoint

1 The only real difference is the method of their production.

Exam question

Physics is an evolving subject and so there are many possible answers to this question. Suggestions include:
Forces and motion You could discuss the ideas of Aristotle, Galileo, Newton and Einstein.
Energy and heat You could compare the caloric and kinetic theories. You could describe the work of Rumford and Joule. Brownian motion could be included. You could bring this timeline up to date by mentioning the quest to reach 0 K.
Waves and particles You could explain that Newton's particle model of light was preferred to Huygens' wave model until after Young's experiment. You could describe the roles of Einstein, Bohr and de Broglie in the development of the wave–particle duality theory.

The *nature of radioactivity*, the *development of particle physics* or a discussion of *advances in astronomy* could all form the basis of an excellent answer to this question.

Examiner's secrets

This type of question, requiring knowledge that is developed over an entire course, is called *synoptic*. Synoptic questions are a feature of all A2 courses.

Reflection and refraction

Checkpoints

1 The Moon does not produce any light. We can only see it if light from stars, e.g. the Sun, bounces off it and into our eyes. By day, our view of the sky is dominated by the Sun so the Moon is not visible.
2 Both would still slow down, but they would travel straight on without bending.
3 It would bend away from the normal.

Exam question

(a) (i) 3.40°, (ii) 10.14°
(b) The frequency of the light in both media is 7.5×10^{14} Hz. The wavelength of violet in the unknown medium is 333 nm.

Examiner's secrets

Get a list of all the formulae and constants that will be provided in your exams, it will help reassure you during revision. With regular revision, you will need to use it less and less.

Total internal reflection and fibre optics

Checkpoints

1 41.8°
2 60.5%
3 It changes direction, slows down and its wavelength decreases.
4 The wavelength is reduced by the same proportion as the speed, i.e. by ²⁄₃ in this case.

Exam question

(a) Try to maintain a balance in essay-style questions that contain multiple parts. In this case you could choose two points relating to technological changes (perhaps in medicine where doctors use endoscopy to see inside the body and in telecommunications where a single optical fibre cable can carry up to half a million telephone calls at the same time).

Keyhole surgery made possible by fibre optics is much simpler than conventional surgery and can often mean that patients spend much less time in hospital which is more cost effective.

The use of fibre optic technology has contributed to cheaper telephone call charges. Fibre optics have opened the way for digital television which will allow many more television channels. This may persuade people to spend even more time watching TV. Cheaper telephone call charges may allow people to keep in closer contact with friends and family living abroad.
(b) Fibre optics can be used to light road signs, they are found in some security fences, the speed of information transfer along fibre optics makes the internet accessible to many people and fibre optics can even be used for decoration!
(c) See the diagrams on page 104.

> **Examiner's secrets**
>
> Read each question carefully. It would be easy to do part (a) of this question and leave nothing to write about in part (b).

Diffraction and resolution

Checkpoints

1 This phenomenon would help you!
2 0.017 5°

Exam question

(a) The entrance to harbours acts in the same way as a gap in a ripple-tank experiment. The incoming water waves are diffracted, spreading out in a semicircular pattern. Each time that the resulting semicircular wavefronts reach the shoreline they erode it. It is as if they are taking semicircular 'bites' from the shoreline so that it becomes semicircular. See the upper diagram on page 106.
(b) The metal grid prevents microwaves from escaping. As the gaps in the grid are much smaller than the wavelength of microwaves, they cannot squeeze through. The wavelength of light is much smaller than the grid gaps so light can pass in and out unaffected.

> **Examiner's secrets**
>
> If more time been allocated to this question a more detailed answer would be expected. Explaining how Huygens' construction predicts the formation of curved secondary wavefronts, i.e. diffraction, could extend part (a). Part (b) is similar to something you may have come across at GCSE: hot short-wavelength rays from the Sun can squeeze into greenhouses but as the rays cool, their wavelengths increase and they cannot get out again. Just be careful not to stray too far off the point!

Superposition

Checkpoints

1 A supercrest would be formed.
2 The path difference between wavelets from A and C is a whole wavelength, between A and B it is half a wavelength.

Exam question

(a) The principle of superposition predicts that if a crest is added to a similarly sized trough, zero displacement will result – destructive interference. If we were considering sound waves, this would result in silence. Noise is defined as any unwanted sound. To combat noise pollution, a microphone could be used to sample the unwanted noise. An amplifier could be used to reverse the sound; the result could be called antisound. If the antisound is played over the original, the original can be cancelled.
(b) Establish a line of zero disturbance by moving your hand in a horizontal plane at a certain height. Show that the result of adding a crest (move one hand up) to a trough (move the other hand down) would be zero disturbance (referring back to our original horizontal sweep). Extend the answer by explaining how supercrests and supertroughs could be formed.

> **Examiner's secrets**
>
> Examiners are impressed when candidates can link different parts of their course. Synoptic assessment will make this skill even more important. For example, 'antisound was unsuccessfully tested in the car industry. Instead the engineers chose to avoid materials whose resonant frequencies would be reached during normal use of the car, thus reducing boominess.'

Interference

Checkpoints

1 0.33 m
2 1×10^{-4} m

Action point

4 000

Exam question

(a) Coherence means constant phase difference. To get constructive interference it is essential that the combining waves are coherent – that they always travel in step. For example, if two crests always reach a point at the same time a supercrest will be formed. But if a crest only occasionally arrives with another crest, no clear result will emerge. Even if the two waves are always out of phase by the same amount, say one wave is always ahead of the other by half a wavelength, a clear interference pattern will form. The essential requirement to get a clear interference pattern is that there must be a constant phase difference. Obtaining coherent wave sources can be difficult and so interference patterns are not very common in everyday life.
(b) Laser light is coherent. Light from an ordinary light bulb is incoherent.
(c) Use of $d \sin \theta = n\lambda$

$0.4 \times 10^{-3} \sin \theta = 2 \times 600 \times 10^{-9}$

$\sin \theta = 3 \times 10^{-3}$

$\theta = 1.72°$

> **Examiner's secrets**
>
> Try to minimize spelling mistakes. Some words, like *synchronous*, are often incorrectly spelt. Why not build up a private word bank of difficult spellings?

Standing waves

Checkpoint

1

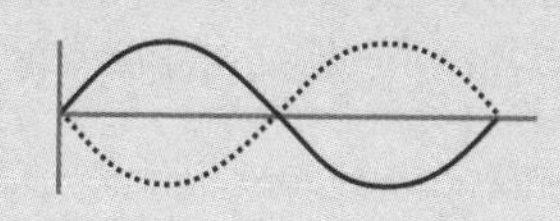

2 Wavelength = 0.64 m, velocity of sound = 327.68 m s^{-1}.

By examining the diagram of sound standing waves on page 113, you can see that the standing-wave pattern extends into the air beyond the top of the glass tube. (The air above the glass tube acts as an extension to the tube itself.) This means that when we measure the length of the glass tube above the water level, we are underestimating the length of the standing-wave pattern.

Exam question

A good answer would be to describe how to measure the wavelength of sound waves as detailed on page 113. This would allow you to mention end corrections as described in the answer to checkpoint 1, see above. Having calculated wavelength and noted the frequency f of the tuning fork, use $v = f\lambda$ to get velocity v.

You could then go on to describe using Kundt's dust tube. Dust is evenly deposited inside a glass tube. One end of the tube is closed and a loudspeaker is attached to the other end.

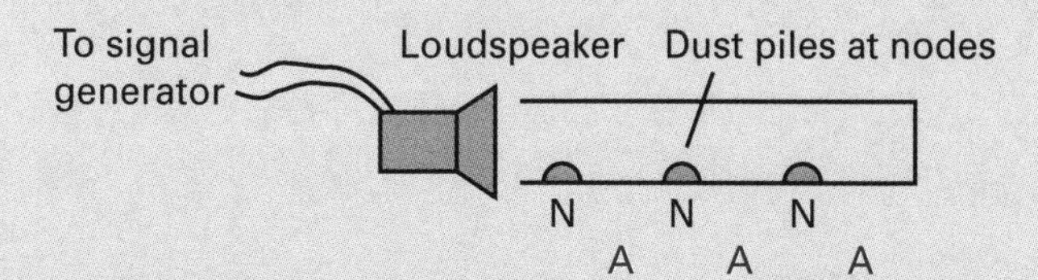

A signal generator is connected to the speaker so that sound waves can be sent down the tube. When a standing-wave pattern is set up, the dust vibrates violently at the antinodes (A). This causes the dust to build up at the nodes (N). The position of the nodes and antinodes can be observed and measured to give an accurate value for the wavelength of the sound waves. The frequency can be read off the signal generator and the velocity can be calculated as before.

Examiner's secrets

Check which experiments are specifically mentioned on your syllabus and learn them!

Planck's constant

Checkpoints

1 Intensity would get very high.
2 $E = hc/\lambda$

Exam question

Classical physics 19th century physics. The most important feature of classical physics is that the laws of Newtonian mechanics are obeyed.
Modern physics Theories that have emerged since around the turn of the 20th century, e.g. quantum theory and the theory of relativity.
Black bodies Hypothetical objects that would be able to absorb all electromagnetic radiation that fell on them.
Discrete Consisting of distinct or separate parts. In quantum theory this word could be used to describe individual packets of energy, or quanta.
Continuous Gradually changing in value in an unbroken series. This word could describe the electromagnetic spectrum in a classical sense.
Quanta The smallest quantities of a physical property, such as energy, that a system can possess according to the quantum theory.
Quantum number One of a set of integers (whole numbers) characterizing energy levels of particles or a system of particles.
Quantum theory A theory regarding the behaviour of physical systems based on Planck's idea that they can possess only certain properties, e.g. energy, in discrete amounts (called quanta).
Standing waves Also called stationary waves. They are waves produced by interference between two waves travelling in opposite directions. Standing waves have unchanging amplitudes at each point along their axes. Electrons have only certain energy levels available to them as they are trapped within standing-wave patterns in the atom.
Energy states Also known as energy levels. One of the states of constant energy in which a system may exist. Each energy state is separated by finite amounts of energy.

Photoelectric effect

Checkpoints

1 Metals that hold onto their electrons more strongly have larger work functions. This means that more energy is required to allow them to escape from the metal. $E = hf$ suggests that energy E is directly proportional to frequency f. As there is a minimum amount of energy that is required to let electrons escape, and as this value varies from metal to metal, there is also a minimum frequency of light required. This also varies from metal to metal.
2 Delocalized electrons.

Exam question

(a) The work function is the minimum amount of energy required to release an electron from a metal. If a photon delivers the work function exactly, it will be able to let the electron reach the surface of the metal. If less than the work function is delivered, not even this will happen, but if the photon delivers more energy than the work function any excess will reappear as kinetic energy as the electron moves away from the surface of the metal. Two electrons could leave the metal with different kinetic energies. This could be because photons of different energies released them or perhaps they were bound more or less tightly to the metal.

(b) Einstein's theory was that each photon delivered its packet of energy to just one electron. If the first photon to strike the metal had sufficient energy, it could release an electron immediately. There was no need to build up enough energy, which would take longer.

(c) Brighter more intense light does not mean that the light photons are more energetic. It simply means that there are more photons. As each electron receives just one photon, more photons mean that more electrons can escape, not that those that do escape will be more energetic. So intensity does not affect the maximum kinetic energy of photoelectrons released from metals in the photoelectric effect.

Examiner's secrets

Try to use the correct scientific terminology wherever possible. When used appropriately, this technique will not only make your answer more professional but it could also save you time. So be specific! Don't say *electron* when you could mean *free electron* or *photoelectron*.

Quantum behaviour

Checkpoints

1 $f = E/h$
$= 3 \times 10^{-19}/6.6 \times 10^{-34}$
$= 4.55 \times 10^{14}$ Hz.

2 Probability is proportional to amplitude squared.
$P_A/P_B = 4^2/1^2 = 16/1$
So $P_A : P_B = 16 : 1$

Exam question

(a) $v^2 = \dfrac{2 \times 16 \times 10^{-17}}{9.11 \times 10^{-31}}$
$v = 1.9 \times 10^7$ m s^{-1}

(b) $f = \dfrac{16 \times 10^{-17}}{6.6 \times 10^{-34}} = 2.4 \times 10^{17}$ Hz

(c) $\lambda = v/f = \dfrac{1.9 \times 10^7}{2.4 \times 10^{17}}$
$= 7.92 \times 10^{-11}$
$= 0.079$ nm

Atomic line spectra

Checkpoints

Flame tests are used to identify chemicals on the basis of the colour they produce on burning. Metal ions can be identified in this way. For example, Na^+ burns with an intense golden yellow colour. A Nichrome or platinum wire is cleaned by repeatedly dipping it in hydrochloric acid, then heating it in a roaring bunsen burner flame. When no colour is given to the flame by the wire we assume that the wire is clean. It is moistened with dilute hydrochloric acid and then used to pick up a sample of the compound, which is then held in a colourless flame for identification.

$f = 2.46 \times 10^{15}$ Hz (remember to convert the energy of the photon in eV into J before using $E = hf$).
Wavelength = 122 nm.

Exam question

(a) and **(b)** Emission spectra show the light emitted by hot gases. Photons are emitted when electrons in hot gases fall from higher to lower energy levels. Only certain energy levels are available to the electrons, so only certain energy falls are possible. The energy levels and possible energy falls are characteristic of the element in question. The energy of the photon is equal to the difference in the two energy levels. The energy (and wavelength) of the emitted photons thus identify the element that produced the light. Absorption spectra are formed when white light emitted from the very hot core of a star passes through cooler outer layers. The atoms in the cooler regions can only absorb photons of particular energy and therefore of particular wavelength. It is the electrons within the target atoms that actually absorb photons. These electrons have only certain energy levels available to them. If the incident photon can deliver exactly the right amount of energy to lift an electron to a higher energy level, it will be absorbed. If it cannot, the photon will not be absorbed. As the energy levels available to the electrons are characteristic of the target atom, by examining which photons are absorbed astronomers can identify which elements are present within the star under observation. Refer to the energy-level diagrams on page 120.

Examiner's secrets

Examiners often get ideas for questions from the science that is reported on the TV, or the radio, or in newspapers. They set the questions well in advance of the date that you will sit the examination so it may well be worthwhile to keep yourself informed of what is happening throughout your course. Astronomy is frequently reported by the media, for example.

De Broglie's equation and atomic models

Checkpoints

1 The wave and particle models complement one another. It is not possible to use only the wave model, or only the particle model, to describe matter or radiation. Both models are required.

2 The Thomson and Rutherford models agree that atoms contain both positive and negative charges. However, the Thomson model has positive and negative charges evenly distributed within the atom. The Rutherford model has the positive charge concentrated in a central nucleus with electrons orbiting around the outside.

Exam question

(a) De Broglie suggested that all particles might have a dual nature. His equation, $\lambda = h/p$, linked a wave property (wavelength λ) with a particle property (momentum p). From this equation we can see that momentum is inversely proportional to wavelength as h is a constant – Planck's constant. So, as momentum increases, wavelength decreases. The momentum of an object that we might see in our everyday life is so big that its associated wavelength is incredibly small. The wavelength of a 100 m sprinter would be about 10^{-36} m for example. To see the sprinter exhibit a wave property, like diffraction, would require him/her to squeeze through a gap of about 10^{-36} m. That's unlikely to happen and so we are unlikely to observe matter behaving as waves in everyday life.

Without an understanding of wave–particle duality, one might assume that negatively charged electrons should spiral towards the protons in the nucleus as opposite charges attract. However, electrons have a wavelength as they can be considered to be waves. Squeezing an electron in towards the nucleus would shorten its wavelength but this would make the electron too energetic for the protons to hold! (Energy is proportional to $1/\lambda^2$.)

See page 123, for a diagram of the Rutherford–Bohr atomic model. The corresponding energy-level diagram looks like this:

Energy levels	Orbits
E_3	$n = 4$
E_2	$n = 3$
E_1	$n = 2$
E_0	$n = 1$ Ground state

(Note the diagram on page 123 shows E_0 and E_1 only.)

(b) $\frac{1}{2}mv^2 = 1.6 \times 10^{-19}$ J

$$\therefore v = \sqrt{\frac{3.2 \times 10^{-19}}{9.11 \times 10^{-31}}}$$

$$= 5.9 \times 10^6 \text{ ms}^{-1}$$

$$\lambda = \frac{6.6 \times 10^{-34}}{9.11 \times 10^{-31} \times 5.9 \times 10^6}$$

$$= 1.22 \times 10^{-10} \text{ m}$$

Particle diffraction and probing the atom

Checkpoints

1 λ for X-rays is in the range 10^{-8} to 10^{-13} m. As they can have a similar λ to the spacing between crystals, good diffraction is possible, and so X-rays can be used to study the structure of crystals.

2

Cathode ray tube	mass and charge of electrons.
Electron microscope	structure of viruses.
Cyclotron	structure of nuclei.
Linear accelerator	evidence for quarks.
Synchroton	bottom quark.

Exam question

(a) Electrons can be considered to be waves with a wavelength approximately the same size as the spacing between atoms in a solid (10^{-10} m). This means that electrons can be noticeably diffracted by the spaces between atoms in a metal. As each metal has a different lattice structure, each one will produce an identifiable diffraction pattern.

(b) A fundamental particle is one that is not made from other particles. Atoms are not fundamental particles as they contain electrons, protons and neutrons. Electrons are fundamental particles.

(c) As mentioned above, electrons are fundamental particles. Protons and neutrons are not fundamental particles. A proton contains two up quarks and one down quark while a neutron consists of two down quarks and one up quark. *Up* and *down quarks* are fundamental particles.

Examiner's secrets

If your syllabus has a section or column entitled *suggested activities* rename it as *essential activities*. You must cover everything that is mentioned in the syllabus, as the examiner will consider it all when setting questions.

Fields

Non-contact forces (sometimes called at-a-distance forces) can be described using the concept of a field. In this section, gravitational forces and fields are considered but there are also strong links to magnetic fields. The idea of field strength and potential is also important and is used to explain interactions that occur between objects that have mass and/or charge.

Exam themes

- → Force between masses
- → Satellites and orbits
- → Calculation of *g* for planets of different masses and radii
- → Use of vector algebra to find resultant force and field
- → Field strength between capacitor plates
- → Motion of charged particles
- → Combinations of electrical and gravitational fields
- → Similarities and differences between fields
- → Calculation of potential and potential difference

Topic checklist

○ AS ● A2	AQA/A	AQA/B	CCEA	EDEXCEL	OCR/A	OCR/B	WJEC
Newton's law of universal gravitation	●	●	●	●	●	●	●
Gravitational fields	●	●	●	●	●	●	●
Electric forces and fields	●	●	●	●	●	●	●
Electric potential* and charged particle acceleration	●	●	●	●	●	●	●
Comparisons: gravitational and electric fields	●	●	●	●	●	●	●
Synoptic skills	●	●	●	●	●	●	●

* *Note*: the amount of detail that you are expected to know about electric potential depends on the syllabus that you are following. Make sure that you are familiar with the content of yours and revise accordingly.

Newton's law of universal gravitation

Newton's law of universal gravitation allows you to calculate the force of attraction that occurs between *any* objects that have mass. It was proposed almost 350 years ago after Newton had pondered upon why Kepler's laws of planetary motion worked and also more mundane matters such as what made an apple fall to Earth! He realized that they could both be explained using the same idea.

> *"Truth is ever to be found in the simplicity, and not in the multiplicity and confusion of things."*
>
> Sir Issac Newton

Examiner's secrets

Learn this law – it often comes up and can give you 'cheap' marks in an exam.

Statement of Newton's law of universal gravitation

- Every particle in the Universe attracts every other with a force which is directly proportional to the product of their masses and inversely proportional to the square of their separation.

The law is usually summarized in a diagram like this:

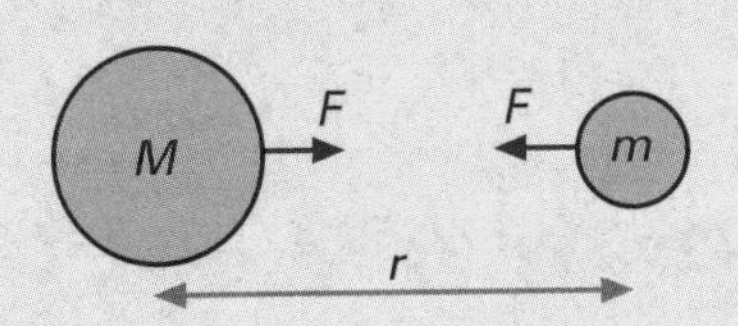

Equation

The law gives this equation:

$$F = G\frac{Mm}{r^2}$$

Where M and m are the two masses, r is the separation, F is the force and G is a constant of proportionality, known as the **universal constant of gravitation** ($= 6.67 \times 10^{-11}$ N m^2 kg^{-2}). Note that:

- the force is always attractive and the range is infinite
- strictly, the equation applies only to point masses, but you can use in all examples you will come across at AS- or A2-level
- r is the distance between the *centres* of the bodies
- G is *very* small – hence gravitational forces are very small, unless one (or both) of the masses is huge, as with a planet

The jargon

Some syllabuses use m_1 and m_2 instead of M and m – get used to yours.

Checkpoint 1

Be able to work out why G has these units – this is another common exam question.

Watch out!

Remember that forces are vector quantities. There is a convention: attractive forces are negative and repulsive forces are positive.

Inverse square laws

Newton's law of universal gravitation is an example of an **inverse-square law**. This simply means that the size of the force is *inversely* proportional to the *square* of the separation of the objects:

- if the separation is doubled (*multiplied* by 2), the force is quartered (*divided* by 2^2)
- if the separation is made 10 times *bigger*, the force is 10^2 (= 100) times *smaller* etc.
- inverse-square laws generate graphs which have a characteristic shape, as shown on the opposite page.

Links

Coulomb's law, see page 136, is another inverse-square law.

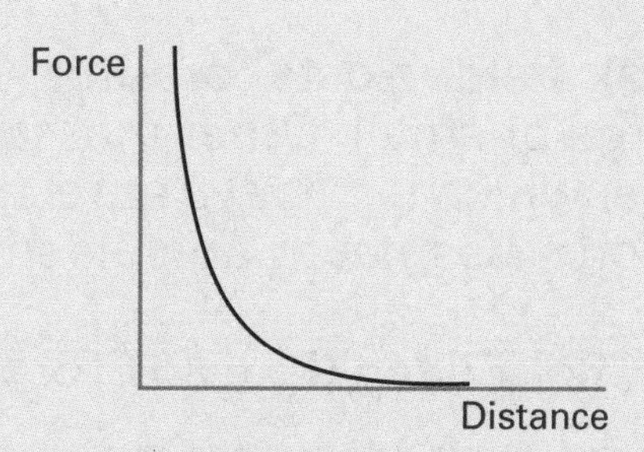

Note With inverse-square laws, the force drops off very rapidly with distance.

Motion of planets and satellites

For an object in orbit, the centripetal force required to keep it moving in a circle is provided by the gravitational force, given by the equation opposite. Thus:

$$G\frac{Mm}{r^2} = m\omega^2 r$$

Where M is the mass of the object *being orbited*, m is the mass, v is the velocity and ω is the angular velocity of the *orbiting* object, r is the radius of the orbit. Note:

- m can easily be eliminated from the equation, showing that the mass of the orbiting object is not relevant
- the equation can be rearranged to apply to particular problems
- in particular, it can generate the equation which relates the *radius* r of the orbit to its *period* T:

$$T^2 = \frac{4\pi^2}{GM}r^3$$

Which takes you back to Kepler's third law and the challenge originally taken up by Newton.

Links

See *circular motion*, pages 30–1.

Action point

It was the motion of the planets that Newton was trying to explain when he formulated his law of universal gravitation – read about Tycho Brahe's observations (and how he lost his nose in a duel!) and the laws that Johannes Kepler worked out to describe them.

Checkpoint 2

Try to derive this equation from the one above it.

Exam questions answers: page 144

1. What is the gravitational pull between a 3 kg mass and a 5 kg mass placed 0.15 m apart? (5 min)
2. Express G in SI *base* units. (5 min)
3. State Newton's law of universal gravitation and explain how it applies to the Earth/Moon system, drawing a labelled diagram to show the forces which act on the two bodies. (5 min)
4. Communication satellites are usually parked in *geostationary* orbits. Explain what is meant by this and show that the height of such an orbit is approximately 36 000 km above the Earth's surface. (15 min)

Gravitational fields

A gravitational field exists around *any* mass, no matter how large or small. Other masses in this region will feel a force, which is *always* attractive: a repulsive gravitational force has never been identified.

Don't forget

Gravitational fields are *created* by *all* masses and *felt* by *all* masses.

Gravitational field lines (lines of force)

Gravitational fields are always drawn using gravitational **field lines**, which show the direction of the force on a mass placed at any point in the field. Around a spherical mass the lines look like this:

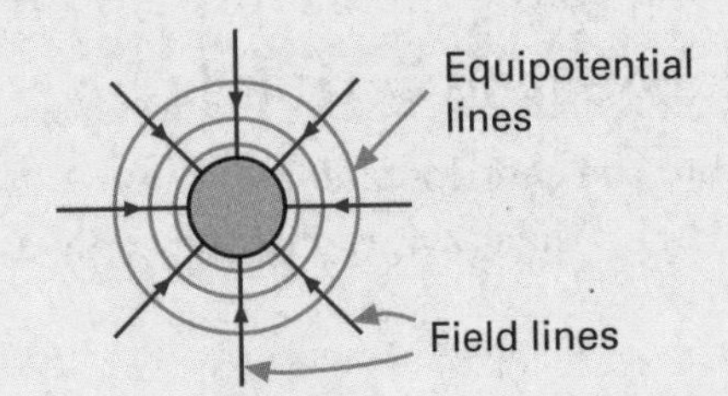

The jargon

The arrows always point towards the object – since the force is always attractive. *Equipotential surfaces* are explained on the opposite page.

Near the Earth's surface, the field is nearly uniform and the lines are evenly spaced:

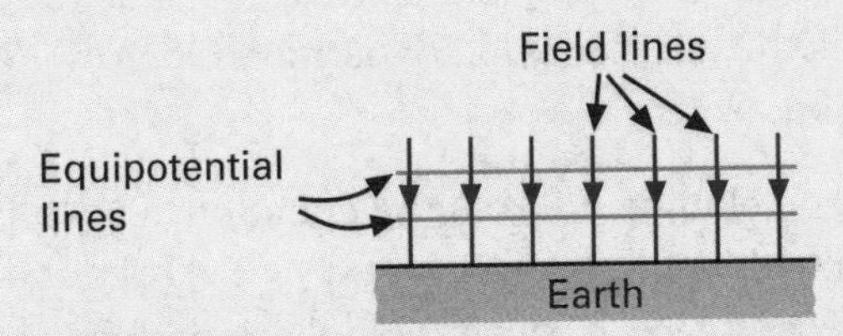

Examiner's secrets

All field lines (gravitational, electric and magnetic) must never touch or cross and they always enter and leave surfaces at right angles to them.

Gravitational field strength *g*

The gravitational field strength g at a point in a field is the force per unit mass on an object placed at that point:

$$g = \frac{F}{m}$$

It follows from Newton's universal law of gravitation that, around a spherical mass M:

$$g = G\frac{M}{r^2}$$

Where r is the distance from the *centre* of the mass. Note that:

- g has units of N kg^{-1}
- g may also be regarded as the acceleration due to gravity (in m s^{-2}) at the point
- g is a *vector* quantity
- the variation of g with distance from a mass is another *inverse-square* relationship

Checkpoint 1

Is M the mass of the object *causing* the field or *feeling* it?

Checkpoint 2

Can you show that N kg^{-1} is identical to m s^{-2}?

Watch out!

You need to take *direction* into account when you combine field strengths.

Earth's gravitational field

- Its strength is $9.81\ \text{N kg}^{-1}$ at the surface of the Earth.
- It falls off with distance above and below the surface as shown in the graph on the opposite page.

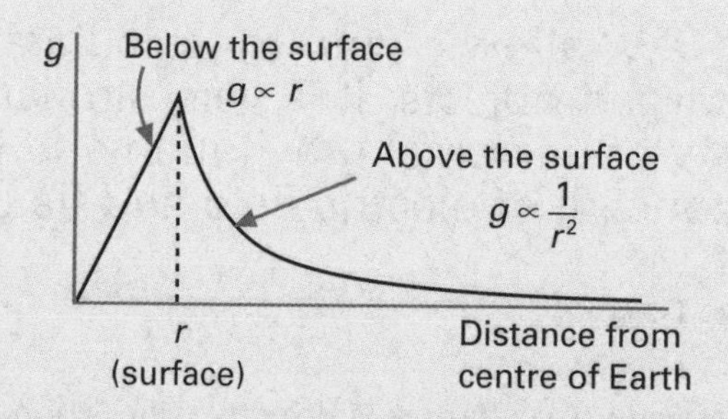

Gravitational potential *U*

When a mass is moved against a gravitational force, *work* is done. This is described using the concept of gravitational potential (symbol U, but V in some text books), defined as the work done in taking unit mass from infinity (∞) to a point in a field.

Thus, at a distance r from a mass M, it can be shown that:

$$U = -G\frac{M}{r}$$

Note:

- → U is a *scalar* quantity
- → U is zero at infinity
- → at all other places, U is negative (since a 'negative' amount of work has to be done to move a mass against an attractive force)
- → the gravitational field strength is equal to the *potential gradient* (the *slope* of a graph of U against r)

Escape velocity

If a mass at the Earth's surface can be given a kinetic energy equal to its gravitational potential energy (= mU), it will escape completely from the Earth's gravitational field (and end up at infinity!). The velocity required to do this is called the **escape velocity**, is not dependent upon the mass and is given by:

$$v = \sqrt{\left(\frac{2GM}{r}\right)} = \sqrt{(2gr)}$$

Equipotential surfaces

Points that are at the same potential lie on **equipotential surfaces**. Around a spherical mass, these are concentric spheres – or circles in two dimensions – see opposite. Equipotential surfaces are *always* perpendicular to field lines.

Checkpoint 3

The linear variation in g below the surface of the Earth is an approximation, based on the (incorrect) assumption of uniform density. Why is the graph this shape *above* the Earth's surface?

Examiner's secrets

Be familiar with your own syllabus: *learn* any definitions necessary (like this one) and try to make sure that you know when to use all the equations.

Checkpoint 4

What would a graph of U against r look like? Would it be different from g against r?

Checkpoint 5

What does each of the quantities in this equation stand for? They have been used frequently in the last four pages.

Examiner's secrets

Very often gravitational potentials are described in terms of gravitational *wells*. To escape from the surface of the Earth, a space craft must climb out of the Earth's potential well.

Exam questions answers: pages 144–5

1 Explain what you understand by escape velocity and use the approximate values of $r_E = 6.4 \times 10^6$ m and $g = 10$ N kg^{-1} to show that the escape velocity from the Earth is close to 11 km s^{-1}. (5 min)

2 It is proposed that a black hole with a mass equal to that of the Earth would have a radius of 1 cm. What is the gravitational field strength at the surface of the black hole?

How far away would you have to go for the gravitational field strength to have the same value as that on the surface of the Earth? What object has approximately this size? (10 min)

Electric forces and fields

Coulomb's law allows you to calculate the forc between charged objects; it is very similar in form t Newton's law of universal gravitation and leads logic ally to the concept of electric field and field strength.

Coulomb's law

- → The force between two charges is directly proportional to the product of the charges and inversely proportional to the square of their separation.

The law is usually summarized in a diagram like this:

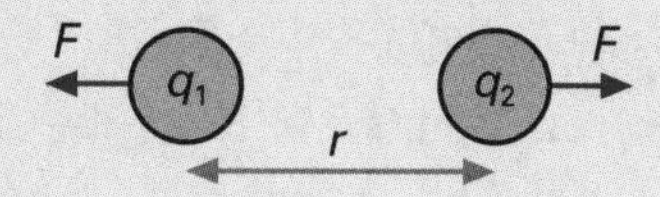

Action point

The form of this equation should be compared with that of *Newton's law of universal gravitation*, pages 132–3. What is the major difference between the two laws?

The law gives this equation:

$$F = k\frac{q_1 q_2}{r^2}$$

Where q_1 and q_2 are the charges and r is their separation. k is a constan of proportionality ($= 9.0 \times 10^9$ N m^2 C^{-2} in a vacuum and effectively the same in air).

Note:

- → if the charges are alike (both positive or both negative), the force is repulsive (as shown above)
- → if the charges are unlike (one positive and one negative), the force is attractive
- → the charges should be small in comparison with their separation
- → r should be measured from the *centre* of the charges

Permittivity of free space

Links

ε_0, ε (and ε_R) are described in *capacitors*, pages 76–7.

If the charges above are in a vacuum, the constant k in the equation is usually expressed as:

$$k = \frac{1}{4\pi\varepsilon_0}$$

Where the 4π is included in order to simplify equations derived from Coulomb's law and ε_0 is known as the permittivity of free space ($= 8.85 \times 10^{-12}$ F m^{-1}). If the medium between the charges is different, then ε_0 is replaced by ε, the permittivity of the medium.

Electric fields

An electric field is a region around a charge in which another charge feels a force. Unlike gravitational fields, the forces can be either attractive or repulsive.

Electric field lines

As with gravitational fields, lines are used to represent electric fields. The shape of common fields is shown on the opposite page.

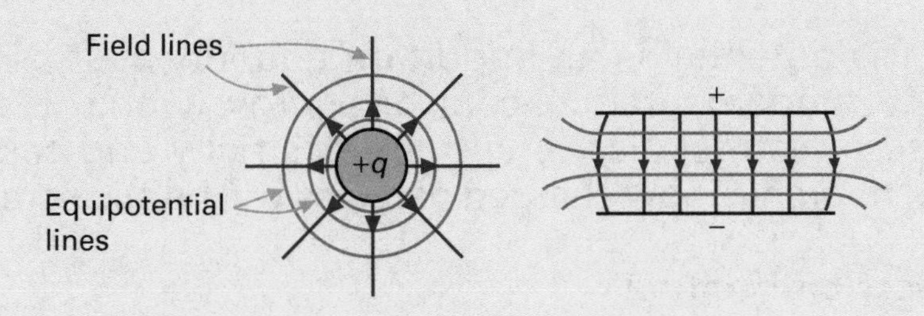

Note:

- at any point, the lines show the direction of the force on a positive charge placed at that point
- as a consequence, field lines around a point positive charge and a point negative charge are in different directions
- the field between parallel plates is approximately uniform

Electric field strength

The electric field strength E at a point in a field is defined as the force on unit charge placed at that point, and so if a charge q feels a force F, then:

$$E = \frac{F}{q}$$

It follows from Coulomb's law that around a point charge Q

$$E = k\frac{Q}{r^2}$$

Note:

- E has units of N C^{-1} (or V m^{-1}) and is a *vector* quantity
- the direction of E is the direction of the force on a positive charge
- inside a hollow conductor, E is zero
- E varies with distance from a charge according to an inverse-square law and therefore the range of E is infinite.

Between parallel plates, the field is uniform and E is given by:

$$E = \frac{V}{d}$$

Where V is the potential difference between the plates and d is their separation – hence the alternative unit for E, above.

Checkpoint 1

Sketch the field lines and equipotential lines around a *negative* charge.

Checkpoint 2

How would you observe electric field lines in the laboratory? (*Hint* One common method uses materials normally found in the kitchen and the garden shed!)

The jargon

Uniform means constant in both magnitude *and* direction because the electric field is a vector quantity.

Checkpoint 3

What is k in this equation?

Examiner's secrets

Be able to show that N C^{-1} and V m^{-1} are equivalent units.

Exam questions answers: page 145

1 Express ε_0 in SI *base* units. (10 min)

2 Draw a free-body force diagram for a tiny particle carrying a charge equal to that of two electrons which is held stationary in an electric field between parallel plates 5 mm apart and with a PD of 1 MV between them. Show the polarity of the plates on your diagram.

If $g = 9.81$ N kg^{-1}, calculate the mass of the particle. What assumptions have you made? What practical problems might be associated with this arrangement? (15 min)

Electric potential and charged particle acceleratio

Electric potential is an important quantity as it brings in the fundamental concept of energy (work). It also relates electrostatics to current electricity and can be used to predict how charged particles will be accelerated

Examiner's secrets

Potential difference (voltage) features strongly in current electricity – if you understand potential you will be well on the way to understanding all of electricity.

Electric potential

If a charge is moved in an electric field, then work is done/energy is converted.

The potential V at a point in a field is the work done W per unit charge q in taking *positive* charge from infinity (where the force, field and potential are all zero) to that point, i.e. $V = W/q$.

Around a spherical conductor, carrying charge Q:

Examiner's secrets

Make sure that you convert distances to metres before you use this equation.

$$V = k\frac{Q}{r}$$

Where r is the distance of the point from the (centre of) the charge Q. Note:

- ➜ V is a *scalar* quantity
- ➜ the electric field strength is equal in magnitude to the potential gradient
- ➜ positive charges move *down* a potential gradient (from *high* to *low* potential)
- ➜ negative charges move *up* a potential gradient (from *low* to *high* potential)
- ➜ the relationship between V and distance is inversely proportional

Checkpoint 1

Sketch a graph showing the variation of V with r.

Equipotential surfaces

Points that are at the same potential lie on equipotential surfaces. Around a spherical charge, these are concentric spheres – or circles in two dimensions (see the diagrams on pages 137 and 145). Equipotential surfaces are *always* perpendicular to field lines.

Motion of charged particles

Acceleration

The idea of potential can be used to describe the acceleration of charged particles. It follows from the definition of the volt (see page 62) that the kinetic energy gained by a particle carrying a charge q when accelerated through a potential difference of V volts is given by:

Checkpoint 2

What is the definition of a volt?

$$\tfrac{1}{2}mv^2 = qV$$

Where m is the mass of the particle and v its subsequent speed.

This equation can be used to calculate the speed of an electron as it emerges from an electron gun, as used in televisions, for instance. The *deflection* of these electrons across the screen is also caused by electric fields.

Deflection

Charged particles are deflected by electric fields and it follows from the definition of electric field strength (see page 137) that the force F on a particle carrying a charge q in a field of strength E is given by:

$$F = Eq$$

If the field is provided by a pair of parallel plates the shape of the path is as shown below and the equation becomes:

$$F = \frac{Vq}{d}$$

Where V is the potential difference between the plates and d their separation.

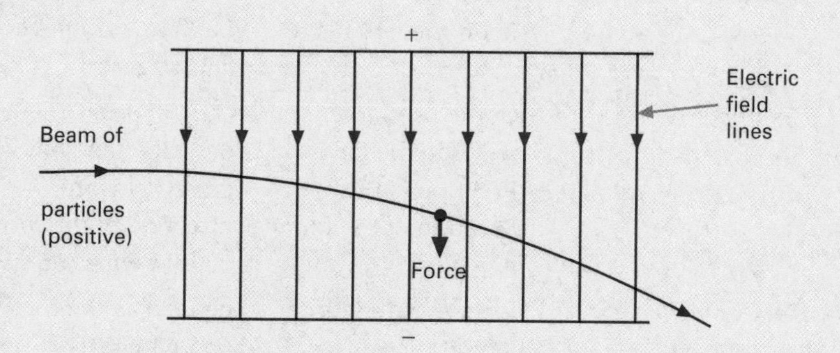

Note:

- the force is always parallel to the field lines
- the path of the particles is a parabola while in the electric field
- the size of the force does not depend on the speed of the particles
- deflection only occurs when the particles are within the field – elsewhere they travel in straight lines
- positive and negative charges are deflected in opposite directions
- we ignore the effects of gravity

Exam questions

answers: pages 145–6

1 A point charge of +4 nC is brought from infinity to a point 30 mm away from a charge of +8 nC. How much work is done? How much more work must be done to bring the charges 10 mm closer together? (10 min)

2 An electron (charge 1.6×10^{-19} C) is accelerated through a potential difference of 2.5 kV and enters an electric field that is perpendicular to the direction of motion of the electron. How fast are the electrons travelling when they enter the field?

Explain why this horizontal velocity does not change as it passes through the field.

If the plates are 5 cm long, for what length of time does the electron stay in the field?

If the field strength is 10 kV m^{-1}, how big is the deflecting force on the electron? (15 min)

Links

Charged particles are also deflected by magnetic fields (see pages 78–9), but *only if they are moving.*

Check the net

As well as televisions, ink-jet printers also work on this principle. Visit www.waythingswork.com/

Checkpoint 3

How is the shape of the path followed by charged particles different in a magnetic field?

Action point

The path of a charged particle in an electric field is a parabola. If you consider the vertical and horizontal motions separately, you should be able to prove that the particle gains KE.

Examiner's secrets

There is an opportunity for examiners to get you to compare the path of a moving charge in an electric field with that in a magnetic field. Make sure you are ready for it!

Comparisons: gravitational and electric fields

There are numerous similarities between gravitational and electric fields – and some important differences. The comparison is easiest to make when presented in the form of a table.

Action point

For each of these similarities, say whether there is an equivalent effect for a magnetic field.

Similarities

	Gravitational property	*Electrical property*
For the force to act . . .	contact not needed – force acts at a distance	contact not needed – force acts at a distance
Range of field	infinite – size of force decreases as distance increases, but in theory never falls to zero	infinite – size of force decreases as distance increases, but in theory never falls to zero
Field lines	can be used to describe the field – direction is the direction of the force on a *mass*	can be used to describe the field – direction is the direction of the force on a *positive charge*
Field strength	defined as the force on unit mass: $g = \frac{F}{m}$	defined as the force on unit charge: $E = \frac{F}{q}$
Force between two objects	given by Newton's law of universal gravitation: $F = G\frac{Mm}{r^2}$ (inverse-square law)	given by Coulomb's law: $F = k\frac{Qq}{r^2}$ (inverse-square law)
Potential	can be used to describe work done in moving masses: $U = -\frac{GM}{r}$ (inverse law)	can be used to describe work done in moving charges: $V = k\frac{Q}{r}$ (inverse law)
Potential energy	gravitational potential energy = mU	electrical potential energy = qV
Relationship between potential and field strength	field strength is (negative of) potential gradient: $g = -\frac{dU}{dr}$	field strength is (negative of) potential gradient: $E = -\frac{dV}{dr}$
Kinetic energy	calculated from: $\frac{1}{2}mv^2 = mU$	calculated from: $\frac{1}{2}mv^2 = qV$

Test yourself

What are the units of field strength? Is it a scalar or vector quantity?

Test yourself

What are the units of potential? Is it a scalar or vector quantity?

Checkpoint 1

How is this expression for gravitational potential energy related to the more usual version – namely *mgh*?

Examiner's secrets

Questions comparing gravitational and electric forces and fields come up frequently. Learn as much as you can of the information in these tables.

Differences

	Gravitational property	*Electrical property*
Origin	produced by and act upon *masses*	produced by and act upon *charges*
Effect	cause attraction *only*	can cause attraction *or* repulsion
Shielding	not possible – no material has been found that is able to shield gravitational forces	possible – shielding possible using devices such as Faraday cages
Comparative size (This difference is brought out in Question 2, below – try it!)	insignificant unless one (or both) of masses is huge	much bigger

Exam questions

answers: page 146

1 List two similarities and two differences between gravitational and electric fields. (5 min)

2 An electron has a mass of 9.11×10^{-31} kg and a charge of 1.60×10^{-19} C. Calculate the gravitational and the electrical force on two electrons 1.00×10^{-10} m apart in a vacuum.

Calculate the ratio of the two forces and comment on your answer.

Other than the magnitude, state one other difference between these forces. (15 min)

3 What is meant by a field line or line of force? Explain in terms of both gravitational and electric fields, stating how the field lines differ in these cases. (10 min)

Synoptic skills

All A-level specifications (a modern name for th syllabus) must include a minimum of 15% synopti assessment. It applies only if you go on to study sub jects to A2, but it covers AS work too. Synoptic assess ment involves the drawing together of knowledge an understanding from different areas of the course Some synoptic skills are taught through case studies others via making comparisons (analogies).

Synoptic skills

Watch out!

Synoptic papers usually contain questions about other areas of the course. So make sure you check the assessment details of *your* specification.

In theory, synoptic skills should be:

- based on parts of the specification common to all students (including AS)
- the same for all students regardless of specification
- based on a range of different types of question

In practice, synoptic assessment in physics may ask you to:

- make connections between different areas of physics
- apply knowledge and understanding in new situations
- analyse and interpret data
- read a passage of text and answer questions based on it
- produce an extended piece of writing
- answer structured questions with connected calculations

Progressing from GCSE

When you first begin studying A-levels, especially physics, it can sometimes be a large jump up from GCSE level. In fact, you may find that what you learned for your GCSEs may not exactly fit with what you are now expected to know for A-level. The science you needed for GCSE was not wrong; it's just that it was simplified to make it easier to understand. At A-level, you study subjects to a much greater depth. You will probably find it is the same if you go on to study physics at university.

"Only connect! . . . Only connect the prose and the passion and both will be exalted . . ."

E M Forster

Making connections

Studying A-level physics can seem rather daunting at first and students often complain that they don't know how things fit together. Making connections between different areas of physics will take time, but it does happen and usually just in time for the synoptic assessment at the end of A2!

There are lots of ways you can make connections between different areas of physics and these connections will help you revise for your examinations too.

Mind maps

There are different types of so-called *mind maps* and all can be effective in studying and revising. A simple mind map, or *spider diagram*, can be just a series of ideas with a few links between them. *Concept maps* are similar, but the connection between ideas has much more meaning. Begin by *brainstorming* to bring up ideas within a particular topic, then link the ideas together. If you can label the connections with words or phrases it will help you remember how the ideas go together.

Card notes

Make up notes on cards using bullet points. If you get into the habit of doing this as soon as you start your AS course, you will have an invaluable set of memory cards or *aide mémoire*. By the time you get to take your A2 examinations you will have a small library of cards from which to revise.

Don't forget

Keep your notes well organized and indexed. You can't afford to waste time looking for notes near to the examination.

Data analysis

Working scientists communicate many of their ideas through tables and graphs. Physics concepts and laws come alive through measured data. Usually they are produced from your own experiments in class. Being able to analyse unfamiliar data is an important skill. The synoptic examination may ask you to read off data from graphs, process raw data or plot graphs in order to draw conclusions.

Background reading

Reading scientific journals and articles from the Internet is an extremely good way of keeping up to date with current developments in physics. You will certainly become more scientifically literate by reading around the subject.

Action point

Scientific American, *New Scientist* and *Physics Review* are all good sources of background reading.

Practical work

Making measurements is an important aspect of physics and you should keep a clear and up-to-date record of the practical work you do in class. Synoptic questions may well link directly to expected experimental work. You should develop a clear and concise way of recalling experimental arrangements, especially details of how to reduce errors.

Synoptic questions often deal with the differences between data collected on a small scale in the laboratory and that obtained on an industrial scale.

Examiner's secrets

Many synoptic questions are set using articles from popular scientific journals. So, you never know how useful your background reading will be!

Answers
Fields

Newton's law of universal gravitation

Checkpoints

1 The equation can be rearranged to give:

$$G = \frac{Fr^2}{Mm}$$

from which it is apparent that the units of G must be $\mathrm{N\,m^2\,kg^{-2}}$.

2 The derivation uses the relationship $\omega = 2\pi/T$:

$$\frac{GMm}{r^2} = m\omega^2 r$$

$$r^3\omega^3 = GM$$

$$r^3\left(\frac{2\pi}{T}\right)^2 = GM$$

$$r^3 = \frac{GM}{4\pi^2}T^2 \text{ or } T^2 = \frac{4\pi^2}{GM}r^3$$

Exam questions

1 $F = \dfrac{6.67 \times 10^{-11} \times 3 \times 5}{0.15^2} = 4.45 \times 10^{-8}\ \mathrm{N}$

2 By rearranging Newton's equation:

$$F = \frac{GMm}{r^2}$$

G can be shown to be given by:

$$G = \frac{Fr^2}{Mm}$$

and, for this equation to be homogeneous, the units of G must be the units of the right-hand side.
These are:

$$\frac{\mathrm{N\,m^2}}{\mathrm{kg^2}}$$

or $\mathrm{N\,m^2\,kg^{-2}}$.
But these are not yet *base* units and N must be replaced by its base equivalent ($\mathrm{kg\,m\,s^{-2}}$), giving units for G of $\mathrm{kg^{-1}\,m^3\,s^{-2}}$.

Examiner's secrets

With this sort of question, always adopt the approach above – namely, rearrange the equation to make the quantity that you are asked to find the *subject* of the equation, then sort out the units.

3 Statement of law as in spread.
For the Earth/Moon system:

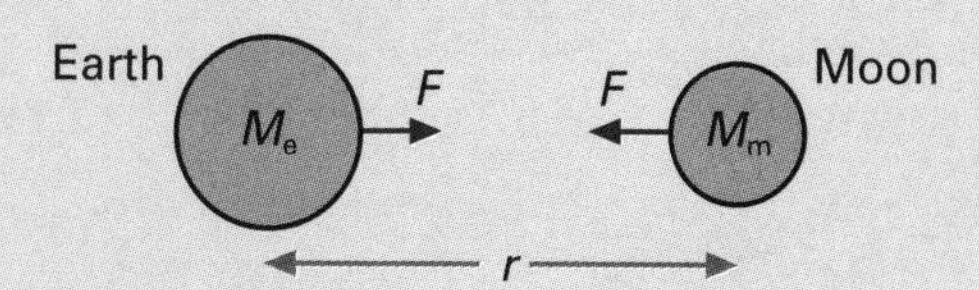

Here the forces labelled F are gravitational forces and are equal and opposite, given by:

$$F = \frac{GM_eM_m}{r^2}$$

4 For a geostationary orbit, the period of the rotation must be exactly the same as that of the Earth – namely, 24 hours. This allows satellites to be *parked* in such an orbit, where they will stay directly above the same point on the Earth's surface, though this point must be directly above the equator. The height of such an orbit can be calculated by equating the centripetal force to the gravitational force at that height. If the height from the centre of the Earth is r, this gives:

$$m\omega^2 r = \frac{GMm}{r^2}$$

Where ω is the angular velocity associated with the rotation. Note that the m can be divided from both sides, showing that the mass of the satellite is not relevant.
Since $\omega = 2\pi/T$ (where T is the period):

$$\frac{4\pi^2}{T^2}r = \frac{GM}{r^2} \qquad r = \sqrt[3]{\left(\frac{GMT^2}{4\pi^2}\right)}$$

Substituting values of $G = 6.67 \times 10^{-11}\ \mathrm{N\,m^2\,kg^{-2}}$, $T = 24$ hours $= 8.64 \times 10^4$ s and $M_e = 6.0 \times 10^{24}$ kg, gives $r \approx 42\,000$ km and, since the radius of the Earth is 6 400 km the orbit is 36 000 km above the surface of the Earth.

Gravitational fields

Checkpoints

1 M is the mass of the object *causing* the field.

2 $\mathrm{N\,kg^{-1}} = (\mathrm{kg\,m\,s^{-2}})\,\mathrm{kg^{-1}} = \mathrm{m\,s^{-2}}$.

3 The graph is the shape shown because the relationship between g and distance is an inverse square.

4 The graph of U against r looks like this:

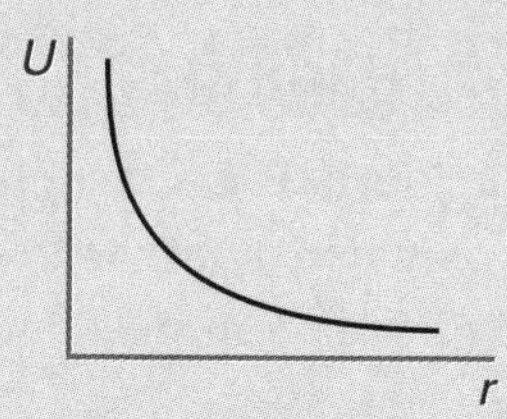

It is different from g against r because it is an inverse relationship, rather than an inverse square.

5 G is the universal constant of gravitation, M is the mass of the Earth (or other planet), g is the gravitational field strength, and r is the radius of the Earth (or other planet).

Exam questions

1 The escape velocity is the minimum velocity that an object must acquire in order to escape completely from a gravitational field.
Using:

$$v = \sqrt{(2gr)}$$

gives:

$$v = \sqrt{(2 \times 10\ \mathrm{N\,kg^{-1}} \times 6.4 \times 10^6\ \mathrm{m})}$$
$$\simeq 11 \times 10^3\ \mathrm{m\,s^{-1}} \simeq 11\ \mathrm{km\,s^{-1}}$$

2 Using:

$$g = \frac{GM}{r^2}$$

gives:

$$g = \frac{6.67 \times 10^{-11}\ \mathrm{N\,m^2\,kg^{-2}} \times 6.0 \times 10^{24}\ \mathrm{kg}}{(1 \times 10^{-2})^2\ \mathrm{m^2}}$$
$$= 4.0 \times 10^{18}\ \mathrm{N\,kg^{-1}}$$

The equation above can be rearranged to give:

$$r^2 = \frac{GM}{g} \quad \text{or} \quad r = \sqrt{\left(\frac{GM}{g}\right)}$$

and for $g = 9.81\ \text{N kg}^{-1}$:

$$r = \sqrt{\left(\frac{6.67 \times 10^{-11}\ \text{N m}^2\,\text{kg}^{-2} \times 6.0 \times 10^{24}\ \text{kg}}{9.81\ \text{N kg}^{-1}}\right)}$$

$$= 6.4 \times 10^6\ \text{m}$$

This is very similar to the radius of the Earth.

Electric forces and fields

Checkpoints

1 Field lines and equipotential surfaces around a negative charge look like this:

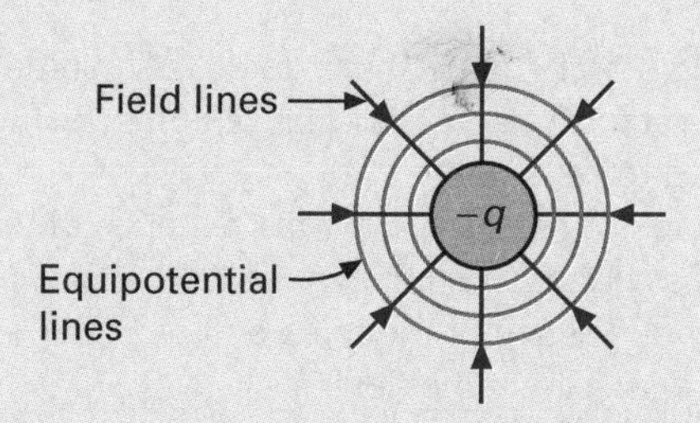

2 To observe electric field lines in the laboratory you could use a high-voltage supply connected to electrodes separated by castor oil; if either grass seed or semolina is sprinkled into the castor oil, the seeds/grains align themselves along field lines. Alternatively, conducting paper can be used to plot equipotential surfaces.

3 k in the equation is $1/(4\pi\varepsilon_0)$.

Exam questions

1 The equation for Coulomb's law:

$$F = \frac{1}{4\pi\varepsilon_0} \times \frac{Qq}{r^2}$$

can be rearranged to give:

$$\varepsilon_0 = \frac{Qq}{4\pi r^2 F}$$

and hence the units of ε_0 must be

$\text{C}^2\ (\text{kg m s}^{-2})^{-1}\ \text{m}^{-2}$

or $\text{C}^2\ \text{kg}^{-1}\ \text{m}^{-1}\ \text{s}^2\ \text{m}^{-2}$

or $\text{kg}^{-1}\ \text{m}^{-3}\ \text{s}^2\ \text{C}^2$

2 Here we have

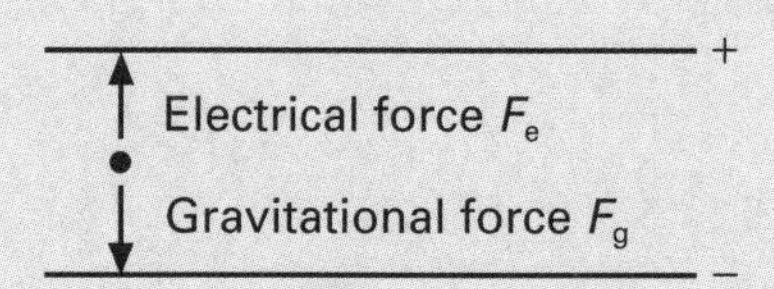

Equating the electrical and gravitational forces gives:

$$\frac{Vq}{d} = mg$$

Hence:

$$m = \frac{Vq}{gd} = \frac{1 \times 10^6\ \text{V} \times 3.2 \times 10^{-19}\ \text{C}}{9.81\ \text{N kg}^{-1} \times 5 \times 10^{-3}\ \text{m}}$$

$$= 6.5 \times 10^{-12}\ \text{kg}$$

It has been assumed that the upthrust on the particle can be ignored.

The problem associated with this arrangement is that the electrical field strength between the plates is huge ($2 \times 10^8\ \text{V m}^{-1}$). This is greater than the breakdown potential for air and therefore the air would begin to conduct.

Electric potential and charged particle acceleration

Checkpoints

1 V varies with r according to an inverse relationship, like this:

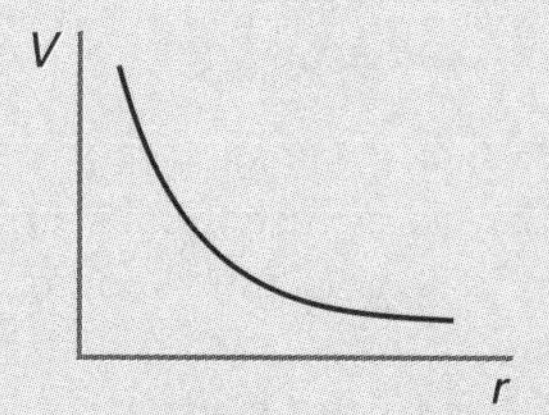

2 If 1 J of energy is converted when 1 C of charge flows between two points, then the potential difference between those points is 1 V.

3 In a magnetic field, moving charged particles follow a *circular* path, since the direction of the force is given by the left-hand rule and is always perpendicular to the direction of motion.

Exam questions

1 The work done in bringing unit charge from infinity to a point a distance r from a point charge is the potential V at that point. To bring a charge q:

work done $= Vq$

$$= \frac{8 \times 10^{-9}\ \text{C}}{4\pi\varepsilon_0 \times 30 \times 10^{-3}\ \text{m}} \times 4 \times 10^{-9}\ \text{C}$$

$$= 9.59 \times 10^{-6}\ \text{J}$$

To take the same charge to a distance of 20 mm:

$$\text{work done} = \frac{8 \times 10^{-9}\ \text{C}}{4\pi\varepsilon_0 \times 20 \times 10^{-3}\ \text{m}} \times 4 \times 10^{-9}\ \text{C}$$

$$= 1.44 \times 10^{-5}\ \text{J}$$

Hence, additional work $= 4.79 \times 10^{-6}\ \text{J}$

2 The kinetic energy of the electron is equal to the work done by the potential difference:

$$\tfrac{1}{2}mv^2 = qV$$

Hence,

$$v = \sqrt{\left(\frac{2qV}{m}\right)} = \sqrt{\left(\frac{2 \times 1.6 \times 10^{-19}\ \text{C} \times 2.5 \times 10^3\ \text{V}}{9.11 \times 10^{-31}\ \text{kg}}\right)}$$

$$= 2.96 \times 10^7\ \text{m s}^{-1}$$

Here the force is always vertical (in the direction of the field lines) and so the horizontal velocity is unaffected and the time will be given by

time = distance/velocity

$$= \frac{5 \times 10^{-2}\ \text{m}}{2.96 \times 10^7\ \text{m s}^{-1}}$$

$$= 1.69 \times 10^{-9}\ \text{s}$$

The force on a charged particle q in a field of strength E is given by:

$F = Eq$

$= 10\,000\ \text{N C}^{-1} \times 1.6 \times 10^{-19}\ \text{C}$

$= 1.6 \times 10^{-15}\ \text{N}$

Comparisons: gravitational and electric fields

Checkpoint

1 The change in gravitational potential energy $\Delta(\text{GPE})$ is given by:

$\Delta(\text{GPE}) = m\Delta U$

Since $U = gr$ and, over short distances, g can be assumed to be constant, this becomes:

mg (difference in height)

and is usually expressed as mgh.

Exam questions

1 *Similarities* Can be represented using field lines; field strength is inversely proportional to the square of the distance etc (see table in spread)

Differences Gravitational fields act on masses, electric fields act on charges; gravitational fields cause attractive forces only, electrical fields cause attractive or repulsive forces.

2 Gravitational force is given by:

$$F_g = \frac{GMm}{r^2}$$

$$= \frac{6.67 \times 10^{-11}\ \text{N m}^2\,\text{kg}^{-2} \times 9.11 \times 10^{-31}\ \text{kg} \times 9.11 \times 10^{-31}\ \text{kg}}{(1.00 \times 10^{-10})^2\ \text{m}^2}$$

$= 5.54 \times 10^{-51}\ \text{N}$

Electrical force is given by:

$$F_e = \frac{Qq}{4\pi\varepsilon_0 r^2}$$

$$= \frac{1.60 \times 10^{-19}\ \text{C} \times 1.60 \times 10^{-19}\ \text{C}}{4\pi \times 8.85 \times 10^{-12}\ \text{F m}^{-1} \times (1.00 \times 10^{-10})^2\ \text{m}^2}$$

$= 2.30 \times 10^{-8}\ \text{N}$

Hence, ratio:

$F_g/F_e = 2.41 \times 10^{-43}$

showing that the gravitational force is a minute fraction of the electrical force. Gravitational force is attractive but electrical force is repulsive.

3 Gravitational field lines show the direction of the force on *unit mass* placed at a point in the field. Electric field lines show the direction of the force on *unit positive charge* placed at a point in the field.

Options

The commonest eight options are summarized in this section: nuclear and particle physics, astrophysics, medical physics, radiation and risk, electronics, materials, telecommunications and turning points in physics. Check your syllabus for details.

Exam themes

- → *Nuclear and particle physics* The standard model, antiparticles, combinations of fundamental particles, fundamental forces and the interactions between particles, particle accelerators.
- → *Astrophysics* Measuring astronomical distances, stellar surface temperatures and spectra, imaging methods, the Big Bang theory (and evidence for it), the fate of the Universe (and what it depends on).
- → *Medical physics* Diagnosis, treatment, the eye and seeing, the ear and hearing.
- → *Radiation and risk* The effects of radiation dose, the risk to health and the probability of long-term damage.
- → *Electronics* Measuring devices, electronic devices and their uses.
- → *Materials* Macroscopic behaviour of metals, ceramics and polymers and its interpretation on an atomic level, and the electrical, magnetic and optical properties of materials.
- → *Telecommunications* Analogue and digital signals, amplitude, frequency and pulse code modulation, time division multiplexing, AM receivers, sending signals by wires, optical fibres and radio waves.
- → *Turning points in physics* Einstein's theory of special relativity, electromagnetic waves, the discovery of the electron and wave particle duality.

Topic checklist

○ AS ● A2	AQA/A	AQA/B	CCEA	EDEXCEL	OCR/A	OCR/B	WJEC
Nuclear and particle physics 1	○*	○*		○	●		
Nuclear and particle physics 2	○*			○	●		●*
Astrophysics 1	●			○	●	●*	
Astrophysics 2	●			○	●	●*	
Astrophysics 3	●			○	●	●*	
Medical and health physics 1	●		●*		●		
Medical and health physics 2	●	○*	●*	○	●		
Radiation and risk	●	○*	●*	○	●	●*	
Electronics 1	●		●*				
Electronics 2	●						
Electronics 3	●						
Materials 1	○*	●*	○	○	●	○*	○*
Materials 2			○	○	●		○*
Materials 3					●		
Telecommunications 1		○*			●		
Telecommunications 2		○*			●		
Turning points in physics 1	●						
Turning points in physics 2	●						

* material found in core module, not an option

Nuclear and particle physics 1

"Three quarks for Muster Mark!"

James Joyce

Atoms were originally thought to be fundamental particles (the word atom comes from the Greek word *atomos* meaning indivisible) until the electron was discovered in 1897. Currently the electron is thought to be one of the 12 fundamental particles which, with their 12 antiparticles, form the standard model of particle physics.

Check the net

You'll find the particle adventure on www.cpepweb.org/

The standard model

The 12 fundamental particles of the standard model form two groups called **leptons** and **quarks**. Each group consists of three pairs of particles, known as the three generations of particles.

Leptons

1	*2*	*3*	*Charge*
electron e^-	muon μ^-	tau τ^-	$-e$
electron neutrino ν_e	muon neutrino ν_μ	tau neutrino ν_τ	zero

Quarks

1	*2*	*3*	*Charge*
up u	charm c	top t	$+{}^2/_3e$
down d	strange s	bottom b	$-{}^1/_3e$

Order of increasing mass →

Matter and antimatter

Each particle has a corresponding antimatter particle or antiparticle. An antiparticle is identical to its particle except that it has the opposite charge. So an antielectron has the same mass as an electron but it has a charge of $+e$. Antiparticles are usually denoted by having a line over the symbol, so an antielectron would be $\bar{e}$. An antielectron also has its own name: the **positron**.

Action point

Draw up two tables like the ones above listing the antiparticles with their corresponding charges.

Matter and energy

When a particle meets with its antiparticle, the two are annihilated and leave behind a photon of gamma radiation. This is an example of matter being a form of energy. The amount of energy released is given by Einstein's equation $E = mc^2$, where c is the speed of light.

The opposite can also occur: gamma radiation can disappear to produce a particle/antiparticle pair.

Checkpoint 1

What energy of photon is required to create an electron–positron pair?

Hadrons, baryons and mesons

Whereas leptons can be found singly, quarks can only exist combined with other quarks. These combinations of quarks are called **hadrons**. A **baryon** is a hadron made from three quarks. For example, a proton consists of two up quarks and a down quark (uud) and a neutron is one up quark and two downs (udd).

Mesons are hadrons consisting of two quarks – a quark and its antiquark. For example, the pi-mesons or pions in the table below.

Pions

Name	*Symbol*	*Quarks*	*Charge*
pi zero	π^0	$u\bar{u}$	zero
pi zero	π^0	$d\bar{d}$	zero
pi plus	π^+	$u\bar{d}$	$+e$
pi minus	π^-	$d\bar{u}$	$-e$

Forces

There are four fundamental forces. The first two weaken with distance, but extend to infinity.

1. Gravity forms planets, stars and galaxies.
2. The electromagnetic force keeps atoms together.

The next two have a range within the size of the atomic nucleus.

3. The weak force is responsible for radioactive decay.
4. The strong force acts between quarks and keeps protons inside the nucleus in spite of their electrostatic repulsion.

Exchange bosons

- The action of forces can be explained by the theory of exchange particles, called **bosons**.
- When two electrons interact, a photon is exchanged.
- The exchange particle for the gravitational force, the **graviton**, has yet to be discovered.
- **Feynman diagrams** (see exam question) are used to show exchange of particles as a force is produced.

Force	*Particles affected*	*Boson*
gravitational	all particles with mass	gravitons
weak	quarks and leptons	W and Z particles
electromagnetic	all particles with charge	photons
strong	quarks and hadrons	gluons for quarks, mesons for hadrons

Checkpoint 2

Show that the combinations of quarks for the proton and neutron give the correct amount of charge.

Checkpoint 3

Show that the combinations of quarks for the pions give the correct amount of charge.

Links

See *gravitational fields*, pages 134–5; *electric forces and field*, pages 136–7; *nuclear instability*, pages 44–5.

Checkpoint 4

Explain what the following groups of particles are and give an example of each one: hadrons, mesons, baryons, leptons, quarks, antiparticles, bosons.

Exam question answer: page 184

An electron and a positron can annihilate by either of the following mechanisms:

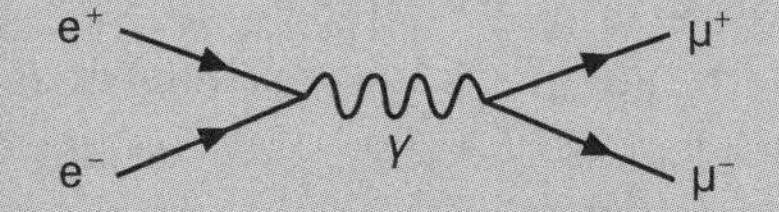

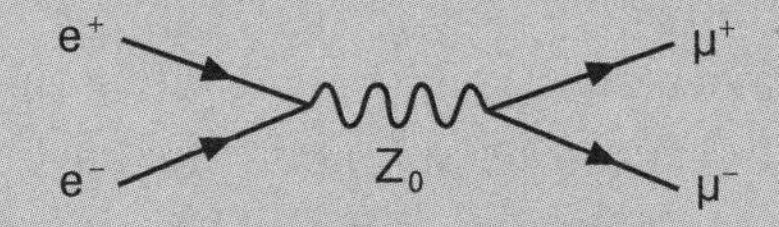

(a) Which of the fundamental interactions is represented by each figure?

(b) Draw another diagram to illustrate the exchange of a π^+ between a neutron and a proton. (5 min)

Nuclear and particle physics 2

The weak force causes one type of particle to change into another. For example, if an electron interacts with a muon neutrino, this could result in a muon and an electron neutrino.

$$e^- + \nu_\mu = \mu^- + \nu_e$$

Stability of hadrons

- Hadrons are thought to be unstable and decay into other particles.
- Neutrons and protons inside the nucleus are relatively stable.
- A free neutron has a half-life of about 15 minutes.
- Free protons are much more stable and their half-life is thought to be of the order of 10^{32} years.

Beta decay

Neutrons decay to produce a proton, an electron and an antineutrino. This can happen in the nucleus of an atom that has a high neutron to proton ratio. It is an example of the weak force causing beta decay as the electron is emitted from the nucleus as a beta particle (β^-). The antineutrino is also emitted while the proton remains in the nucleus.

Using the quark model, this means that a neutron (udd) has become a proton (uud) so a down quark has changed into an up quark.

A different kind of beta decay occurs in nuclei that have too few neutrons. Positron decay is the mirror image of beta decay. Here a proton becomes a neutron while a positron (also known as β^+) and a neutrino are emitted.

Checkpoint 1

Use the quark model to describe the changes that must take place when positron decay occurs.

The jargon

Beta capture is when one of the orbiting electrons is captured by the nucleus and combines with a proton to produce a neutron.

Baryon number, strangeness and conservation

You are familiar with the conservation of momentum in particle interactions. Charge and mass number are also conserved along with other properties, two of which are the baryon number and the strangeness.

Baryon number

All baryons are assigned a baryon number of +1 or –1 while all mesons have a baryon number of 0. For example, protons and neutrons have baryon numbers of +1 while their antiparticles have baryon numbers of –1. In this interaction, a proton and an antiproton interact to produce pi-plus and pi-minus mesons: $p + \bar{p} = \pi^+ + \pi^-$. So –1 + 1 = 0 on the left equals 0 on the right.

Strangeness

Strangeness is a property given to the strange quark. This is said to have a strangeness of –1 while the antistrange quark has a strangeness of +1. Protons and neutrons have a strangeness of 0 since they do not contain strange quarks. The omega baryon (Ω^-) is a cluster of three strange quarks so it has a strangeness of –3.

Checkpoint 2

Apply conservation of charge, baryon number and strangeness to beta decay.

Watch out!

Electrons, positrons and neutrinos are all leptons. Leptons have a lepton number of +1 if they are *real* and –1 if they are antimatter. Lepton numbers must be conserved in particle reactions.

The jargon

The properties or *flavours* of the other quarks (i.e. charm, topness and bottomness) are also conserved!

Particle accelerators

Rutherford used α-particles to probe the structure of the atom. Particle physicists require particles with much more energy in order to probe the structure of the nucleus. This is because:

- the α particles did not have enough energy to reach the nucleus because of the electrostatic repulsion
- energy can be converted to mass, so the greater the energy, the more massive the particles that can be produced
- the more energetic the particle, the smaller the detail that can be resolved

Check the net

The web site of the European Laboratory for Particle Physics (and the home of the world wide web) has information about its particle accelerators at www.cern.ch/

Linear accelerator

- Charged particles travel along a straight evacuated tube.
- Cylindrical electrodes connected to a high-frequency AC supply change polarity when the particles reach the end of an electrode.
- The particles are accelerated across the gap.
- The electrodes increase in length as the particles' speed increases.
- To get high-energy particles, the accelerators have to be very long.

Checkpoint 3

Explain why the tubes that the charged particles travel along must be evacuated.

Cyclotron

- A more compact method uses a magnetic field to make the particles move in a circle.
- Instead of moving particles colliding into a stationary target, oppositely charged particles can be accelerated in the opposite direction increasing the energy of the collision.
- The particles spiral outwards between two semicircular metal D-shaped electrodes connected to a high-frequency AC supply.
- The frequency of the supply is given by $f = Bq/2\pi m$.
- The energy of the particles is limited by the theory of relativity – as they approach the speed of light, their mass increases so they would no longer reach the gaps as the PD changed.

Checkpoint 4

For a particle in a cyclotron, show that if its speed increases, the radius of the circular path increases but the period remains constant.

Synchrotron

- Here previously accelerated particles travel around the same circular path.
- As they are accelerated more, their mass increases.
- The strength of the magnetic field in the deflecting magnets is increased to keep the radius of their path the same.

Exam question answer: page 184

The omega minus (Ω^-), a particle with strangeness –3, was identified in an experiment involving an interaction between a K^- meson of strangeness –1 and a proton: $K^- + p = \Omega^- + K^+ + K^0$.

Is the Ω^- a baryon or a meson? Give *two* reasons for your answer. (5 min)

Astrophysics 1

Checkpoint 1

How many light-years are there in a parsec?

Checkpoint 2

(a) How many light-minutes in a light-year?
(b) How many minutes does it take light to reach us from the Sun?

Checkpoint 3

The angle subtended by a person's eyes at a distant gate post is 1.2°. Given that the person's eyes are 10 cm apart, calculate the distance to the post.

Checkpoint 4

Calculate the Sun's luminosity if the intensity of solar radiation reaching the surface of the atmosphere (the *solar constant*) is 1 360 W m^{-2}.

Links

See *Planck's constant*, pages 114–15. The search for a mathematical description of the black-body spectrum first led Max Planck to treat electromagnetic radiation as quanta – discrete chunks (photons) of energy.

The further we look, the smaller we seem. The distance and time scales dealt with here are astronomical!

- *Astronomical Unit* (AU) The average distance from the Earth to the Sun. 1 AU = 1.496×10^{11} m.
- *Parsec* (pc) The distance to a star that subtends an angle of 1 second at an arc of length 1 AU (see diagram below). 1 pc = 3.086×10^{16} m.
- *Light-year* (ly) The distance travelled by a ray of light in 1 year. 1 ly = 9.46×10^{15} m.

Using parallax to measure distances to nearby stars

(1° = 60 minutes = 3 600 seconds)

1 second
1pc
1AU
Centre of Sun
Centre of Earth

Earth
Parallel angle in second
1AU
Sun
Sta
Distance in parsecs

Parallax works only for nearby stars (up to about 100 pc).

Using the inverse-square law to measure greater distances

The intensity of light from a star obeys an inverse-square law:

$$I = L/(4\pi R^2)$$

Where I is intensity in W m^{-2}, L is power output (or *luminosity*) in watts and $4\pi R^2$ is the area the radiation is spread over at a distance of R.

The black-body spectrum

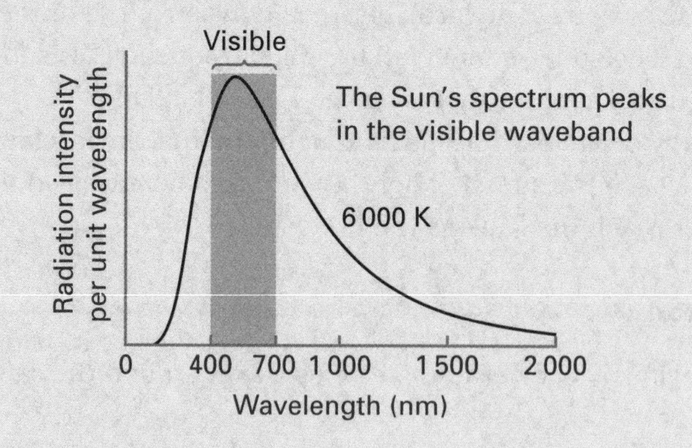

Wien's displacement law

The position of the emission peak in a black-body spectrum is given by:

$$\lambda_{max} T = 2.89 \times 10^{-3}$$

T is surface temperature in kelvins and λ_{max} is given in metres. The higher the temperature, the shorter the peak wavelength.

The Stefan–Boltzmann law (applied to stars)

A star's luminosity L (in watts) is given by:

$$L = 4\pi r^2 \sigma T^4$$

Where $4\pi r^2$ is the surface area of the star (of radius r), σ is the Stefan–Boltzmann constant and T is the star's surface temperature (K).

$$\sigma = 5.67 \times 10^{-8}\ \mathrm{W\,m^{-2}\,K^{-4}}$$

Using magnitudes to work out distances

Apparent magnitude m is a measure of how bright a star *appears* from the Earth. **Absolute magnitude** M is a measure of how bright a star actually *is*. (M is the magnitude a star would be if viewed from a distance of 10 pc.)

- Both m and M are measured on *logarithmic scales*.
- The brightest stars have the most negative magnitudes, the dimmest stars have the most positive magnitudes!
- The difference between m and M tells us how far away the star is:

  $m - M = 5 \log (d/10)$

  Where d is the distance in parsecs. Rearranging the equation, we get:

  $d = 10 \times 10^{(m-M)/5}$
- This technique works well up to distances of around 10 Mpc.

Standard candles

Stars with well known behaviour that can be identified in distant galaxies are called **standard candles**. *Cepheid variable stars* pulsate periodically, with readily measured changes in brightness. The period of the pulsation is directly related to the star's absolute magnitude M, and so the distance to the star and its galaxy can be calculated. Some types of supernova can also be used as standard candles.

- The greater the distance being measured, the greater its uncertainty.

Stellar absorption spectra

Dark lines in a star's emission spectrum are the result of absorption of radiation (at specific wavelengths) by excitable atoms in the cooler outer layers of stellar gas. These line spectra can be used:

- to identify elements present (giving clues to composition and age)
- to measure speeds of relative motion (of distant stars and galaxies)
- to detect spinning stars and binary stars

The Doppler effect

If a star is moving away from us, the light it emits is stretched out to longer wavelengths (*red shift*). If a star is moving towards us, the waves are bunched closer together (*blue shift*).

- *Radar* (using microwaves) can be used to measure the speeds and rates of spin of planets (*radar astronomy*).
- Line spectra from spinning stars are broadened due to red and blue shift. The degree of spread tells us the star's spin rate!

Examiner's secrets

Check your syllabus to see how much you need to know about magnitudes of stars.

Watch out!

Magnitudes are a (difficult) relic from earlier days of astronomy. Since stars have been painstakingly classified by magnitude, we have to learn how to use the data. $m \approx 1$ for the brightest stars visible with the naked eye, $m \approx 6$ for the dimmest stars visible with the naked eye.

Checkpoint 5

What are the sources of uncertainty in using absolute and apparent magnitudes to calculate distances to far-off stars? Why do these uncertainties increase with distance?

Links

See *atomic line spectra*, pages 120–1.

Checkpoint 6

(a) Why are absorption lines so wavelength specific?
(b) Why do absorption lines from spinning stars spread?

Exam questions

answers: page 184

1 Sirius is 2.7 pc away. (a) How long does light from Sirius take to get here? (b) If Sirius' apparent magnitude is –1.46, what is its absolute magnitude? (10 min)

2 The star Betelgeuse has an apparent magnitude of +0.50 and an absolute magnitude of –5.0. How far away is it? (5 min)

Astrophysics 2

Links

See *Newton's law of universal gravitation*, pages 132–3.

Links

See *binding energy and mass defect*, pages 50–1 for more information on energy release by nuclear fusion.

Checkpoint 1

Where would you expect to find most of the hydrogen in a giant star – in the core, or around the edges? Explain.

Checkpoint 2

(a) Why do collapsing stars heat up (ignoring fusion)?
(b) Why must white dwarfs fade?

Checkpoint 3

The (negative) gravitational potential at the surface of a sphere of mass M and radius R is GM/R (J kg^{-1}). Derive the equation for the radius of a dark star ($R = 2GM/c^2$).

Checkpoint 4

Michell's proposed dark stars were based on the idea of escape velocity. Light emitted would be slowed by gravity until it eventually came to a halt, then returned, accelerating to the star. What is wrong with this idea?

Studying stars is an ancient art; understanding them is very modern and still far from complete.

Birth of a star

Huge gas clouds are stellar nurseries. Gravity pulls gas particles together, heating them as they accelerate inwards. When they reach **ignition temperature** (between 10^6 and 10^7 K), collisions begin to result in nuclear fusion and the star *ignites*. Fusion releases huge amounts of energy, heating up the plasma further and causing a **radiation pressure** which halts the **gravitational collapse**.

Death of a star

When a star begins to run out of hydrogen, radiation pressure drops and the star starts to collapse again under gravity. As it collapses, its core heats up; helium ignites, releasing more energy, raising the pressure and causing the star to expand. It becomes a red giant – a huge, unstable star. Gravity dominates at the core, but radiation pressure sometimes dominates around the edges causing continued expansion. The outer layers of a dying red giant may spread out to form an expanding **planetary nebula**; the exhausted core collapses.

White dwarfs, neutron stars and black holes

The fate of a star depends on its mass. The biggest stars (>8 solar masses) explode as **super novae** which can briefly outshine entire galaxies. Smaller stars live longer and die in a less spectacular fashion. They slough off their outer layers more slowly, or not at all.

- *Small stellar cores* (<1.4 solar masses) will collapse to form **white dwarfs** which slowly fade.
- *Medium-sized cores* (1.4 to 2.5 solar masses) will collapse further, forming **neutron stars** – entire stars as dense as atomic nuclei.
- *Large stellar cores* (>2.5 solar masses) undergo complete, perpetual gravitational collapse, forming **black holes**.

(*Quantum physics effects* Electron degeneracy and neutron degeneracy protect white dwarfs and neutron stars, respectively, from further collapse!)

Black holes

John Michell suggested (in 1783!) that if a star's *escape velocity* was greater than the speed of light, then no light could escape it! The radius of such a *dark star* is given (using Newton's law of gravitation) by:

$$R = 2GM/c^2 \qquad (G = 6.67 \times 10^{-11}\ \text{N m}^2\ \text{kg}^{-2})$$

This approach is unjustified as Newton's law breaks down in strong gravitational fields, but it turns out that the critical radius of a black hole is given by exactly the same formula when relativity is properly accounted for! Karl Schwarzschild was first to solve Einstein's equation of general relativity to give the above solution. R is therefore known as the *Schwarzschild radius*.

Hertzsprung–Russell diagram 

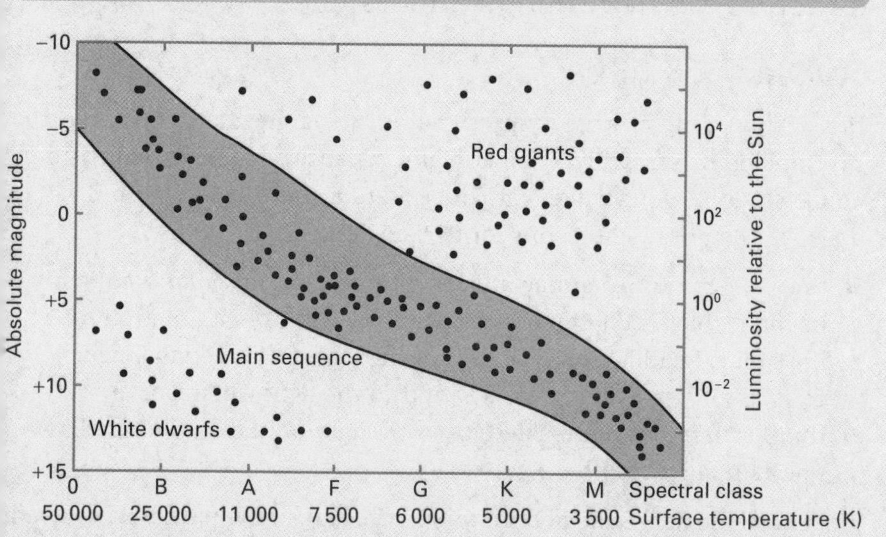

This diagram shows that some hot stars are unusually dim (white dwarfs) and some cool stars are unusually bright (red giants).

How to get the best images of stars 

- *High resolution* Two stars cannot be resolved unless their angular separation (in radians) is greater than λ/D, where λ is the wavelength of light being viewed and D is the diameter of the objective lens or mirror. This is known as the Rayleigh criterion.
- *Long exposure times* Photography and tracking devices . . .
- *High photon capture efficiency*: . . . using a charge coupled device – a light-sensitive microchip used in video and digital cameras which has high photon capture efficiency (typically 70%).
- *Minimal atmospheric interference.*

Telescopes

The angular magnification M of a **refracting telescope** depends on the focal lengths of the objective and eyepiece ($M = f_o/f_e$).

- Powerful refracting telescopes must be very long (large f_o).
- Large mirrors are easier to make than large lenses, and so most powerful optical telescopes are **reflecting telescopes**.

Radio astronomy gives different information on stars, galaxies and interstellar gases. Radio astronomers were first to detect **pulsars**.

Atmospheric transmission of electromagnetic radiation

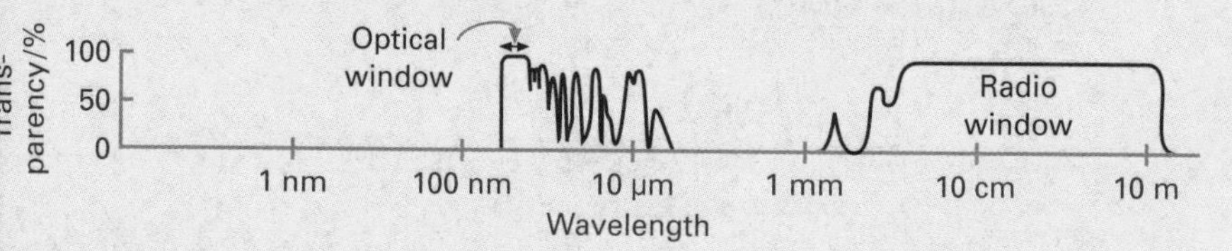

Exam question answer: page 185

Satellite-based X-ray detectors, UV, light, IR and microwave telescopes have all been used to great effect. Why are there few calls for radio telescopes to be launched? Give two reasons. (10 min)

Examiner's secrets

The H–R diagram is a favourite topic for exam questions. Be sure you can trace the life of a typical star (our Sun) from forming a *proto star* through its main sequence life to its *red-giant* phase and subsequent death as a slowly fading *white dwarf.*

Checkpoint 5

Explain why the unusually dim hot stars must be dwarfs and why the unusually bright cool stars must be giants.

Checkpoint 6

What size of radio dish would be needed to give the same angular resolution as a refracting telescope with a 75 mm objective? (Assume the wavelengths of light and radio waves being used are 0.5 µm and 5 m, respectively.) Use your answer to explain why arrays of radio telescopes are often used for detailed radio astronomy.

Check the net

Visit NASA's web site for the latest on satellite launches and discoveries: www.nasa.gov

Checkpoint 7

(a) Calculate the Schwarzschild radius of (i) the Earth, (ii) the Sun, (iii) a 5 solar-mass star (mass of Earth = 6.0×10^{24} kg; mass of Sun = 2.0×10^{30} kg).
(b) Explain why the Earth and Sun will not form black holes.

Astrophysics 3

Cosmology is the study of the entire Universe – it[s] structure, its beginnings and its fate!

Olbers' paradox

Hans Olbers (and others) realized that in an infinite, static Universe, th[e] sky should burn brightly, even at night, because any line of sight you choose to look along will eventually lead to a star – a light-source. It doesn't! The possible reasons for dark nights are:

- the Universe is expanding at such a rate that it has a *dark horizon* beyond which light can never reach us (light from a point on the horizon is heading towards us, but the journey is getting longer – by 3.00×10^8 m every second – and so the light will never arrive!)
- the Universe has not existed for ever. Light from distant stars has not had time to get here yet

Most cosmologists believe both to be true, and so the paradox is solve[d]

Gravity won't allow a static Universe!

Russian physicist Alexandr Friedmann solved Einstein's equations of general relativity and showed that they predict either a constantly expanding Universe (an *open* Universe) or a Universe which first expands, then collapses under gravity (a *closed* Universe). The fate of the Universe depends on its rate of expansion (the Hubble constant) and its density. Both factors are difficult to measure accurately.

Cosmological red shift

Line spectra from distant stars are all **red shifted**, showing that distant stars are receding from us as the Universe expands. Recession speeds are given by the Doppler equations:

$$\Delta\lambda/\lambda = v/c \qquad \text{and} \qquad \Delta f/f = -v/c$$

Relative increase in wavelength = recession velocity/speed of light.

Hubble's law

The speed of recession of a star v is directly proportional to its distanc[e] away d: $v = Hd$. This was discovered by Edwin Hubble in the 1920s. H is the **Hubble constant**. H's value is between 50 and 90 km s^{-1} Mpc^{-1}

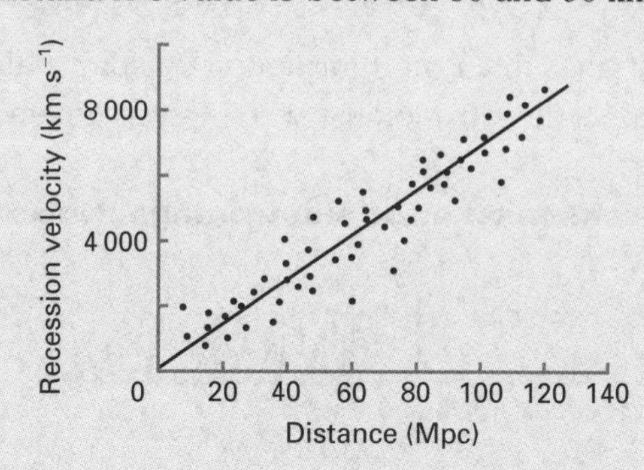

- A large value for H implies a rapidly expanding, young Universe.
- A small value for H implies a slowly expanding, old Universe.
- $1/H$ gives an estimate of the maximum age of the Universe.
- A value of 65 km s^{-1} Mpc^{-1} for H implies the maximum age of the Universe is 15 billion years.

Checkpoint 1

A third possible solution to Olbers' paradox is that the Universe is finite and small enough for us to see to its edges, in which case the blackness is really the void at the edge of the Universe! Why should we discount this conclusion?

Checkpoint 2

To estimate the age of the Universe, convert H's units into s^{-1}, then y^{-1}, then find the reciprocal. Calculate the range of maximum ages the Universe could have corresponding to the range of values of H.

Checkpoint 3

(a) Do you think H has remained constant since the big bang? Explain.
(b) Why is $1/H$ a *maximum* value for the age of the Universe?

Watch out!

The Hubble law does not imply that we are in the centre of the Universe! The same thing would be observed from any position in the Universe.

Cosmological principle

On a big scale, the Universe would look pretty much the same from any position. Hubble's law would hold anywhere.

The Universe can be modelled as the surface of a balloon. As it inflates, the stars separate. Hubble's law holds from every viewpoint. The expansion of space also accounts for cosmological red shift and the stretching of the remnant radiation from the big bang.

Checkpoint 4

Outer stars (only) in galaxies move too fast. Their motion can be explained by the presence of *dark matter* in the outer realms of galaxies. Explain why this dark matter must lie far from the centre of the galaxy, but within the orbit of these fast outer stars. (Difficult.)

Fate of the Universe and the omega factor

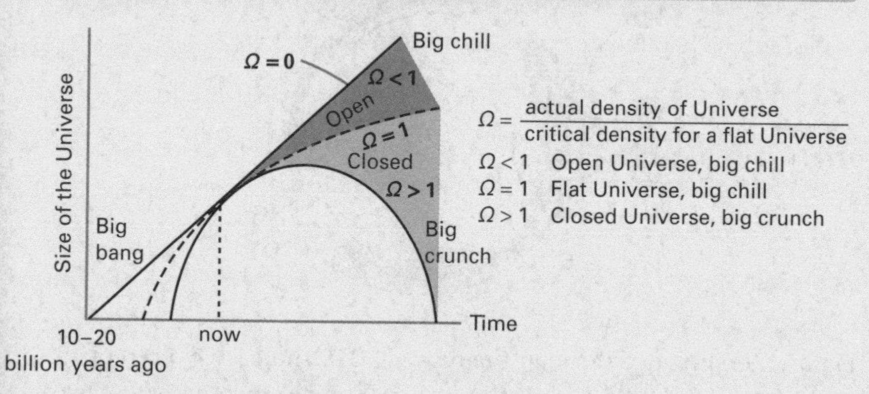

$$\Omega = \frac{\text{actual density of Universe}}{\text{critical density for a flat Universe}}$$

$\Omega < 1$ Open Universe, big chill
$\Omega = 1$ Flat Universe, big chill
$\Omega > 1$ Closed Universe, big crunch

The critical density is about 5 hydrogen atoms per m^3. In a flat Universe, the density is exactly sufficient to eventually (when time reaches infinity!) halt the Universe's expansion. Many cosmologists believe the universe is flat.

Action point

Find out more about why many cosmologists, particularly those with a background in particle physics, favour a flat Universe.

The Big Bang theory

The Universe was very hot in the beginning; the radiation pressure was immense. The Universe expanded and cooled rapidly. When its temperature had fallen to around 3 000 K (after about 300 000 years), the radiation pressure could no longer prevent the formation of atoms. Matter condensed and the Universe cleared; light could at last travel freely. Typical radiation at 3 000 K might have a wavelength of 1 μm. *Remnant microwave background radiation* (MBR) still bathes us from all directions, with a black-body temperature of 2.7 K and a peak wavelength of around 1 mm. Since clearing, the Universe has expanded and cooled by a factor of around 1 000.

Checkpoint 5

Use Wien's law (page 152) to check the peak wavelengths of black-body radiation at 3 000 K and at 2.7 K. What are the corresponding regions of the electromagnetic spectrum?

Evidence for the Big Bang theory

1. The Universe is expanding.
2. Remnant MBR.
3. The proportion of helium in old stars matches the amount the theory predicts would have been created in the big bang.
4. Radio astronomy shows the distant (old) Universe is different from the near, present Universe.

Examiner's secrets

Many syllabuses require you to give evidence that the Universe is expanding. Much of the research evidence changes constantly. Make sure you keep up to date – read scientific journals and stay in touch!

Exam question answer: page 185

The hydrogen absorption spectrum from quasar 3C273 is red shifted by 15%. (a) Calculate how fast it is receding from us. (b) Assuming $H = 65$ km s^{-1} Mpc, work out how far away it is. Convert your answer to light-years. (15 min)

Medical and health physics 1

Physicists and physicians are involved in the scienc of nature. Physicists can help doctors in examinatior diagnosis and treatment, but this partnership has see tragedy. For example, doctors began to use X-ray shortly after their discovery, but before their tru nature was understood. Many doctors died as a result

Checkpoint 1

Explain the similarities between the eye and a closed circuit TV system.

The eye and seeing

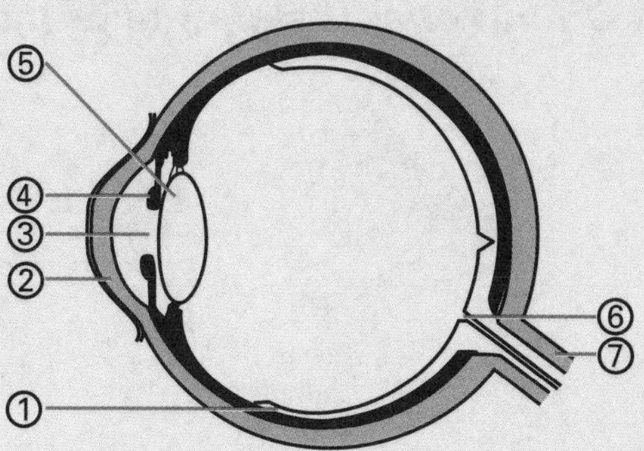

Test yourself

Study the diagram and notes on the structure and function of the eye. Now cover all but the diagram. Write down the name of each labelled part and explain its function(s). Repeat this exercise after a week has passed.

Light enters the eye through the *pupil* ③. This hole is covered by a transparent protective layer, the *cornea* ②. Light then enters a clear solution of salts, the *aqueous humour*. (The aqueous and vitreous humours provide the eye with nutrients and use liquid pressure to kee the eye's shape.) The *iris* ④ covers more or less of the pupil to control the amount of light entering the eye. About two thirds of the refraction and *focusing* of the light occurs at the cornea ②. The *lens* ⑤ makes fine adjustments to the focusing of light onto the *retina* ①. The *ciliary muscles* contract to make the lens thinner and relax to let it get fatter, allowing us to see both distant and near objects. This automatic focusing is called *accommodation*. Light next passes through the *vitreous humour* before forming an image on the retina – a layer of light-sensitive cells (*rods* and *cones*). Rods cannot distinguish different colours, but can pick up lower light levels than the colour-sensitive cones. Rods are useful by night, cones by day. The optic nerve ⑦ carries electrical impulses from the retina to the brain.

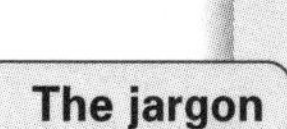

The jargon

The minimum distance an object can be from the eye, while still clearly visible, is called the *near point*. The *far point* is the furthest position of clear vision.

Checkpoint 2

What would happen if light from an object fell on the part of the eye labelled ⑥ in the diagram?

Eye defects and their correction

- *Myopia (short sight)* The eye cannot focus clearly on distant objects. Light from them is focused *short of* (before) the retina. A *concave* or *diverging* lens can be used to correct this problem.
- *Hypermetropia (long sight)* Light from objects near the eye is brought to a focus beyond the retina so that distant objects are not seen clearly. A *convex* or *converging* lens is used to treat this condition.
- *Presbyopia (old sight)* With age the lens becomes stiffer and the ciliary muscles weaken. Bifocal lenses become necessary.
- *Astigmatism* Images are seen clearly in only one plane, e.g. the vertical plane. This can be corrected with lenses shaped to compensate for the cornea not being spherically curved. They have a cylindrical feature to their shape.

Watch out!

It is always worthwhile to use the correct scientific terms, but make sure that you spell them correctly!

The ear and hearing

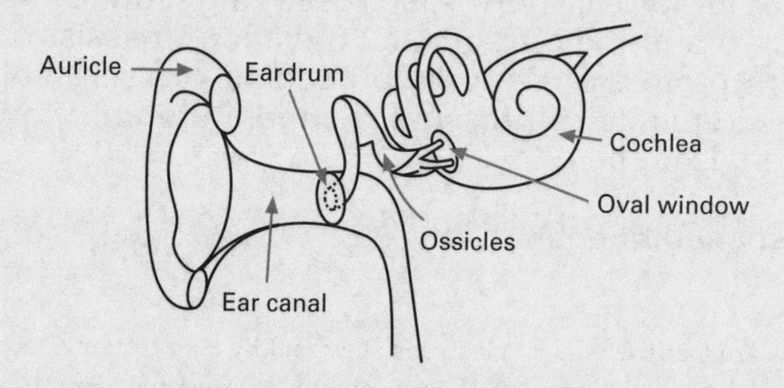

Sound waves reach the ear and are funnelled into the ear canal. They then travel along the ear canal to the eardrum, making it vibrate. This makes the ear ossicles, and then the oval window, vibrate too. The vibrating oval window forces fluid in the cochlea to move. Then tiny, sensory hairs in the cochlea vibrate, sending nerve impulses to the brain where they are interpreted as sounds.

The ear's response

Sounds can be distinguished by their pitch (frequency dependent), loudness (intensity and frequency dependent) and/or quality (determined by the range of frequencies and relative amplitudes within the sound).

- A sound's intensity I is defined as power per unit area ($W\,m^{-2}$).
- The minimum detectable (threshold) intensity $I_0 \approx 10^{-12}\,W\,m^2$.
- The highest safe intensity is about $100\,W\,m^{-2}$.
- The same sound, played to two people, can seem louder to one person than the other. Loudness is an individual response to intensity and frequency. A single frequency note of intensity I has loudness $L = 10\log_{10}(I/I_0)$, where L is in decibels (dB).
- The dBA scale is more sophisticated than the dB scale as it takes account of both intensity and frequency.
- Attenuation is the decrease in intensity of a wave as it travels.
- Threshold intensity I_0 and the ear's ability to respond to different frequencies (frequency response) suffer with age and noise damage.

Exam question

answer: page 185

Use these graphs to describe the ear's response to frequency and intensity. (20 min)

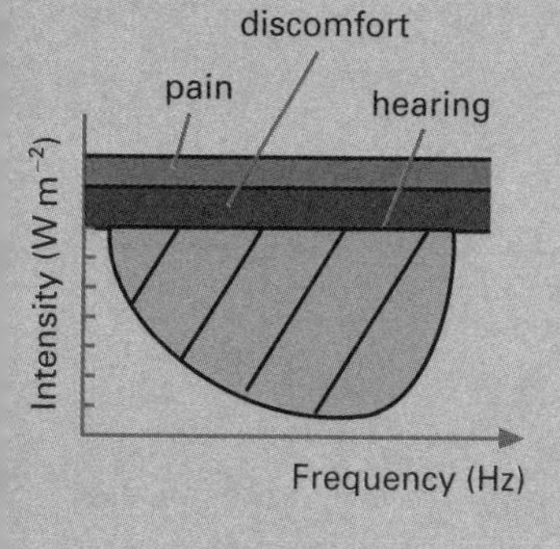

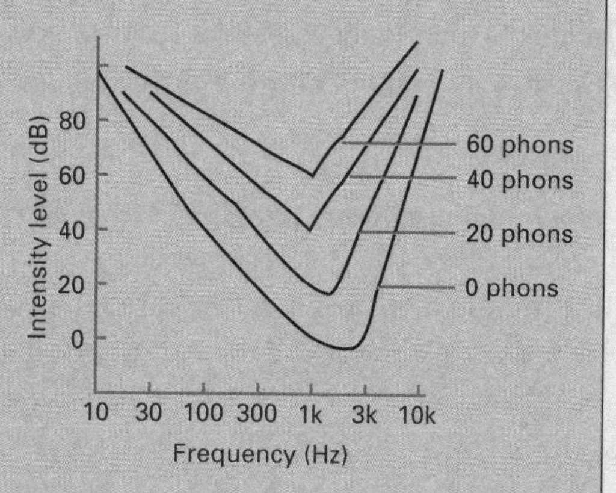

Speed learning

How would you eat an elephant? Answer – in small bits! The same thing applies for revision. To get it all done, do it in small bits!

Speed learning

Sometimes, the silliest things can stick in your head. Change the very serious ear diagram shown here into something much more light hearted. The auricle, or pinna, that funnels sound into the ear could be a filter funnel, the ear canal could have a narrow boat on it . . .

Checkpoint 3

Why does your own voice sound so strange when you hear it on an audio or video tape recording?

Checkpoint 4

After attending a loud music concert you hear ringing in your ears. What does this mean?

Medical and health physics 2

Medical physics began with the use of X-rays an radiotherapy. It now also uses ultrasound, ligh infrared and radio-frequency radiation to assist othe health professionals to help patients. It also involve measurements of pressure, temperature and flow c materials in the body.

Medical diagnosis

X-rays

X-rays are made when fast electrons slow rapidly after hitting a target. X-rays can be detected when they react with photographic film. As they pass through the patient, different body tissues absorb different amounts of their energy. Dense structures, e.g. bones, absorb most energy, and so cast dark shadows on a photographic film. The energy c an X-ray beam falls off exponentially as absorber thickness x increases. If soft tissue, e.g. the stomach, is to be X-rayed, a contrast medium (such as barium sulphate) can be used. The barium enters the stomach absorbs X-rays and improves contrast.

Conventional X-ray images are longitudinal (up and down) only. **Computed tomography** (CT) scans allow any cross-section to be imaged. An X-ray tube is moved around the patient. Very narrow X-ray beams pass through the patient and are detected by an array of severa hundred scintillation counters. Information from the scan is then collected, reconstructed and displayed by a computer.

The jargon

Exponential changes occur when the next value in a sequence depends upon the previous one. E.g. the number of bacteria that will be present in a colony tomorrow depends upon how many are there today. The amount of energy that will be able to travel deeper than 50 cm inside a human body depends upon how much energy got as far as 50 cm.

Ultrasound

Unlike X-rays, **ultrasound** is non-ionizing. It is believed to be harmless. Ultrasound waves (longitudinal pressure waves) can pass through human tissue and are reflected at the boundaries between body parts. The reflections are used to create images of the patient. Ultrasound waves are made when a high-frequency alternating voltage is applied tc a special piezo-crystal, e.g. lead zirconate–lead titanate, that changes shape when an electric field is applied to it.

When an ultrasound echo returns to the scanner, the surface of the crystal is squeezed so that the crystal then creates an electric potentia difference. This is piezoelectricity. Information from the scan can be displayed on a cathode-ray oscilloscope (an A-scan) or on a TV screen (a B-scan). As ultrasound waves are reflected at the boundary between materials of different density, a coupling gel or oil is applied to the patient's skin to let the ultrasound waves enter the body by matching the acoustic impedance of the skin.

The jargon

Ultrasound waves have a frequency above the limit of normal human hearing i.e. greater than 20 kHz.

The jargon

Acoustic impedance describes how difficult it is for sound waves to pass through a certain medium. Acoustic impedance is the product of the medium's density times the velocity of sound in that medium.

MRI

Magnetic resonance imaging (MRI) maps the position of protons in th human body. Hydrogen nuclei each contain a proton so MRI maps the position of hydrogen nuclei. As the human body is 70% water (H_2O) anc 20% fat (CH_3 and CH_2), MRI is ideal for imaging soft tissue.

Nuclei, such as hydrogen nuclei in the human body, spin! As they are also charged (containing protons), they act like magnets. If they are

The jargon

Magnetic resonance imaging (MRI) was once called nuclear magnetic resonance (NMR) imaging. The N in NMR referred to the nuclei being scanned. The name was changed because of fear associated with the word nuclear.

laced within another large magnetic field, these tiny magnets line up, ust as a compass lines up within the Earth's magnetic field.

The hydrogen nuclei can line up in one of two ways: either parallel r antiparallel to the external magnetic field. If a second external nagnetic field, this time constantly changing direction, is superimposed n top of the first, the hydrogen nuclei can be forced to flip from arallel to antiparallel and vice versa. As the hydrogen nuclei flip, they ake energy from the external oscillating magnetic field. This change in nergy is detected to produce an image – more energy will be drained t points where there is a high concentration of hydrogen nuclei.

ndoscopy

ndoscopes are light pipes used to look inside the body. Light is sent own one optical fibre, it reflects off the structure under investigation nd returns along a second optical fibre towards the doctor's eyes. ight bounces off the inside of optical fibres by total internal reflection TIR). TIR means that the light pipe does not have to be straight (so it an be pushed along cavities that are not straight, like blood vessels).

adioisotopes

adioactive isotopes (radioisotopes) can trace the passage of ubstances through the body, e.g. doctors can watch iodine-123 pass hrough the kidneys to ensure they are not blocked. Other applications nclude: testing thyroid function, checking blood flow through the brain nd measuring blood volume. The half-life of the radioisotope has to e long enough to allow measurements to be taken but short enough o that it does not linger! These radioisotopes emit gamma rays, then ecay into stable daughter products. Gamma cameras are used to etect the radiation coming from radioisotopes.

Medical treatment

adiotherapy uses X-rays, gamma rays or electrons to kill cancer cells. he radiation prevents cell division (which can cause cell death) by ttacking the DNA within the cancer cell's nucleus. There are three ypes of treatment. A beam of radiation can be directed at the patient rom the outside, a radioactive source can be placed into, or alongside, he tumour, or a radioactive liquid which is then taken up by the umour can be injected into, or swallowed by, the patient.

Early radiotherapy involved using ordinary X-ray machines at lightly higher energies than for diagnosis. More energetic (between .17 and 1.33 MeV) gamma ray photons, from cobalt-60, then became vailable to treat deeper tumours. Today, linear accelerators that can roduce X-rays of up to 25 MeV are preferred. They can also produce lectrons of varying energies to treat surface tumours.

Exam question

answer: page 186

(a) Construct a table to compare the main medical imaging techniques.

(b) Describe how ionizing radiation is used in radiotherapy. (25 min)

Check the net

For more on MRI, go to siemens.de/med/e/gg/mr/mr.html

Checkpoint 1

Light reflected back to the doctor travels along a bundle of fibres. What would happen to the image if the fibres became twisted over one another?

Checkpoint 2

By referring to relative ionizing powers and ranges, explain why gamma rays are more useful tracers than alpha or beta particles.

Speed learning

Visualizing a process often allows one to remember it. E.g. if a deep-seated tumour is to be attacked by external radiation, careful planning can avoid unnecessary damage to surrounding healthy tissue. Several beams can be directed at the tumour from different directions, rather like several needles piercing an apple. The needles are directed towards the rotten core of the apple. Each individual needle inflicts minimum damage as it travels towards the core. But maximum damage is inflicted where all the needles meet, at the core!

Radiation and risk

"O Roentgen, then
the news is true,
And not a trick of idle
humour,
That bids us each
beware of you,
And of your grim and
graveyard humour."

G. E. M. Jauncey

Links

See more ionizing radiation on pages 46–7.

X-rays were discovered in 1895 and within a year they were used in a popular fairground attraction. A mass x-ray screening programme was started in the 1940s to diagnose tuberculosis, a disease that claimed many lives. If you bought a new pair of shoes in the 1950s it was very likely that you would have had an x-ray of your feet. Comfort considerations are not now worth the risk when it comes to shoes, but the risk is worth taking to cure toothache or to see broken bones. So what exactly *is* risk?

X-ray films

Bones and teeth appear white on an x-ray film because these materials absorb x-rays. Cavities and soft tissue are quite transparent to x-rays – they go straight through, turning x-ray film black. Different materials absorb x-rays by different amounts and in the case of teeth and bones it is this that gives the contrast in the image.

Half-thickness

Knowledge of the way materials absorb x-rays is used to shield them to reduce the risk to people.

A given thickness of any material reduces the intensity of x-rays by a constant fraction. This means that the intensity decreases exponentially with increasing thickness.

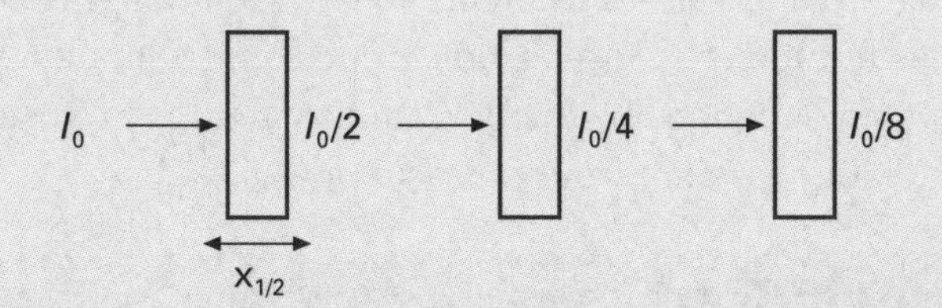

Half-thickness ($x_{1/2}$) is defined as the thickness of absorber needed to reduce the intensity of x-rays by half.

The table below shows typical half-thickness for 200 keV x-rays.

Concrete	Lead
25 mm	0.5 mm

Watch out!

Half-thickness is also known as *half-value thickness.*

Don't forget

Half-thickness applies to gamma rays too.

When you have an x-ray these days the nurse or dentist will keep well away from the x-ray machine. It's nothing personal, but they may even leave the room! In either case they will be shielded from the harmful x-rays by standing behind leaded glass.

This reduces the dose of x-rays they get, but what about the patient? Shouldn't they be protected too? Well, the answer is yes, but the difference is that a dentist or nurse may carry out many x-ray procedures every day. In fact, there are strict limitations on the allowed dose to medical staff and patients too.

It is worth remembering that when it comes to many medical procedures there is always a conflict between risk and benefit.

Checkpoint 1

If the intensity of an x-ray beam is I_0, what would be the intensity penetrating:
(a) 50 mm of concrete?
(b) 1.5 mm of lead?

Absorbed dose (Gy)

The amount of radiation absorbed by the human body is measured in units called **grays** (Gy) where 1 Gy is equivalent to 1 Jkg^{-1}. This seems simple enough, but the energy absorbed is not all there is to it.

It turns out that the gray is not good enough in predicting the potential consequences of exposure to radiation. The risk must also take into account the *type* of radiation involved.

Quality factor

The table below gives the quality factor of different ionizing radiation.

Radiation	Quality factor
alpha	20
neutrons	10
gamma	1
x-rays	1
beta	1

The quality factor takes into account the variation in sensitivity of body tissue to different types of radiation. The table shows that exposure to alpha radiation carries the greatest potential risk, whereas x-rays, gamma rays and beta radiation are potentially the least harmful.

Dose equivalent (Sv)

The dose equivalent is the absorbed dose multiplied by the quality factor. To distinguish it from absorbed dose, it has the unit **sieverts** (Sv). So, 2 Gy from an alpha source leads to an absorbed dose of 40 Sv (2 × 20). For x-rays, an equivalent absorbed dose gives rise to only 2 Sv (2 × 1).

Risk = probability × consequences

Risk from radiation can often be difficult to quantify because it is by no means certain that the potential effects of exposure will actually turn out to be life threatening. The concept of risk combines the probability of an event with the consequences of that event.

The current estimate of risk due to radiation dose is 5% per sievert. This means that 5 in 100 can be expected to contract a fatal radiation-induced cancer.

Watch out!

It is a popular myth that gamma rays are more damaging than alpha particles. This is not actually the case. Gamma rays, because they are uncharged, generally pass straight through the body. On the other hand, alpha particles are strongly ionizing and are absorbed easily in the skin, doing lots of damage.

Examiner's secrets

Dose equivalent in Sv is often just known as **dose**. To distinguish it from **absorbed dose** (Gy) you must learn that the units are different.

Checkpoint 2

Calculate the dose equivalent of a combination of doses totally 2 mGy, from a source that emits 10% alpha particles and 90% gamma rays.

Checkpoint 3

Background dose equivalent is about 2 millisieverts (2 mSv). How many people out of a population of 1 million may be expected to die from background radiation?

Exam question answer: page 186

A new x-ray cancer-screening programme is suggested. Every person over the age of 35 is to have a single chest x-ray that gives them a dose of 50 × background. How many deaths in a population of 20 million may be expected from this programme? Is the risk worth it? (5 min)

Electronics 1

The discovery of the transistor in 1947 has led to the miniaturization of electronic circuits and to the vast increase in electronic devices, such as mobile phones, computers, CD players, that are common in everyday life.

Electrical measurements

Digital or analogue meters can be used for electrical measurements.

Digital	*Analogue*
→ high input resistance so draws negligible current	→ resistance depends on range, less than that of a digital meter
→ cheap and robust	→ expensive and delicate
→ resistance scale is the same as for voltage and current	→ non-linear resistance scale
→ needs a power supply	→ only needs a power supply for measuring resistance
→ can do other things; e.g. measure capacitance, test transistors	→ can be used to observe values that change

Cathode-ray oscilloscopes

A CRO is a high-resistance voltmeter.

- → The voltage can be calculated from the height of the trace if the number of volts per scale division (voltage sensitivity) is known.
- → The current can be calculated using $I = V/R$ if the voltage is measured across a known resistor.
- → The time for a cycle of a periodic voltage can be measured from the time-base setting (the number of seconds per division), and its frequency calculated using $f = 1/T$.

Checkpoint 1

Work out (a) the peak voltage, (b) the frequency of the AC mains signal shown on the CRO screen if the voltage sensitivity is 50 V div^{-1} and the time-base setting is 2 ms div^{-1}.

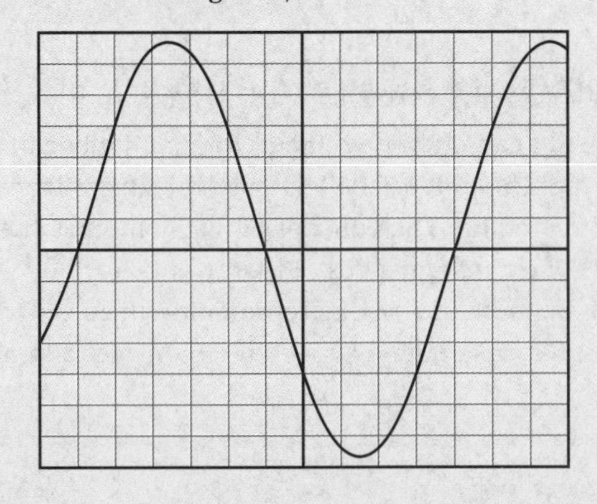

Capacitors

Links

See *capacitors*, pages 76–7.

Discharging

- → A capacitor discharging through a resistor has an **exponential decay curve** for the charge (and voltage and current).
- → The **time constant** is CR seconds and is the time taken for the charge to drop to $1/e$ (37%) of its initial value.

Checkpoint 2

Show that CR has the units of time.

- In 5*CR* seconds the charge will have dropped to 0.67% and the capacitor is said to be effectively discharged.

Capacitors and AC

When AC is applied to a capacitor, the current depends on the frequency as well as the capacitance. The resistance to AC, or **capacitative reactance**, is given by $X_C = 1/(2\pi fC)$ and is measured in ohms. It varies with frequency, as shown.

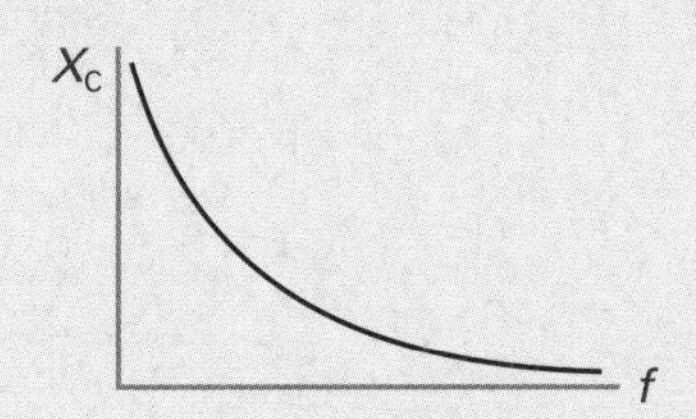

RC filters

- A **high-pass filter** lets high frequencies through and reduces low frequencies. A **low-pass filter** lets low frequencies through.
- The circuit consists of a resistor in series with a capacitor as a potential divider.
- At high frequencies the reactance is low so the voltage across the resistance is larger than the voltage across the capacitor.
- At low frequencies the reactance is high so the voltage across the capacitor is the larger.
- A high-pass filter takes the output across the resistance and a low-pass filter takes the output across the capacitance.

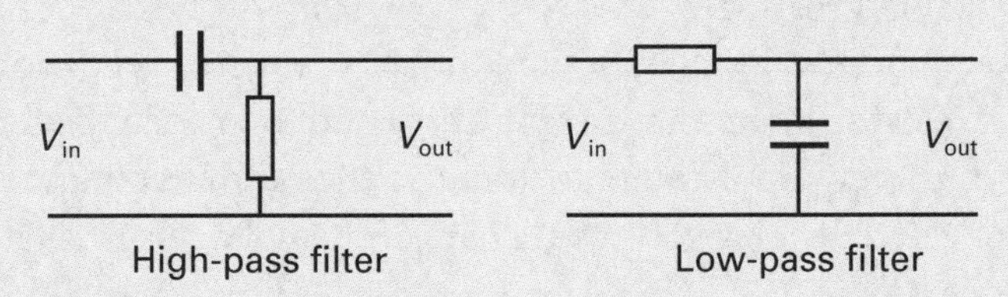

Square waves and *RC* circuits

When a square wave is fed into a capacitor and resistor in series, the voltage across the capacitor will increase then decrease exponentially. The voltage across the resistor will vary as the current in the circuit so it will be large in one direction and decrease exponentially as the capacitor charges up, then large in the opposite direction and decrease exponentially as the capacitor discharges.

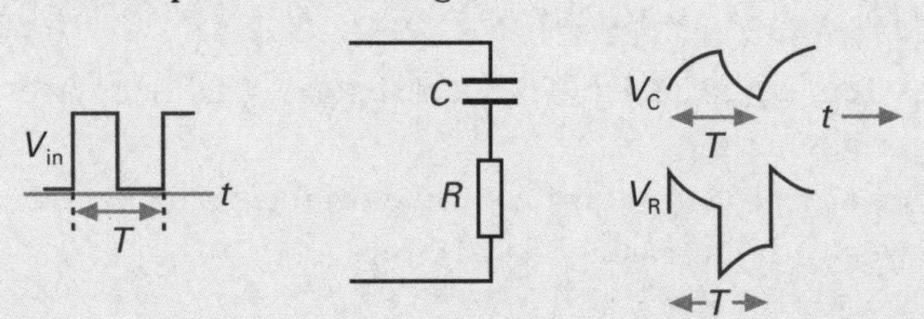

Exam question

answer: page 186

A square waveform of period *T* is fed into an *RC* circuit. Sketch graphs to show how the voltages across *R* and *C* vary (i) if $RC \gg T$ and (ii) if $RC \ll T$. (10 min)

Jargon

Electrolytic capacitors are *polarized* and have to be connected the correct way round in a circuit, unlike unpolarized ones.

Because a perfect insulator does not exist, capacitors conduct a little. Ceramic capacitors have a small *leakage current*.

Polypropylene capacitors have a high *working voltage* – that is, they can withstand a high voltage before being damaged.

Capacitance changes with temperature. Polycarbonate capacitors have a greater *temperature range* than cheaper polyester types.

Checkpoint 3

Calculate the capacitative reactance of a capacitor with a capacitance of 47 μF if the frequency of the AC is 50 kHz.

Checkpoint 4

A fully charged 2 200 μF capacitor is discharged through a 180 kΩ resistor. Calculate the time constant for the circuit.

Electronics 2

There are many different types of electronic component, each with its individual characteristics. Selections of these make up the circuits that operate electronic devices.

The jargon

Current passes when the anode of a diode is about 0.7 V more positive than the cathode. It is said to be *forward biased*. When it is *reverse biased* (the cathode is more positive than the anode) there is only a small *reverse current*. The *reverse breakdown voltage* is the maximum reverse voltage before the diode is damaged. The latter could be 115 V for a typical germanium diode or 5.1 V for a typical Zener diode.

Diodes

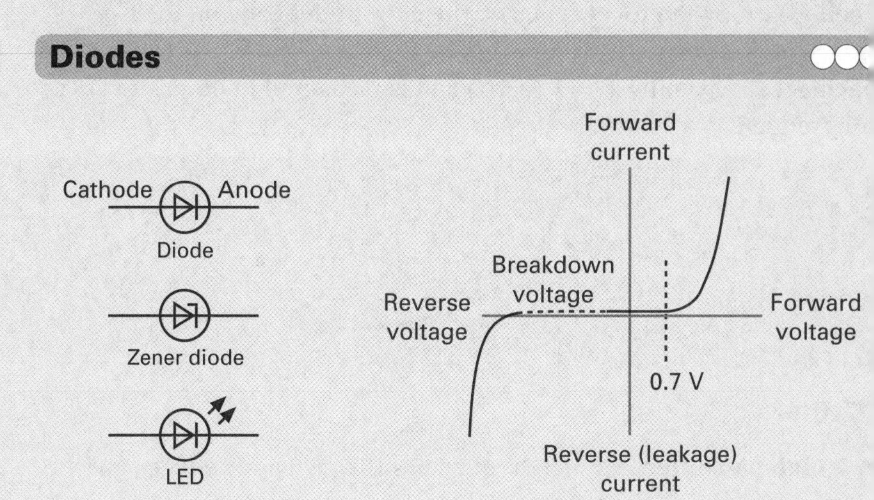

Links

See *alternating currents*, pages 74–5 for more on rectification.

Links

See *capacitors*, pages 76–7.

A diode allows current to pass only one way through it. This one-way conduction property of diodes is used to convert AC to DC in rectifier circuits. **Half-wave rectification** uses a single diode that cuts off half the power.

Full-wave rectification uses four diodes as a bridge rectifier. Both halves of the AC signal flow through the load resistor in the same direction. A large-value capacitor makes the output smoother. The initial pulse charges up the capacitor which then discharges slowly through the load resistor.

A **Zener diode** is designed to operate at its reverse breakdown voltage. A resistor limits the current so it is not damaged. Zener diodes are used in power supply units to stabilize the output voltage. Even though the current through it may change, the voltage across it remains constant.

A **light-emitting diode** emits light when it conducts in forward bias. The reverse breakdown voltage is about 5 V so care needs be taken to connect them correctly.

Diodes are sensitive to light but are covered in a light-proof package. A **photodiode** has a transparent covering, and so becomes a light sensor.

Checkpoint 1

A typical forward voltage for light to be emitted from an LED is 1.7 V and the maximum forward current is 20 mA. A series resistor is needed to protect the LED. For a supply of 5 V, calculate the PD across the resistor and the value of its resistance.

Resistive transducers

A **light-dependent resistor** has a resistance that decreases with increasing light intensity.

A **negative temperature coefficient thermistor** has a resistance that decreases with increasing temperature.

Junction transistors 

A transistor has three terminals called the **base**, the **emitter** and the **collector**. There are two types: npn and pnp. An npn transistor lets current flow from the collector to the emitter when the base is 0.7 V *more positive* than the emitter. A pnp transistor lets current flow from the emitter to the collector when the base is 0.7 V *more negative* than the emitter. Transistors can be used as switches that also amplify, since a small base current will switch on a larger collector current.

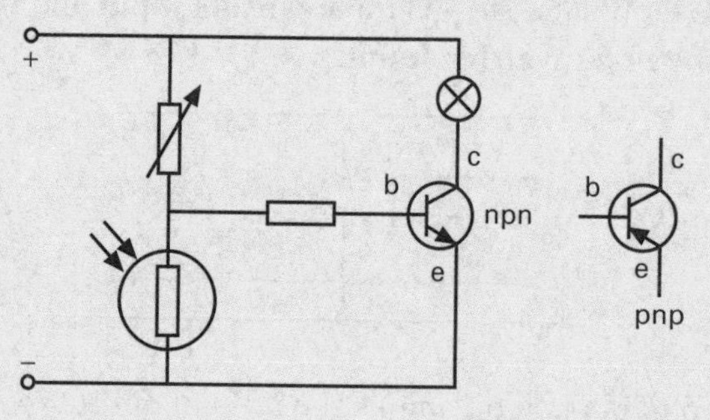

- A light-sensitive switch has an LDR and resistor forming a potential divider.
- In the dark, the LDR's resistance is high and most of the supply voltage is across it.
- The base of the transistor will be more positive than the emitter.
- It switches on causing a current through the lamp.
- In the light, the LDR's resistance drops.
- When the base voltage drops low enough the transistor switches off and the lamp goes out.

Checkpoint 2

Explain how the light-sensitive switch will operate if the positions of the LDR and the resistor are interchanged.

Checkpoint 3

What is the advantage of having a variable resistor in series with the LDR?

Mechanical switches 

A **relay** is an electromagnetic switch that uses a small current to switch larger currents on and off.

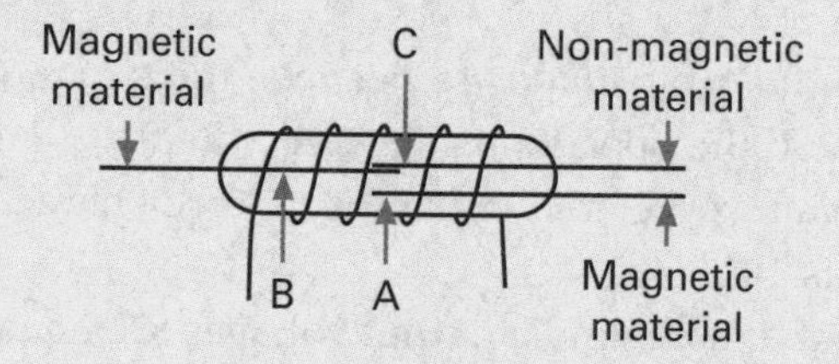

In a **reed relay**, A and B are not touching – they are normally open (NO). B and C are touching or normally closed (NC). A small current through the coil magnetizes A and B so they attract and make contact. When a relay switches off electromagnetic induction causes a big voltage across the coil that could damage the circuit connected to it. A diode connected across the coil prevents this.

Exam question answer: pages 186–7

(a) Sketch a circuit diagram of a temperature switch circuit that switches on a warning light when the temperature gets too high.

(b) Explain how the circuit operates. (15 min)

Electronics 3

An amplifier should produce an exact but larger copy of an input signal.

Operational amplifiers

An **operational amplifier** or **op amp** is an amplifier circuit contained on a single integrated-circuit chip.

Most op amps require a voltage supply that gives $+V_S$, 0 V and $-V_S$. The power supply and connections to it are often not shown on circuit diagrams. The input labelled – is the **inverting input** and the input labelled + is the **non-inverting input**.

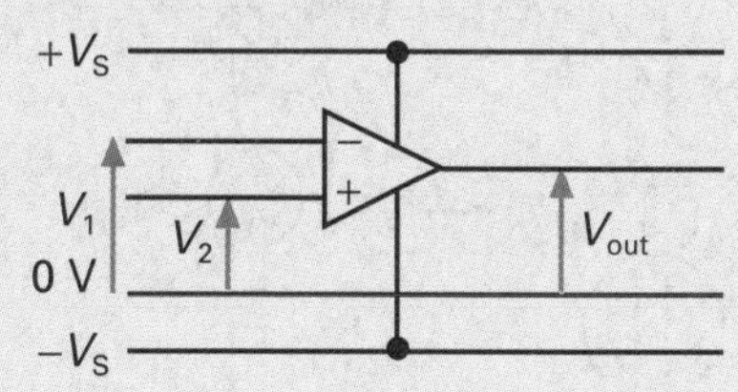

The **open-loop gain** of an op amp is the ratio of output to input voltage when it has no feedback. The **closed-loop gain** is the ratio of output to input voltage when the op amp has feedback.

An ideal op amp has the following characteristics:

- → infinite open-loop voltage gain
- → infinite input impedance so that it draws no current
- → zero output impedance so that the load has maximum current
- → very large bandwidth.

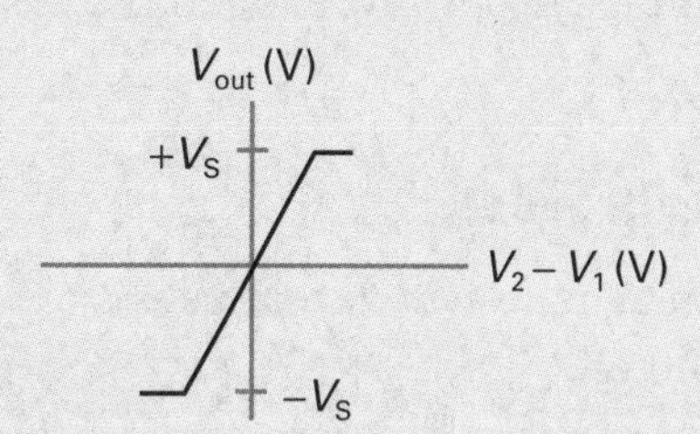

The op amp amplifies the *difference* between the two input voltages V_1 and V_2, and so if the open-loop gain is A_{OL} then $V_{out} = A_{OL}(V_2 - V_1)$. The open-loop gain is the gradient of the voltage characteristic graph. A typical value is 10^5.

The maximum and minimum output voltages are equal to $+V_S$ and $-V_S$ in theory although in practice they are slightly lower than this. In order for the output to be a copy of the input then the op amp should operate on the linear part of the graph outside of which it has become saturated.

A voltage comparator

A **voltage comparator** uses an op amp in open-loop mode.
If $V_2 > V_1$ then $V_{out} = +V_S$ and if $V_1 > V_2$ then $V_{out} = -V_S$
as long as the op amp is saturated.

The inverting amplifier

An op amp is not used as an amplifier without feedback since it takes only a small voltage to become saturated. Feedback reduces the gain but increases the bandwidth.

The jargon

Feedback is when some of the output is fed back into the input.

Negative feedback is when the output is in antiphase with the input so the feedback reduces the input.

Positive feedback occurs when the output is in phase with the input so the input is reinforced.

The *voltage gain* is the ratio of the output voltage to the input voltage.

The *power gain* of an amplifier can be very large so a logarithmic ratio is used: power gain in decibels (dB) = $10\log(P_{out}/P_{in})$.

The *bandwidth* is defined as the frequency range over which the gain does not fall by more than 3 dB from the value in the middle of the band. The voltage gain is proportional to the square root of the power gain so the bandwidth is the frequency range over which the voltage does not fall below $1/\sqrt{2}$ of its maximum value.

The total opposition to AC of a circuit containing, for example, both resistance and capacitance is called the *impedance* Z which is given by $Z = V_{RMS}/I_{RMS}$ and is measured in ohms.

Checkpoint 1

Show that if the open-loop gain is 100 000, then an op amp becomes saturated if $V_2 - V_1$ exceeds ±150 μV for a 15 V supply voltage.

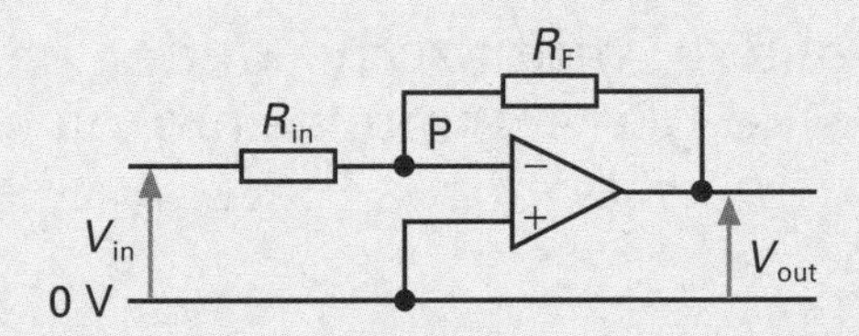

n *ideal op amp* takes *no current* so the same current *I* flows through e input resistor R_{in} and the feedback resistor R_F. An ideal op amp also as infinite gain which makes the two inputs at essentially the *same otential.*

P is also at 0 V since the non-inverting input is earthed. The voltage cross R_{in} must be $V_{in} = IR_{in}$ and the voltage across R_F must be $-V_{out} = IR_F$ aking the closed-loop gain, $V_{out}/V_{in} = -R_F/R_{in}$.

Jargon

Point P is called a *virtual earth* since in practice the op amp will take a small current so P will not be exactly 0 V.

Checkpoint 2

If R_{in} of an inverting amplifier has a value of 10 kΩ, what value should be used for R_F to give a voltage gain of −100? Calculate the output voltage when the input voltage is 10 mV.

Summing amplifier

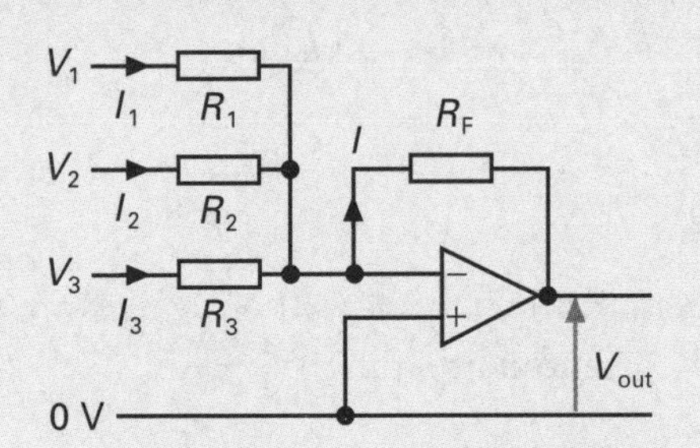

ere $I = I_1 + I_2 + I_3$ so $-V_{out}/R_F = V_1/R_1 + V_2/R_2 + V_3/R_3$.
$R_F = R_1 = R_2 = R_3$ then $V_{out} = -(V_1 + V_2 + V_3)$.

Checkpoint 3

R_{in} of a non-inverting amplifier has a value of 10 kΩ and R_F is 40 kΩ. What is the voltage gain? Calculate the output voltage when the input voltage is 1 V.

Non-inverting amplifier

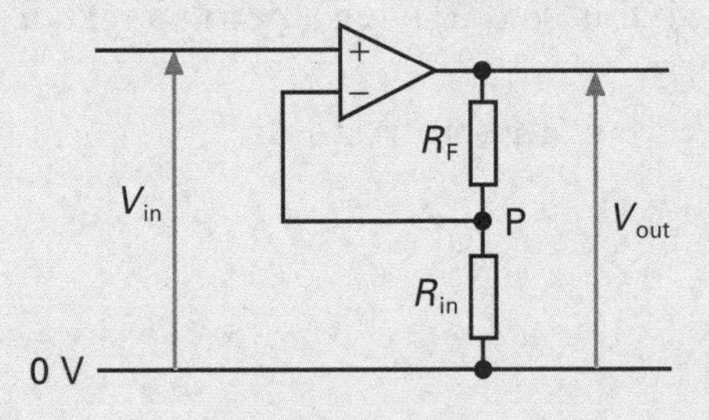

must be at 0 V (V_{in}) and the two resistors act like a potential divider. o $V_{out}/R_F + R_{in} = V_{in}/R_{in}$ giving $V_{out}/V_{in} = 1 + R_F/R_{in}$.

Exam question answer: page 187

The diagram shows a non-inverting amplifier.

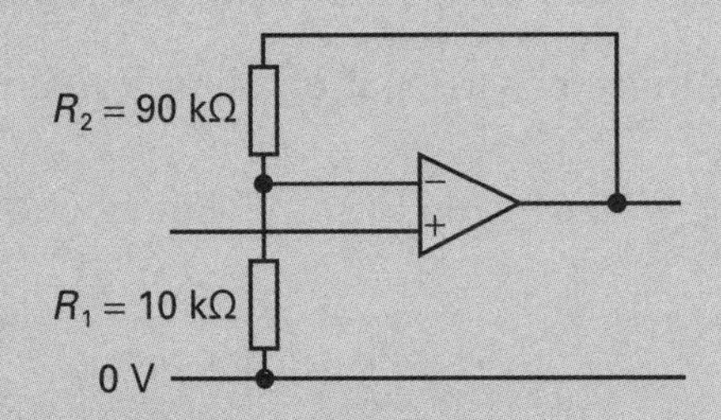

Explain the term *non-inverting amplifier.*

The closed-loop voltage gain of a non-inverting amplifier is given by the formula $1 + R_2/R_1$. Calculate the gain of the amplifier shown.

Show how this expression for the closed-loop voltage gain is derived. (10 min)

Materials 1

Materials behave differently under stress. Whe dropped, a wine glass shatters into pieces, a rubbe ball deforms then bounces back and a metal ca dents!

Links

See *stress, strain and Hooke's law*, pages 26–7.

Checkpoint 1

Write down the definitions of stress, strain and Young's modulus of a material. Give the units of each.

The jargon

The *strength* or *ultimate tensile stress* (UTS) of a material is the greatest stress it undergoes before breaking.

The *yield stress* is the stress when the material begins plastic behaviour.

The *breaking stress* is the stress at the breaking point.

Checkpoint 2

High-carbon steel is more rigid, or stiffer, than mild steel. It has a greater UTS and yield stress and undergoes less plastic strain. Sketch a graph to show how their stress–strain graphs will differ.

Watch out!

The energy stored by an elastic material is the area under the force–extension graph. The area under the stress–strain graph gives the energy stored *per unit volume* of the material.

Stress–strain graphs for different materials

Metal wires in tension

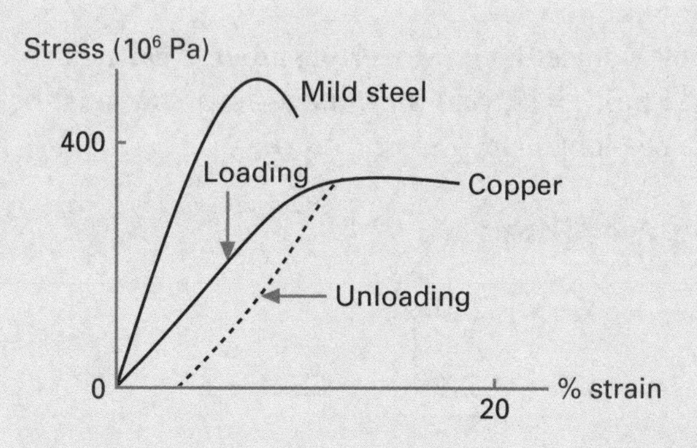

Metals are **polycrystalline** materials – their atoms are arranged as lots of bits of randomly oriented crystals.

- → The wire is **elastic** for small strains (about 0.1%) then it becomes **plastic**.
- → The **limit of proportionality** is the point after which the line is no longer straight.
- → After the **elastic limit** or **yield point**, the wire no longer returns to i original length but is said to have a **permanent set**.

Freshly drawn glass fibre in tension

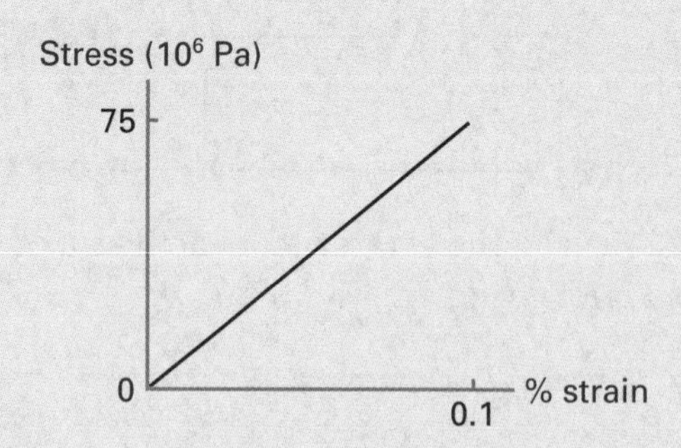

Glass is **amorphous** – its molecules have no regular order.

- → The fibre is quite strong – it requires a large force to break it.
- → It is elastic.
- → It breaks cleanly with a brittle fracture without showing plastic behaviour.

Rubber band in tension

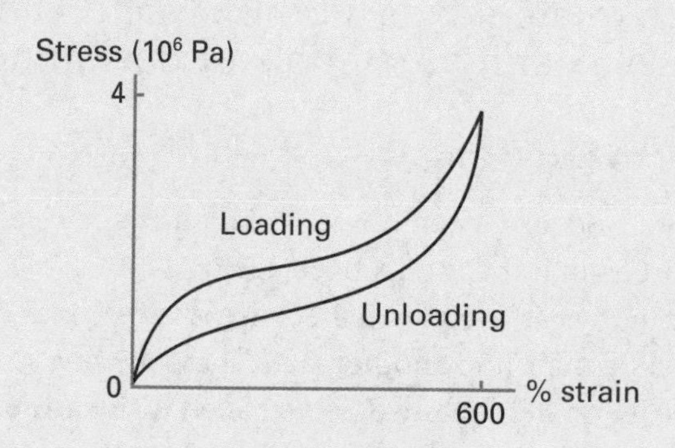

Rubber is a polymer – it consists of long-chain molecules, hydrocarbons in this case.

- Rubber is elastic for very large strains.
- It stretches easily at first but then becomes **stiffer**.
- The unloading line does not coincide with the loading line.

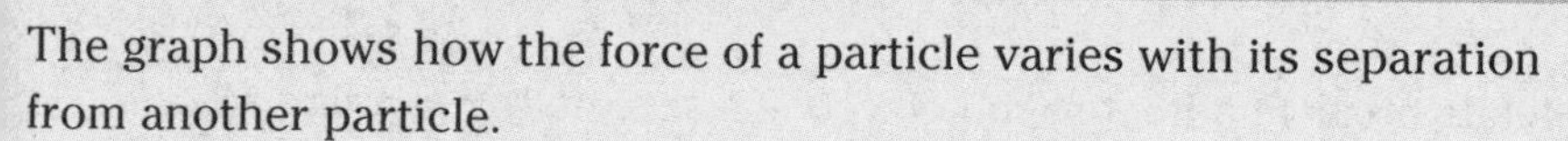

Forces between particles

The graph shows how the force of a particle varies with its separation from another particle.

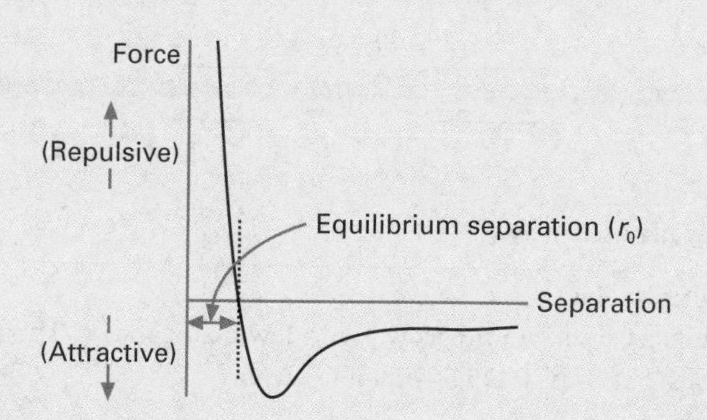

- The **equilibrium separation** r_0 is when the force between particles is *zero*.
- When the particles are closer than r_0 the force is repulsive and when they are further apart the force is attractive.
- The linear nature of the graph near the equilibrium separation means that the material obeys **Hooke's law** when it is in **tension** or **compression**.

Checkpoint 3

The following word pairs have opposite meanings. Give the meaning of each one: stiff/flexible, strong/weak, brittle/tough, elastic/plastic.

Examiner's secrets

You should be able to give examples of materials which are, for example, *weak and brittle* or *strong and tough.*

Checkpoint 4

Which force is largest when the particles are close together and which is largest when they are far apart?

Exam question answer: page 187

(a) Describe how you would measure Young's modulus of copper using a copper wire.

(b) Explain why it is necessary to use a long thin wire.

(c) Two wires, X and Y, are made from the same material. Wire X is three times as long as Y and has twice the diameter of Y. When a load is suspended from X the wire extends by 8 mm. How much will wire Y extend with the same load? (15 min)

Materials 2

Scientists now understand how the properties of a material are related to its atomic structure, and so they can improve existing materials or design new ones.

Behaviour of metals 

Elastic behaviour occurs when the stretched metal is released and the interatomic forces pull the atoms back again.

Crystalline materials are plastic because of **slip** – planes of atoms (**slip planes**) slide over one another. This happens in a single crystal.

Metals usually contain many defects called **dislocations**. This is an extra layer of atoms in the crystal structure. The dislocation moves easily under tension or compression, giving the material a permanent change in shape.

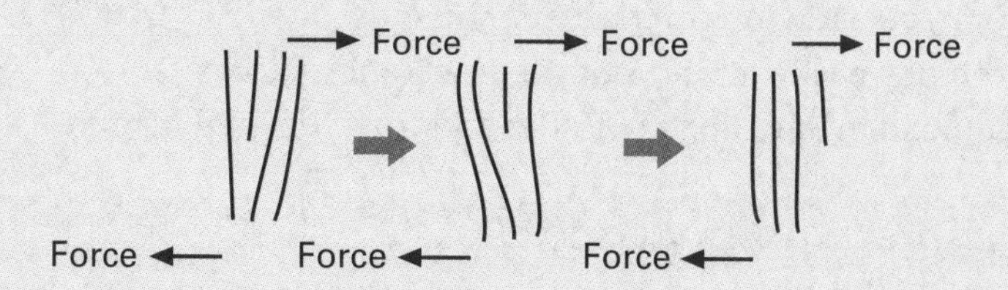

Metals break in a ductile *cup and cone* fracture after forming a *neck*.

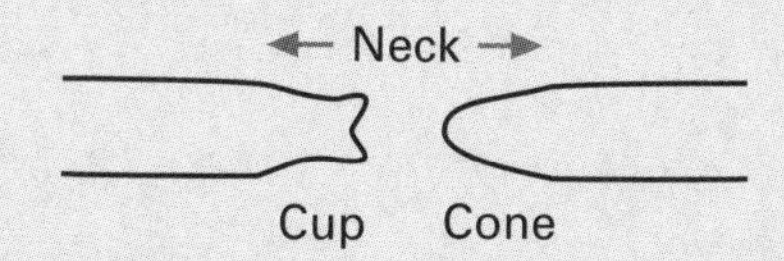

Reducing plastic flow

- Remove all defects
  - Single crystals can be grown with few dislocations. These are strong but impractically small in size.
- Add defects
  - Work hardening (or cold working) introduces more dislocations which jam up and prevent plastic flow.
  - Dislocations cannot move past the edge or **grain boundary** of a crystal. Heat treatment can add grain boundaries reducing plastic flow. **Quenched** steel has been heated then cooled suddenly so only tiny crystals have time to form. It is hard and brittle.
  - Added impurity atoms, such as the carbon atoms in steel, occupy the spaces between the iron atoms and make it harder for dislocations to move.

Behaviour of glass 

There can be no slip with glass, because there is no crystal structure. Glass fails because of tiny scratches in its surface. In tension, the stress at the tip of a scratch or crack is very large and this causes the crack to spread through the material, making a clean break.

The jargon

In a regular arrangement, each metal atom has six nearest neighbours. This is known as *close packing*.

Point defects are where an atom is missing at a site, or where an impurity atom replaces one of the metal atoms, or squeezes in the gaps of a regular array.

Checkpoint 1

Explain why it is more important to know the yield stress of a metal than its breaking stress if it is to be used for a load-bearing structure, such as the cables in a suspension bridge.

The jargon

A metal which undergoes considerable plastic deformation before breaking, like copper, is said to be *ductile*.

Plastic deformation also allows a metal to be *malleable*, that is it can be hammered into shape.

A *hard* metal is resistant to plastic flow.

The jargon

Tempered metal has been quenched then reheated to a lower temperature and cooled slowly. It is tough and elastic.

Annealed metal has been heated then cooled slowly. It is flexible and ductile.

The jargon

Fatigue failure occurs when a metal has been subjected to repeated loading and unloading.

Checkpoint 2

Explain why repeatedly flexing a piece of metal paper clip makes it become stiff and finally break.

Preventing brittle fracture

- Cracks do not propagate in ductile materials since slip occurs which blunts the tip of the crack and reduces stress.
- Pre-stressed or toughened glass is cooled by blowing jets of cold air on the surface. This makes the outside cool and contract while the inside is still molten. When the inside eventually cools and contracts, it holds the outside in compression so it is difficult for cracks to propagate.
- Fibre glass is an example of a composite material. It consists of thin glass fibres glued in a resin called the matrix. A crack in one fibre is stopped at the resin boundary.

Behaviour of rubber

The long-chain molecules of rubber are tangled at room temperature. When a force is applied, the chains untangle and line up. On release, thermal energy causes the chains to become tangled again.

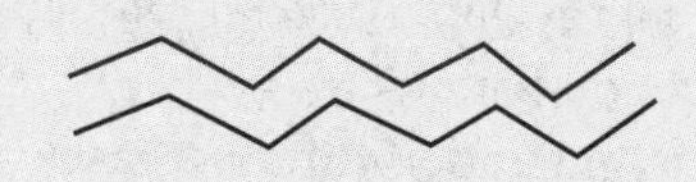

Rubber undergoes elastic hysteresis and this is why the unloading line does not coincide with the loading line. When rubber is stretched and released, not all of the work done in stretching it is recovered. Some of the work done increases its internal energy, making it hotter.

Other polymers

Thermoplastics can be moulded when warm.

- Below their **glass transition temperature**, they are flexible and rubbery.
- Above this temperature, they are rigid and glassy.
- Near the melting point they are **viscoelastic** – the chains slide past one another – behaving like a viscous liquid, although a rapidly applied stress causes elastic behaviour.
- Perspex is amorphous. It is vacuum moulded to produce crash helmets, for example.
- Polythene and nylon are semicrystalline.

Thermosets do not soften on heating as the long chains are crosslinked.

- They are rigid and brittle.
- Bakelite and melamine are used to make electrical fittings.

The jargon

A *tough* material is resistant to the propagation of cracks and requires a relatively large amount of energy to break it. The work done is the area under the force–extension graph, and so this area will be large for a tough material.

Checkpoint 3

Concrete is amorphous. Explain why it is weak in tension. Pre-stressed concrete has concrete mix poured around metal rods held in tension. When the concrete is set, the rods are released. How does this make the concrete stronger?

The jargon

Creep is when a loaded material gradually extends with time. In rubber this is due to the plastic flow of the molecules. The process of vulcanization prevents plastic flow by causing the chains to be crosslinked by sulphur atoms.

Checkpoint 4

Bouncing putty is a silicone-based polymer. Explain why, when rolled into a ball and dropped, it bounces, but it flows into a thin disc if left alone.

Exam question

answer: page 187

Explain why (a) copper is ductile, (b) glass is brittle and (c) rubber is very elastic. (15 min)

Materials 3

An understanding of the atomic structure of materials has led to improvements not only in mechanical behaviour but also in electrical, magnetic and optical properties.

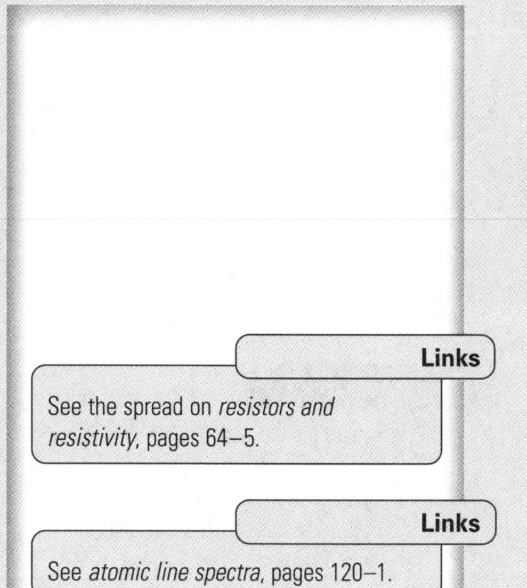

Electrical properties

Conduction

Links

See the spread on *resistors and resistivity*, pages 64–5.

Links

See *atomic line spectra*, pages 120–1.

- **Electrical conductivity** σ is the inverse of resistivity, $\sigma = 1/\rho$. Its units are mho m^{-1} or $(\Omega\ m)^{-1}$.
- In a solid, the outer electrons of neighbouring atoms interact and give rise to ranges of allowed energy levels called bands. The highest band is the **conduction band**, below this is the **valence band**.
- In metals, the valence band and conduction band overlap so there are always electrons that are free to move. An insulator has a wide energy gap between the bands. At room temperature the valence band is full and the conduction band is empty. A semiconductor has a smaller energy gap. At room temperature a few electrons occupy the conduction band.
- An increase in temperature increases atomic vibrations that impede the movement of charges. This causes the resistance of metals to increase with temperature. An increase in temperature increases the number of electrons in the conduction band for semiconductors and insulators (to a lesser effect) so their resistance decreases with temperature.
- A light-dependent resistor (LDR) is made from semiconducting material. An increase in light intensity gives some electrons energy to jump into the conduction band, and so the resistance decreases.
- The **drift speed** of charges in a wire carrying current is given by $v = I/nAq$, where n is the number of charge carriers per unit volume.

Checkpoint 1

How would you expect the drift speeds of the charges in samples of a metal and a semiconductor to compare if they had the same cross-sectional area and carried the same current?

Hall effect

This arises when a thin slice of conductor carries a current at right angles to a magnetic field. The sides of the slice become charged giving rise to the Hall voltage $V_H = Bvd$ where v is the drift speed of the charges and d is the width of the slice. The Hall voltage can be used to compare magnetic field strengths.

Links

See *electromagnetism*, pages 78–9.

Checkpoint 2

Explain why Hall slices are made from semiconducting material rather than metal.

Superconductors

- When some metals are cooled below a certain temperature, the **transition temperature**, their resistance becomes zero.
- A current in such a cooled superconductor continues indefinitely without a power supply.
- Superconducting magnets are used to make charged particles move in circular paths in particle accelerators.

Magnetic properties

Hysteresis loops show how the field strength changes when a magnetic material is placed in an applied magnetic field.

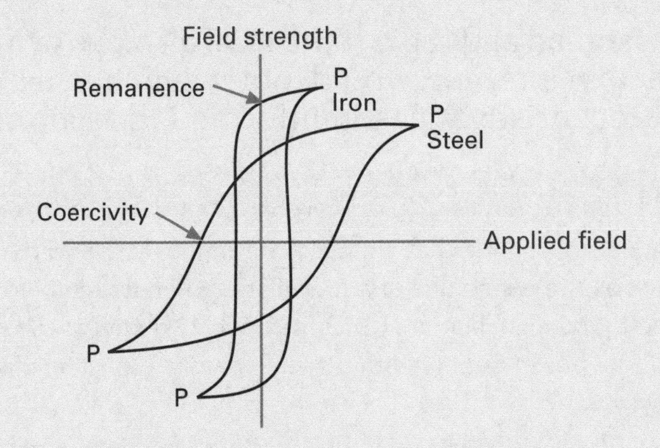

When the applied field is zero, the material retains some magnetism.
- A reverse field is needed to demagnetize the material.
- The area inside the loop represents the energy needed to remagnetize the material, which becomes thermal energy.

omains

Iolecular dipoles line up inside the material forming areas called **omains**. In a non-magnetized material, the magnetic fields of the omains cancel. Changing the applied field causes the domains to ealign.

Optical properties

- The energy of a photon of visible light is less than the energy gap in an insulator, and so the photon is not absorbed: the insulator is transparent. Metals are opaque as photons of visible light can raise electrons from the valence to the conduction band.
- Impurity atoms in glass add levels in the energy gap causing specific wavelengths to be absorbed.
- Minute changes in density give rise to Rayleigh scattering, where the amount of light scattered is inversely proportional to λ^4, and so long wavelengths are scattered least.
- LEDs and lasers can be used to send light signals down optical fibres. LEDs are cheap but have low powers and a spread of wavelengths. Lasers are more expensive but have higher powers and a smaller spectral spread giving less dispersion.

The jargon

At P the material has become *magnetically saturated.*

The *remanent flux density* is the field that remains when the applied field is removed.

The *coercive field* is the reverse field needed to demagnetize the material.

A *soft* magnetic material, like iron, has a high remanence and a low coercivity because it is easy to realign the domains.

A *hard* magnetic material, like steel, has a low remanence and a high coercivity because a lot of energy is needed to realign the domain walls.

A magnetic material becomes non-magnetic above its *Curie temperature* as the thermal vibrations disrupt the alignment of the dipoles.

Checkpoint 3

Explain why heating or repeated hammering of a magnet will cause it to lose its magnetism.

Links

See *atomic line spectra*, pages 120–1.

Checkpoint 4

The voltage across an LED when it *just* emits light of wavelength 600 nm is 2.1 V. What value does this give for Planck's constant?

Exam question answer: pages 187–8

(a) Sketch the hysteresis loops for both hard and soft magnetic materials.

(b) Use your graph to explain what is meant by coercivity, and how this differs for both types of material.

(c) What is the significance of the area enclosed by each hysteresis loop?

(d) Give an example of a situation where a magnetic material with a hysteresis loop of small area would be used. (15 min)

Telecommunications 1

The widespread and rapid transfer of all types of informa-tion via the internet would not have been possible without recent advances in information technology.

Signals

An **analogue** signal is one that changes continuously over a range of values, such as the resistance of a thermistor when its temperature changes. A **digital** signal is one which has only two states – a lamp connected to a power supply with a switch can be either on or off but not in between.

Checkpoint 1

Give an example of:
(a) an analogue device
(b) a digital device.

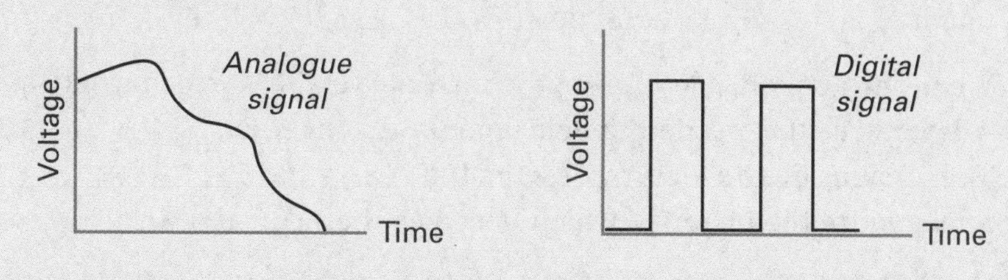

Principles of communication

Sound and TV signals are carried long distances using radio waves.

Bandwidth means a range of frequencies. A higher bandwidth carries more information more quickly. Telephone requires a lower bandwidth (4 kHz) than music (15 kHz) which requires a lower bandwidth than colour TV (6 MHz).

Amplitude modulation is where the lower-frequency analogue signal is used to vary the amplitude of the high-frequency carrier wave. AM is simple, but is affected by unwanted signals or **noise** – as this alters the amplitude of the carrier wave.

The jargon

The *signal to noise ratio* is the proportional amount of signal compared with noise added and it gives an idea of how distinguishable the signal will be.

All complex waves can be broken down into a combination of sine waves of different frequencies. An AM wave sending signal frequencies from 50 Hz to 4.5 kHz on a carrier wave of frequency 100 kHz has a bandwidth of 9 kHz consisting of the carrier frequency plus two sidebands.

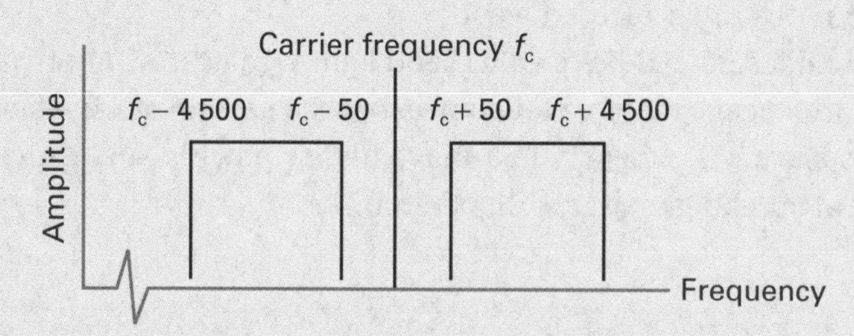

Checkpoint 2

Sketch waves to show:
(a) (i) a quiet, (ii) a loud signal on an AM wave.
(b) (i) a quiet, (ii) a loud signal on an FM wave.

Frequency modulation is where the signal wave alters the frequency of the carrier wave. FM is more complicated to implement and it has a higher bandwidth of 180 kHz. Although noise can still affect the amplitude of an FM wave, the signal wave is unaffected.

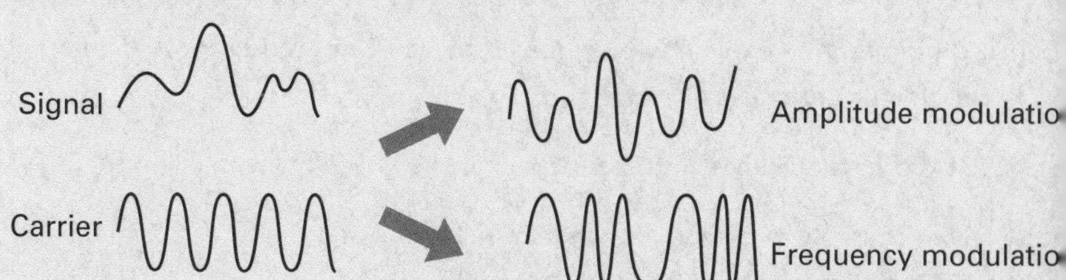

Pulse-code modulation is where an analogue signal is converted to a digital one using an analogue to digital converter (ADC). Distortion to a digital signal can easily be eliminated.

1 **Sampling** First the voltage of the analogue signal is taken at regular intervals. The sampling frequency has to be at least twice the highest frequency in the signal.
2 **Quantization** Next the voltage range is divided into a set of levels and the voltage value is given a number determined by the nearest level below it.
3 **Encoding** Finally the quantization level is converted to a binary number and transmitted as a series of pulses.

An error occurs with quantization since the voltages are rounded down. The error is greatest with smallest signals. **Companding** makes the levels closer together for smaller signals which reduces the error. At the receiving end a digital to analogue converter (DAC) reverses the process.

1 The binary number is converted back into a base ten number.
2 This changing digital signal is then smoothed by a capacitor.

Time-division multiplexing (TDM)

Many digital signals can be carried down the same communication link by TDM. The pulses are divided into time segments, compressed, speeded up and slotted between other signals. They are reassembled at the other end.

AM radio receiver

An AM receiver is less complicated than an FM one.

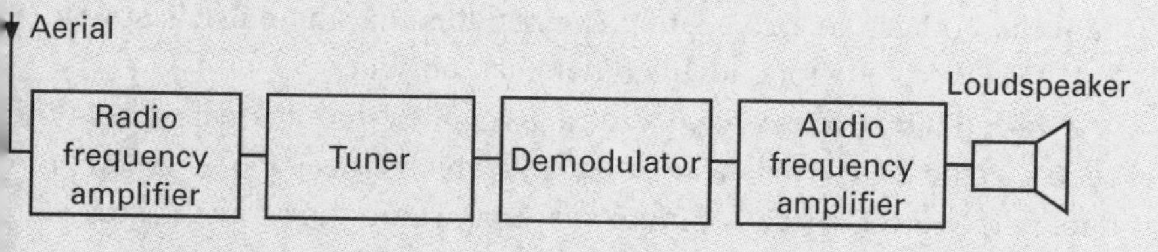

1 An aerial picks up lots of radio frequencies.
2 A variable capacitor in the tuning circuit selects one frequency.
3 A diode in the demodulator lets only the positive signal through and a capacitor blocks the high frequencies.
4 A loudspeaker responds to the average current.

Checkpoint 3

A human's full audible range is 20 kHz. What is the lowest sampling rate that should be used for transmitting music?

The jargon

Since digital signals can have only two states, *on* coded as 1 and *off* coded as 0, numbers have to be represented on the *binary* or base two system. In binary 110 has 1 in the fours column, 1 in the twos column and zero units making the base ten number 6. A *four-bit binary number* uses four binary digits to represent the number, so 0110 would represent the number 6. Four bits can represent 2^4 or 16 different numbers whereas eight bits can represent 256.

Exam questions

answers: page 188

1 Modulation and multiplexing are two processes associated with radio communication. Explain the meaning of each of these processes and why they are necessary. (8 min)

2 Discuss the relative advantages and disadvantages of pulse-code modulation over amplitude modulation for the communication of information. (11 min)

Telecommunications 2

Today data can be sent as electrical pulses along metal cables, as electromagnetic waves through air or space, or as flashes of light down optical fibres.

Communication channels

Wire pairs

A pair of metal wires was the original channel for transmitting communications signals. Wires carrying AC have changing fields around them which induces currents in other wires. This crosslinking means that other conversations can be heard faintly. Twisting the wires can reduce the effect.

Loss of energy caused by the wire's resistance means the signal has to be amplified at regular intervals. Wire pairs have a narrow bandwidth. They are still the most common way of carrying telephone signals to homes from local exchanges.

Links

See *electromagnetic induction*, pages 80–1.

The jargon

The amplifier used for analogue signals is called a *repeater*. Repeaters amplify the noise as well. Energy loss causes digital signals to become smaller and distorted. They are actually reproduced by a *regenerator* which removes the noise.

Coaxial cable

Coaxial cable has a braided metal wire cylinder surrounding the inner wire, with insulation between them. It has a bandwidth of 10^9 Hz. It is used to carry UHF (ultra-high-frequency) signals from aerials to televisions. It was used to carry trunk calls but is being replaced by optical fibres and microwaves.

Optical fibres

Step-index fibres have a glass core surrounded by cladding of lower-refractive-index glass. The change in refractive index is to ensure total internal reflection. Digital signals suffer from **pulse broadening** when travelling down step-index fibres. Some rays follow a much more zig-zag path than others so arrive later. The result is that the pulse is spread out. This limits the rate at which data can be sent.

Graded-index fibres have a refractive index that decreases from the centre to the edge. This means that rays which zig-zag more, spend more time in the lower-refractive-index material where they travel faster.

Monomode fibres are extremely narrow which reduces the number of different paths that can be taken.

Optical fibres have the following advantages over metal cables:

- glass is cheaper than metal
- optical fibres that carry the same number of signals as metal wires are thinner and lighter
- there is less attenuation, or energy loss, so fewer regenerators are needed to restore the pulses
- there is no crosslinking and signals are not affected by electrical interference
- security is better as optical circuits cannot be tapped into

Links

See *total internal reflection and fibre optics*, pages 104–5.

Checkpoint 1

In a particular digital telecommunication system, the signal to noise ratio must not be less than 21 dB.

(a) If the noise power is 10^{-20} W, what is the minimum power that the transmitted signal can have?

(b) If the attenuation is 2.0 dB km^{-1} and the initial signal power is 10^{-3} W, what is the maximum distance the signal can travel before it is regenerated?

Radio waves and microwaves

The table shows how parts of the electromagnetic spectrum are used in telecommunications.

Wavelengths	*Classification*	*Use*	*Frequencies*
100–1 km	low frequency long wave	AM radio	3 kHz–300 kHz
1 km–100 m	medium frequency medium wave	AM radio	300 kHz–3 MHz
100 m–10 m	high frequency short wave	amateur radio police	3 MHz–30 MHz
10 m–1 m	very high frequency	FM radio	30 MHz–300 MHz
1 m–10 cm	ultra high frequency	TV cellphones	300 MHz–3 GHz
10 cm–1 cm	microwaves	satellites trunk lines	3 GHz–30 GHz
100 μm–10 μm	infrared	optical fibres	3 THz–30 THz

Radio-frequency currents oscillating along metal aerials broadcast radio waves. They are picked up most efficiently by aerials tuned to a particular frequency. The simplest tuned aerial is the **dipole** aerial made from two lengths of metal each a quarter of the wavelength long. Dipole aerials are used for VHF and UHF bands but would be impractical for LF and MF.

Long-distance radio communication, originally thought to be restricted by the curvature of the Earth, is possible by:

→ **ground waves** that follow the surface of the Earth provided this is a good conductor

→ **sky waves** that bounce off a reflecting layer of the atmosphere (the ionosphere) and the Earth's surface.

Microwave networks have a very large bandwidth and are used for TV, telephones and data transfer. They are propagated by line of sight between repeater stations. Parabolic aerials are used to focus the microwaves and to collect enough energy.

Geostationary satellites have 24 hour orbits above the equator, and so stay above the same part of the Earth. They receive microwave signals from one place and transmit to another so communication coverage of most of the world is possible.

Checkpoint 2

Explain why FM radio signals are transmitted in the VHF band.

Checkpoint 3

Work out the length of the dipole aerial needed for a 198 kHz radio wave.

The jargon

In *cellular* mobile phone systems, the country is divided into areas called cells, each containing a microwave transmitter/receiver. As users travel from one cell to the next, signals are automatically transferred to the new cell's transmitter/receiver.

Exam question

answer: page 188

Discuss the reasons for, and advantages of, changing from a metal-cable transmission system to an optical-fibre transmission system. (7 min)

Turning points in physics 1

Check the net

For more on the discovery of the electron, go to iop.org/Physics/Electron/section 2/discovery.html

The jargon

Electricity comes from the Greek word for amber, *elektron*.

The jargon

Cathode-ray tubes are similar to modern TV tubes. They contain gas at low pressures. Cathode-ray particles (electrons) are emitted from a cathode (negative terminal) and accelerated towards an anode (positive terminal).

The jargon

Thermionic emission is a process in which metals emit electrons. The metal is heated so that some of its free electrons gain enough kinetic energy to leave its surface.

Checkpoint 1

Thomson calculated e/m to be -1.76×10^{11} C kg^{-1}. The specific charge Q/M for hydrogen ions was known to be 9.65×10^{7} C kg^{-1}. Thomson assumed that the charge on an electron e was equal and opposite to the charge on a hydrogen ion Q. Explain what this meant about the mass of an electron compared with the mass of a hydrogen ion.

Checkpoint 2

As $Bev = mv^2/r$, prove that $e/m = v/Br$.

Can you imagine life without electricity? You might take electricity for granted but your grandparents, or greatgrandparents, were probably very excited when 'the electric' came to their home. The discovery of the electron shed new light on the structure of the atom and much more!

The discovery of the electron

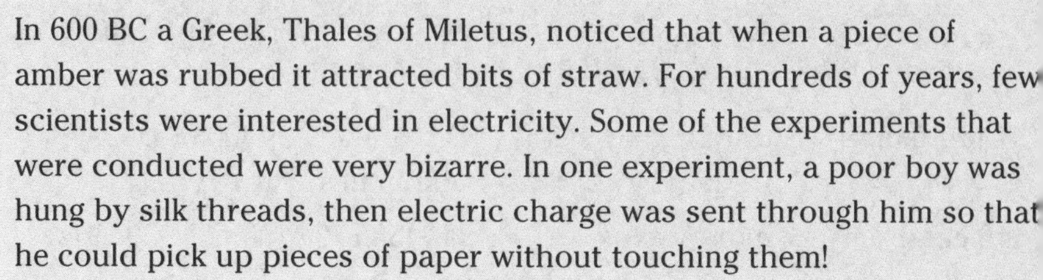

In 600 BC a Greek, Thales of Miletus, noticed that when a piece of amber was rubbed it attracted bits of straw. For hundreds of years, few scientists were interested in electricity. Some of the experiments that were conducted were very bizarre. In one experiment, a poor boy was hung by silk threads, then electric charge was sent through him so that he could pick up pieces of paper without touching them!

Radioactivity was discovered in 1896. This discovery meant that earlier ideas about atoms were wrong. Atoms could be broken up!

J. J. Thomson's discovery of the electron

In 1897 J. J. Thomson was studying particles produced in cathode-ray tubes by a process called thermionic emission. Thomson could not measure the charge and mass of cathode-ray particles (or corpuscles as he called them) separately, but he could change their direction using an electric field and deduced that they had a negative charge. He was also able to bend them using a magnetic field.

Thomson used an electric field to move the corpuscles in one direction and a magnetic field to move them in the opposite direction. He was able to calculate their charge to mass ratio (e/m) and using a value for their charge e taken from electrolysis experiments, Thomson concluded that the corpuscles were much smaller than atoms.

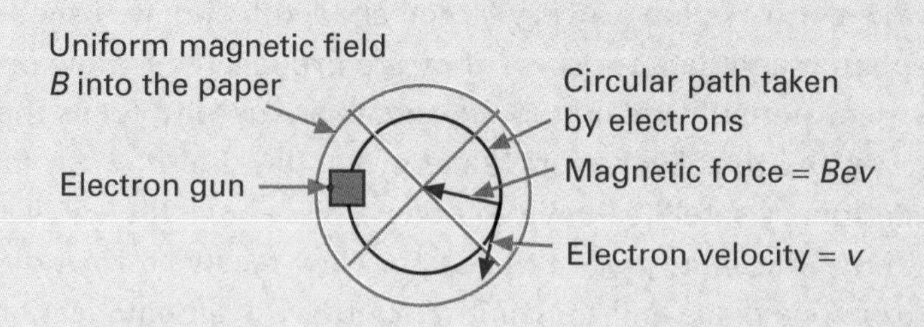

This diagram shows a way to find e/m. A pair of Helmholtz coils produce a uniform magnetic field that makes electrons follow a circular path. A centripetal force (mv^2/r) is provided by, and equal to, the magnetic force (Bev). It can be shown that as $Bev = mv^2/r$, $e/m = v/(Br)$.

Thomson was the first to discover a subatomic particle. Others called it the electron because it had an electric charge. Thomson found that the value of e/m was not affected by the gas contained in the cathode-ray tube. This implied that all atoms contain identical corpuscles, or electrons. J. J. Thomson received the Nobel prize in 1906.

illikan's oil-drop experiment to find a value for *e*

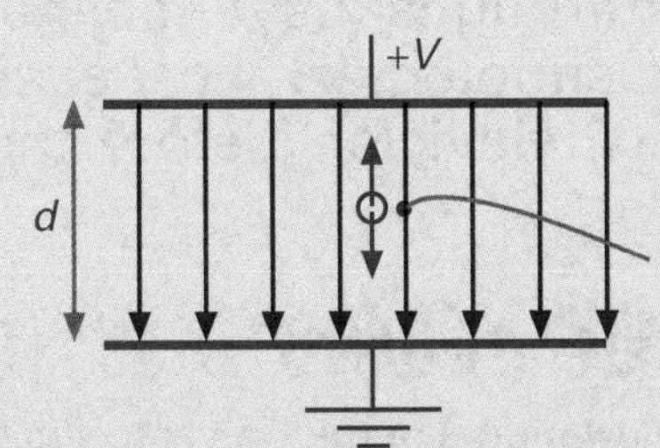

electric force = weight

$\frac{qV}{d} = mg$

small oil drop of mass m
radius r carrying charge q

the equipment is adjusted so that the oil drop remains still,

Force due to electric field = weight

$$qV/d = mg = 4\pi r^3 \rho g/3$$

the charge q on the oil drop of density ρ equals $4\pi r^3 \rho g d/3V$. To
ıd q, the radius of the drop r had to be calculated. To do this the
ectric field was switched off. The drop accelerated at first but then
ached its terminal velocity. At this point the drop's weight equalled
s viscous force $6\pi\eta r v$ (Stokes' law). η is called the coefficient of
scosity of air.

q was always found to be a whole number multiple of -1.6×10^{-19} C,
ıggesting charge is carried in packets of -1.6×10^{-19} C (by electrons).

Wave–particle duality and microscopes

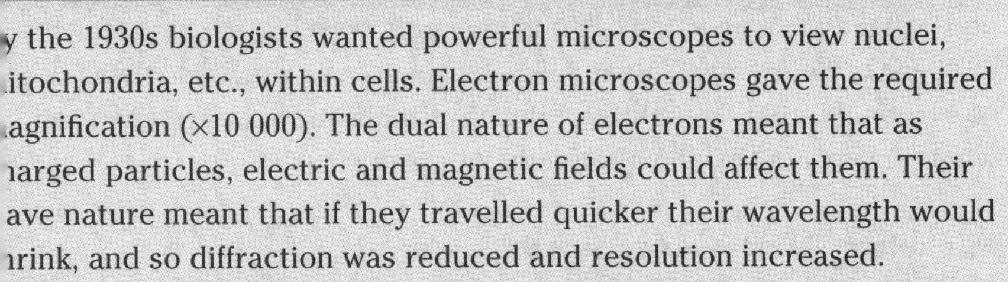

y the 1930s biologists wanted powerful microscopes to view nuclei,
itochondria, etc., within cells. Electron microscopes gave the required
agnification (×10 000). The dual nature of electrons meant that as
ıarged particles, electric and magnetic fields could affect them. Their
ave nature meant that if they travelled quicker their wavelength would
ırink, and so diffraction was reduced and resolution increased.

The **transmission electron microscope** (TEM) was the first electron
ıicroscope. A focused beam of electrons passes through a sample.
lectrons that pass through the sample strike a phosphor screen to
roduce light. Thick or dense areas of the sample allow fewer electrons
ırough so a darker image results.

Scanning tunnelling microscopes (STMs) were invented in 1981.
very fine probe (with a single atom projecting from its end) scans
small area of the sample's surface. The probe is held at a constant
eight (≤1 nm) above the surface. Electrons *tunnel* across the gap.
the probe comes across a raised atom, more electrons can tunnel
cross so the *tunnelling current* increases. Variation in the tunnelling
urrent as the probe moves allows an image to be formed.

Checkpoint 3

As the volume of a sphere = $\frac{4}{3}\pi r^3$, obtain an expression for the weight of a spherical oil drop of density ρ, radius r when acceleration due to gravity = g.

The jargon

Terminal velocity is the constant maximum velocity reached by an object falling under gravity through a fluid.

Check the net

To find out more about electron microscopes visit the Museum of Science, Boston at www.mos.org/

Checkpoint 4

In what ways is a TEM rather like a slide projector?

Exam question answer: page 188

Use $\lambda = h/(2meV)^{1/2}$ to estimate the anode voltage needed to produce wavelengths of the order of the size of atoms. Describe two advantages of an STM over a normal electron microscope. (10 min)

Turning points in physics 2

"It's not that I'm smart, it's just that I stay with problems longer."

Albert Einstein

Our understanding of the Universe has develope over many years. Breakthroughs have opened u previously unforeseen technologies. For exampl who could have imagined television before the di covery of the electron?

Einstein's theory of special relativity

Albert Einstein provided the first new model of the Universe since Newton, two hundred years before. Einstein's contemporary, Charlie Chaplin, said that people cheered him (Chaplin) because everyone understood him, but that people cheered Einstein because no one understood him.

Measuring the speed of light *c*

Galileo was possibly the first scientist to measure c. Since then c has been measured more often, and more accurately, than any other physical constant. Albert Michelson is the person most closely associated with this work. He spent fifty years measuring c. In 1879, Michelson an Edward Morley measured c to be $299\,910\,000\ \text{m s}^{-1} \pm 50\,000\ \text{m s}^{-1}$.

Relativity

If you stand by the side of a road as a car drives past at $50\ \text{km h}^{-1}$, you see it travelling away from you at $50\ \text{km h}^{-1}$. If you jump on a bicycle and pedal after it at $10\ \text{km h}^{-1}$, it is now only travelling away from you at $40\ \text{km h}^{-1}$. It is travelling at $40\ \text{km h}^{-1}$ relative to you. Michelson and Morley found that light always travels at the same speed, no matter whether you are travelling towards it or away from it.

Checkpoint 1

Use the idea of a car overtaking a person cycling along a road, then turning around and driving towards the cyclist, to explain why Michelson and Morley expected the value that they would obtain for c to vary depending on whether light (the car) was travelling towards or away from Earth (the cyclist).

The jargon

In the 19th century, scientists thought that if light is a wave, it must require a medium to travel through. As light can travel from the Sun through the vacuum of space to us, this medium had to be rather special. They called it *ether*. It is now known that ether does not exist.

Michelson–Morley experiment

Michelson and Morley intended to measure the speed of Earth travelling through the *ether*. They assumed that light would travel at a constant speed through the ether. They thought that if the Earth was moving in the same direction as the light they studied, their measurement of c would be less than if the Earth was moving towards the light source.

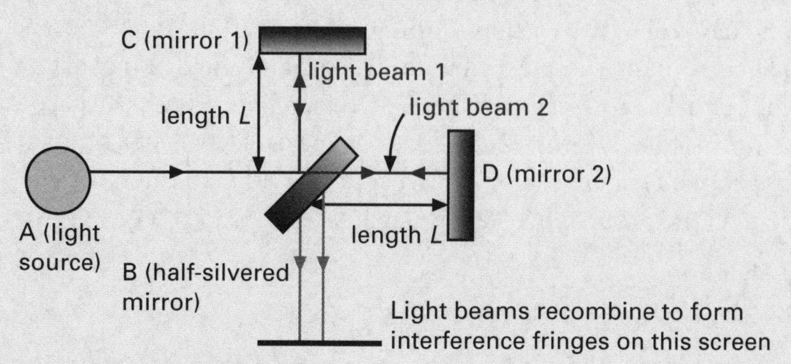

Mirrors 1 and 2 were placed at equal distances L from a half-silvered mirror B. B splits light coming from the light source so that some light, beam 1, travelled towards mirror 1 and some, beam 2, went to mirror 2 After reflection at mirrors 1 and 2, the beams recombined to form interference fringes on the screen. Michelson and Morley expected the two beams to be slightly out of phase when they recombined.

If the equipment (situated on the Earth which they believed to be ıvelling through the ether) was moving in the same direction as light ɔm A, they predicted that the journey time of ray 1 would be slightly ss than that of ray 2. No time difference was ever found; this suggested at the Earth was stationary! Was nature conspiring against them?

nstein's postulates

ɔ one accepted that the Earth was stationary. Einstein solved the oblem by saying that c was always constant. It did not matter hether the Earth was moving towards or away from the light source, c ɔuld always be the same! Michelson and Morley were vindicated. nstein used two basic assumptions to rewrite Newton's laws:

- the speed of light in free space is always constant
- physical laws have the same form in all inertial frames

instein's predictions

Time dilation If the time between two events was measured, an observer moving at speed v would measure a longer time interval t than the time interval t_0 that a stationary observer would:

$$t = \frac{t_0}{\sqrt{(1 - v^2/c^2)}}$$

Length contraction If a rod moves in the same direction as its length it appears shorter, length l, than when it is stationary, length l_0.

$$l = l_0\sqrt{(1 - v^2/c^2)}$$

Relativistic mass The mass of a stationary object m_0 is less than the mass of the same object m moving at speed v.

$$m = \frac{m_0}{\sqrt{(1 - v^2/c^2)}}$$

Energy and mass Einstein showed that energy and mass were interchangeable in his most famous equation: $E = mc^2$.

he theory of special relativity has since been confirmed by experiment.

lectromagnetic waves

axwell predicted the existence of electromagnetic waves in 1864. He ıowed that a changing current in a wire creates electromagnetic waves ıat spread out from the wire with speed $c = 1/\sqrt{(\mu_0\varepsilon_0)}$, where μ_0 and ε_0 e constants.

Hertz is generally given the credit for discovering radio waves in 388. However, he was only able to interpret his observation, that a ɔark between two spheres sent out an electromagnetic wave that ɔuld be picked up some way away, because of Maxwell's earlier work. arconi began experimenting with radio waves in 1894. In 1901 he ansmitted across the Atlantic, revolutionizing communications.

Exam question

answer: page 188

Muons are particles that disintegrate after an average lifetime of 2.2×10^{-6} s. They are created at the top of the atmosphere, some 10 km up. How far could they travel at the speed of light? Bearing this answer in mind, explain why they can be found at ground level. Illustrate your answer with mathematics. (15 min)

"A complete conspiracy is itself a law of nature."

Poincaré

"Newton was the greatest genius who ever lived, and the most fortunate, for there cannot be more than once a system of the world to establish."

J. L. Lagrange (1736–1813)

Action point

Confirm that at speeds other than speeds approaching c, Newton's laws of motion are still valid. Do this by substituting $v = 0.1c$ in each of the first three equations shown opposite. You should find that $t \approx t_0$, $l \approx l_0$ and $m \approx m_0$.

Checkpoint 2

How much energy is released when a uranium nucleus of mass $4.046\,865\,3 \times 10^{-25}$ kg decays to produce a thorium nucleus of mass $3.978\,741\,2 \times 10^{-25}$ kg and an alpha particle of mass $6.804\,42 \times 10^{-27}$ kg?

Answers
Options

Nuclear and particle physics 1

Checkpoints

1 $E = mc^2 = 2 \times 9.11 \times 10^{-31} \times (3 \times 10^8)^2 = 1.64 \times 10^{-13}$ J.

2 uud $= +\frac{2}{3} + \frac{2}{3} - \frac{1}{3} = +1$. udd $= +\frac{2}{3} - \frac{1}{3} - \frac{1}{3} = 0$

3 $u\bar{u} = +\frac{2}{3} - \frac{2}{3} = 0$
$d\bar{d} = -\frac{1}{3} + \frac{1}{3} = 0$
$u\bar{d} = +\frac{2}{3} + \frac{1}{3} = +e$
$d\bar{u} = -\frac{1}{3} - \frac{2}{3} = -e$

4 *Hadrons* particles made from combinations of quarks, e.g. pion, proton.
Meson made from 2 quarks, e.g. pion.
Baryon made from 3 quarks, e.g. proton.
Lepton fundamental particle, e.g. electron.
Quark fundamental particle, e.g. up quark.
Antiparticle antimatter particle with opposite charge to its fundamental particle, e.g. positron.
Bosons particles which are exchanged when forces act, e.g. photon.

Exam question

(a) Left figure: electromagnetic. Right figure: weak.
(b)

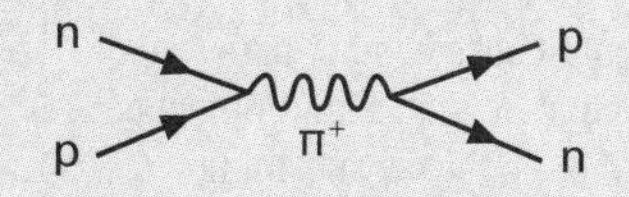

or

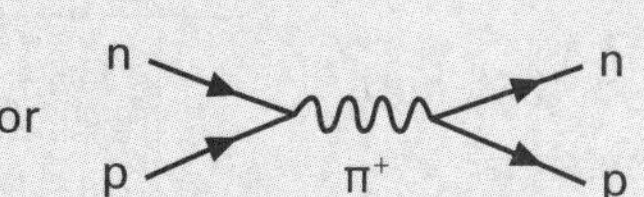

Nuclear and particle physics 2

Checkpoints

1 An up quark has changed into a down quark.

2 n = p + e + v
charge 0 = +1 − 1 + 0 = 0
baryon number 1 = 1 + 0 + 0 = 1
strangeness 0 = 0 + 0 + 0 = 0

3 The particles would collide with any gas particles in the tube and lose their energy.

4 Using $F = Bev = mv^2/r$ then $v/r = Be/m$ so $v \propto r$.
T = distance/speed $= 2\pi r/v = 2\pi m/Be$ = a constant.

Exam question

It is a baryon as the baryon number has to be conserved, and as it has strangeness of −3 it must be made of three quarks.

Astrophysics 1

Checkpoints

1 3.26 ly = 1 pc.

2 (a) 525 600; (b) 8.3 minutes.

3 θ (in radians) = arc length/radius. 1.2° is a small angle. $\theta = 1.2 \times 2\pi/360° = 2.09 \times 10^{-2}$ rad. Distance to post = eye separation/θ in radians = 4.8 m.

4 $L = 3.825 \times 10^{26}$ W.

5 The dimmer the star, the fewer photons captured in a given time, increasing relative errors in the measuremen of apparent magnitude. To estimate absolute magnitude you need a good estimate of surface temperature. A good clear spectrum requires as much light as possible. Observation of dimmer stars is also beset by problems of light pollution and atmospheric turbulence. The furthest stars tend to be dimmest. Use of standard candles depends on these stars behaving as we expect them to. We can only see the brightest very distant star (e.g. quasars). Our understanding of them is incomplete etc.!

6 (a) Each absorption line corresponds to a specific quantu leap in electron energy level. (b) Light from one side of a spinning star is red shifted (because that side is moving away from us); light from the other side is blue shifted. The relative velocity varies over the surface of the star, with maximum Doppler shifts at the edges. The faster the spin, the bigger the spread (of each and every absorption line).

Exam questions

1 (a) 2.7 × 3.26 = 8.80 years.
(b) $-1.46 - M = 5 \log(2.7/10)$ so
$M = -1.46 - 5 \log 0.27 = 1.38$.

2 $d = 10 \times 10^{(0.5-5.0)/5} = 10 \times 10^{1.1} = 126$ pc. (Betelgeuse is actually a bit closer than this rough calculation suggests.

Astrophysics 2

Checkpoints

1 Around the edges. Hydrogen is the least dense gas.

2 (a) Because gravitational potential energy is converted to kinetic energy and then heat as the star's imploding atoms accelerate inwards and collide.
(b) Because there is no further energy source; the gravitational collapse has been stopped and the temperature has not been raised sufficiently to cause fusion of the remaining material (or possibly there is no fuel left, since the fusion chain has reached iron).

3 Gravitational potential is the energy necessary to pull eac kg of matter out of the gravity well.

KE per kg $= \frac{1}{2}v^2$. For escape, $\frac{1}{2}v^2 > GM/R$. The limit t speed is c, the speed of light. If $GM/R > \frac{1}{2}c^2$ even light will be trapped. The critical radius is therefore given by $R = 2GM/c^2$.

Note this line of reasoning suggests that although th star would be dark from a great distance, if you travelled close to it, light would get to you, R is not the distance from the star's core to a dark horizon. Real black holes are different. The Schwarzschild radius ($2GM/c^2$) is an absolute cut off. No light escapes it! What goes on insid the horizon cannot be seen (Penrose's cosmic censorshi conjecture).

4 Light does not emerge more slowly from a massive star than from a less massive one! Light's speed through space is constant. Although light is deflected by gravity, its speed is not altered. (Gravity warps space itself!)

5 Luminosity depends on surface temperature *and* surface area. A dim white star must therefore be small and a bright red star must be big.

6 Equal resolution requires that $\lambda/D_{light} = \lambda/D_{radio}$ (5 m). $D_{radio} = 5 \times 0.075/5 \times 10^{-7} = 7.5 \times 10^{5}$ m (750 km!). Radio telescopes generally have poor resolution. Arrays of widely spaced telescopes can be linked to give improved resolution.

7 (a) (i) For the Earth $R_S = 8.9$ mm. (ii) For the Sun, $R_S = 2.95$ km. (iii) For a 5 solar mass star, $R_S = 14.7$ km. (b) The Earth and the Sun are not massive for gravity ever to crush them sufficiently to form black holes. The minimum mass for a black hole is about 2.5 solar masses.

Exam question

The atmosphere is clear in the radio waveband up to 10 m wavelength, so satellite-based telescopes are not essential (the atmosphere is opaque to X-rays and to important IR wavebands, so satellites are essential if these wavebands are to yield any useful information). Also, a radio telescope would need to be big – difficult to launch. (*Note* it may be possible to cheat the Rayleigh criterion by using the motion of the satellite to broaden the base, or by use of several satellites.)

Astrophysics 3

Checkpoints

1 The Universe looks roughly the same in all directions. If the Universe has an edge and we are close enough to see it, then we must be in its centre (which seems unlikely). This breaks the cosmological principle; the Universe would look different from different places. The other solutions already explain Olbers' paradox, so there is no excuse for this one!

2 1 Mpc = 3.086×10^{22} m. 50 km s^{-1} Mpc^{-1} = 5×10^{4} m s^{-1}/ 3.086×10^{22} m = 1.62×10^{-18} s^{-1}. If $H = 50$ km s^{-1} Mpc^{-1}, maximum age of the Universe = $1/1.62 \times 10^{-18} = 6.17 \times 10^{17}$ s = 19.6 billion years. If $H = 90$ km s^{-1} Mpc^{-1}, maximum age of Universe = 10.9 billion years.

3 (a) Gravity is the only significant force to operate at great distances and it is a force of attraction, and so the rate of expansion of the Universe should be slowing down (H should be decreasing). (b) $1/H$ is only a good estimate of the age of the Universe if we can ignore gravity and assume constant H. If H was greater in the past than it is now, the Universe must be younger than the Hubble time ($1/H$). For a flat Universe, the age is around ⅔ of the Hubble time.

4 The speed of a star depends on the mass inside its orbit. If the dark matter was in the centre of the galaxy, *all* stars would orbit anomalously fast, not just the outer ones. Any matter beyond a star does not contribute to the gravitating mass pulling it towards the galaxy's centre.

5 When $T = 3\,000$ K, $\lambda_{max} = 0.96$ μm (infrared). When $T = 2.7$ K, $\lambda_{max} = 1.07$ mm (microwave).

Exam question

(a) At 15% light speed (4.5×10^{7} m s^{-1})
(b) $v = Hd$ so $d = 4.5 \times 10^{4}$ (km)/65 (km s^{-1} Mpc^{-1}), $d = 692$ Mpc = 2.26×10^{9} ly

Medical and health physics 1

Checkpoints

1 Both are stimulated by light, form an image on a screen, transmit information about that image electrically and produce 2-D images.

2 The object would be invisible. Point ⑥ is the blind spot. At the point where the retina disappears down the optic nerve ⑦, there is no light-sensitive layer for light to fall on.

3 Normally you hear your voice after the sound has passed through the bone and other tissues in your head. Therefore your voice sounds very different to everyone else, or on a recording, as this is not then the case.

4 You have already damaged your hearing.

Exam question

Different parts of the ear are affected by sounds of different frequencies. Each part has a different natural frequency and so each part resonates when a certain frequency is heard. For example, the outer ear resonates when notes of around 3 300 Hz are heard and the middle ear responds to sounds between 700 Hz and 1 500 Hz. The individual response of all the regions of the ear allows us to hear sounds between 20 and 20 000 Hz, though this upper limit declines with age.

The graph on the left shows that the minimum threshold for hearing is about 10^{-12} W m^{-2} between about 1 kHz and 3 kHz. It also shows that the intensity thresholds for discomfort and pain are not frequency dependent. 120 dB cause discomfort irrespective of frequency, for example.

The loudness of a sound (in phons) is the intensity level at 1 kHz that has the same loudness as the sound to a normal ear. The graph on the right shows equal loudness curves. Notice that a 40 dB sound at 100 Hz has the same loudness (0 phons) as a 20 dB sound at 300 Hz.

Medical and health physics 2

Checkpoints

1 Each fibre of the fibre optic carries a small part of the overall image to the eyepiece. If the fibres are scrambled the image will be too.

2 Gamma rays have the longest ranges and weakest ionizing powers. They are therefore absorbed least by the patient and reduce the patient's exposure to ionizing radiation.

Exam question

(a)

	Cost	Cost/image	Size	Exposure	Resolution	Contrast	Patient risk	Good points	Poor points
X-ray	£50 k	£50–300	most heavy and fixed	0.01–10 s	0.05 mm	5%. Good for bone-soft tissue	very slight	superb resolution	info on structure only
CT	£500 k	£150	fixed in place	0.2–5 s	0.25 mm	0.3%	high radiation dose	good contrast	radiation fear
γ-camera	£100–400 k	£100–500	fixed in place	5 min	5–15 mm	often very high 1–10%	radiation approx. 50 mSv	good for physiological processes	poor resolution
Ultrasound	£10–15 k	£50–100	many mobile	5–20 min	1–5 mm	excellent for soft tissue	none proven	non-ionizing	can't see through bone
MRI	£0.5–1.5 m +	£200–400	most heavy and fixed	up to 20 min	0.3–1 mm	excellent for soft tissue	low	gives info on blood flow, pH, int. temp	cost

(b) See page 159.

Radiation and risk

Checkpoints

1 (a) 50 mm of concrete is 2 × half-thickness, so the intensity is $I_0/4$.
(b) 1.5 mm of lead is 3 × half-thickness, so the intensity is I_0/

2 Dose equivalent from alpha:
0.10 × 2 mGy × 20 = 4 × 10^{-3} Sv.
Dose equivalent from gamma:
0.90 × 2 mGy × 1 = 1.8 × 10^{-3} Sv.
Total dose = (4 + 1.8) mSv = 5.8 mSv.

3 2 mSv is equivalent to a risk of 5% × 2 × 10^{-3} = 10 × 10^{-3}%
This is a risk of 0.01 in 100 or 100 in 1 million. So 100 people in a population of 1 million may be expected to get cancer.

Exam question

50 × background dose is 10 mSv. This is a risk of 10 × 10^{-3} × 5% = 0.05 %.
So 0.05 in 100, or 5 in 10 000 may get a fatal cancer from this programme. For a population of 20 million, that is 10 000 people. You should consider the consequences of such a programme.

Electronics 1

Checkpoints

1 Peak voltage = 6.8 × 50 = 340 V.
Frequency = 1/(10 × 2 × 10^{-3}) = 50 Hz.

2 CR = charge/voltage × voltage/current
= current × time/current = time.

3 $X_C = 1(2\pi \times 50 \times 10^3 \times 47 \times 10^{-6}) = 0.067\ \Omega$.

4 $CR = 2\,200 \times 10^{-6} \times 180 \times 10^3 = 400$ s.

Exam question

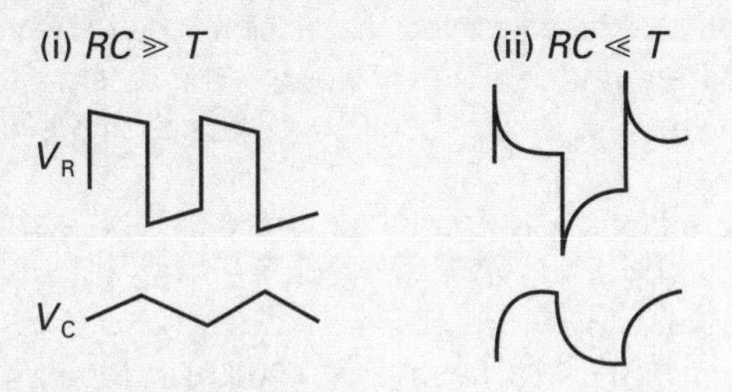

Electronics 2

Checkpoints

1 PD across the resistor = 5 – 1.7 = 3.3 V
$R = V/I = 3.3/20 \times 10^{-3} = 165\ \Omega$.

2 The lamp would come on in the light.

3 So that the circuit could be adjusted easily for the lamp t
light at different light levels.

Exam question

(a)

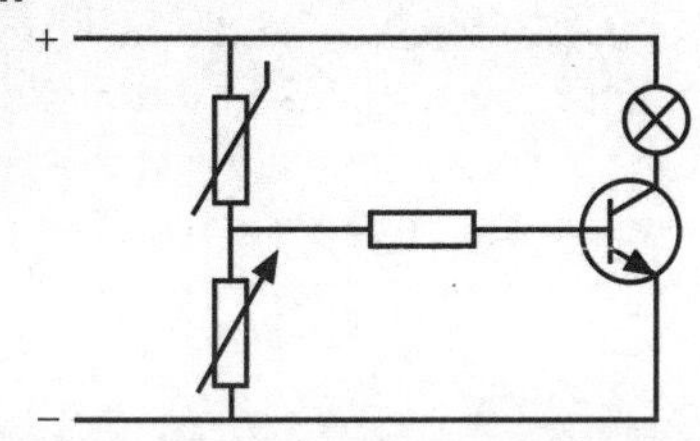

(a) When hot, the thermistor's resistance is low so the base of the transistor is high, and the lamp comes on. When cold, the thermistor has a high resistance so most of the voltage drop is across it and the lamp goes out.

Electronics 3

Checkpoints

Using $V_{out} = A_{OL}(V_2 - V_1)$ gives $V_{out} = \pm 150\ \mu V$
$R_F = A_{CL} \times R_{in} = 1\,000\ k\Omega$, $V_{out} = -1$ V
$A_{CL} = 1 + 40/10 = 5$, $V_{out} = 5$ V

Exam question

non-inverting amplifier has the input signal connected to e non-inverting input so the voltage gain is positive.
$A_{CL} = 10$
ne two inputs of the op amp are at the same potential, and e op amp takes no current so V_{in} equals the votage drop cross R_1 and V_{out} equals the voltage drop across $R_1 + R_2$. So $A_{CL} = V_{out}/V_{in} = (IR_1 + IR_2)/(IR_1) = 1 + R_2/R_1$.

Materials 1

Checkpoints

Stress = force/area in Pa (1 Pa = 1 N m^{-2})
strain = extension/length (a ratio, so no units)
Young's modulus = stress/strain in Pa.

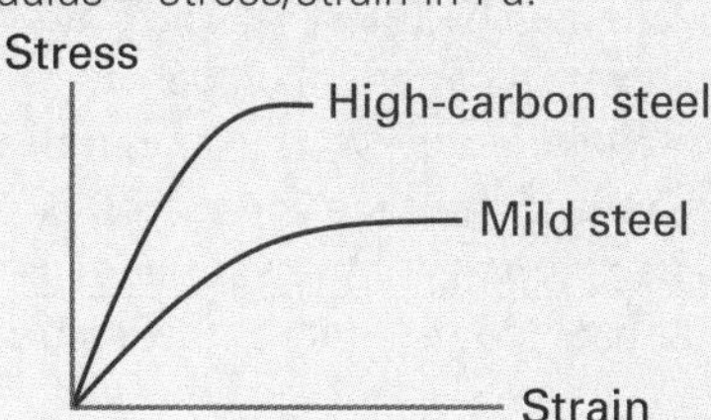

Stiff needs a large force to stretch it. *Flexible* stretches easily.
Strong needs a large force to break it. *Weak* breaks easily
Brittle shatters easily with a clean break. *Tough* does not shatter easily.
Elastic returns to its original shape when the stress is removed. *Plastic* acquires a permanent set.
The repulsive force is largest when the atoms are close together and the attractive force is largest when they are far apart.

Exam question

(a) Apparatus:

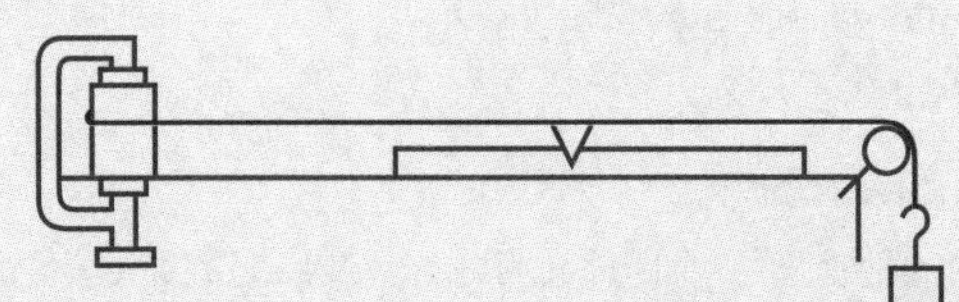

Measure the original length of the wire with a small mass on it to make it taut. Measure the diameter in several places with a micrometer. Add weights to the wire and measure the extension. Plot a graph of force in newtons against extension in metres. Young's modulus of the copper is the gradient of the linear part of the graph multiplied by l/A.

(b) A long thin wire is used so that the extension is long enough to be measured accurately.

(c) Using $e = Fl/EA$, extension is proportional to length, and so Y will have $^1/_3$ of the extension of X. Extension is proportional to 1/area or 1/(diameter)^2 so Y will have four times the extension of X. Together, extension of Y = $^4/_3 \times 8 = 10.7$ mm.

Materials 2

Checkpoints

1. It is important that a load-bearing structure is not permanently deformed after being stressed.
2. When the paper clip is bent backwards and forwards, it is initially flexible as the dislocations move. After a while, they jam up so the material begins to harden and eventually undergoes brittle fracture.
3. Concrete has an amorphous structure so when pulled apart, cracks will propagate. Stressed rods hold the concrete in compression.
4. A sudden stress causes elastic behaviour as the long-chain molecules do not have time to flow, but left under its own weight it behaves like a viscous liquid.

Exam question

(a) Copper is polycrystalline and contains many dislocations that move easily causing plastic flow.

(b) Glass is amorphous and under tension cracks on the surface have great stress at the tip, which causes the crack to run through the material making a clean break.

(c) Rubber consists of long-chain hydrocarbon molecules that are tangled due to the thermal energy. When rubber is stretched, the molecules straighten out.

Examiner's secrets

Examiners will be impressed if you use diagrams to help with your explanations.

Materials 3

Checkpoints

1. Since $v \propto 1/n$ and n is larger for a metal, the drift velocity must be less.
2. Since $V_H \propto v$ and v is larger for a semiconductor, then the Hall voltage will be larger.
3. Both effects will disrupt the alignment of the dipoles in the domains.
4. Using $h = eV/c = eV\lambda/c$, $h = 6.7 \times 10^{-34}$ J s.

Exam question

(a) See the diagram on page 175 *materials 3*.

(b) The coercivity is the reverse field needed to demagnetize a material. It is larger for steel as this is harder to demagnetize than iron.

(c) The area enclosed by the hysteresis loop represents the work done per cycle in reversing the magnetization of a material and it is dissipated as heat in the material.

(d) To reduce energy losses in the core of a transformer which is being remagnetized frequently.

Telecommunications 1

Checkpoints

1 For example (a) an amplifier (b) a calculator.

2 (a) (i)

(ii)

(b) (i)

(ii)

3 40 kHz

Exam questions

1 *Modulation* is where a high-frequency radio wave carrier is altered in some way to contain the information in the audio signal wave which can then be carried long distances. You could explain AM or FM with a diagram too as an example.

Multiplexing is when many signals are sent down the same communications link. You could explain TDM as an example. Multiplexing enables many communications signals to be carried by the same communications link, thus greatly expanding telecommunications networks as well as cutting down the weight of connections in aircraft for example.

2 AM is a simple and inexpensive way of carrying radio signals but it suffers from noise and fading caused by interference by two signals arriving by two different routes. AM has a small bandwidth. PCM is when analogue signals are turned into digital ones before transmission. It is more complicated but even when the signal is degraded it can be correctly decoded at the receiving end. There is some noise caused by quantization errors but companding reduces this. PCM requires a large bandwidth. Many digital signals can be sent down the same link by TDM, they can be fed straight into computers and can have a check digit added.

Telecommunications 2

Checkpoints

1 (a) Using power in dB = $10 \log(P_2/P_1)$, $21 = 10 \log(P_2/10^{-20})$, $P_2 = 1.3 \times 10^{-18}$ W.

(b) power in dB = $10 \log(1.3 \times 10^{-18}/10^{-3}) = -149$ dB at a rate of 2 dB km^{-1} gives 74 km.

2 They have a large bandwidth (180 kHz) and therefore need to be put in a band which can accommodate a number of channels. The bandwidth of the VHF band is 20 MHz.

3 $\lambda = c/f = 3 \times 10^8/198 \times 10^3 = 1\,520$ m so the aerial would be 760 m long.

Exam question

Your answer should include the fact that infrared has a higher bandwidth than microwaves (and so can carry information at a greater rate), cost, weight, attenuation, crosslinking and interference, and security.

Turning points in physics 1

Checkpoints

1 As $e/m = -1.76 \times 10^{11}$ C kg^{-1}, $e = -1.76 \times 10^{11}$ m. As $Q/M = 9.65 \times 10^7$ C kg^{-1}, $Q = 9.65 \times 10^7$ M. Thomson assumed that the charge on an electron e would be equal in size and opposite in charge to the charge on the hydrogen ion Q, i.e. $e = -1.76 \times 10^{11}$ $m = 9.65 \times 10^7$ M. So the mass of an electron m would be very much less than the mass of a hydrogen ion M.

2 $Bev = mv^2/r$. Dividing both sides by Bv gives $e = mv/Br$. Dividing both sides by m gives $e/m = v/Br$.

3 Volume $V = \frac{4}{3}\pi r^3$. Weight $W = mg$. As density, $\rho = m/V$, mass $m = \rho V = \frac{4}{3}\pi r^3 \rho$. $W = mg = \frac{4}{3}\pi r^3 \rho g$.

4 When a projector shines a beam of light through a slide, the light is affected by the slide. The slide 'filters' the light so that only certain parts of the beam pass through certain parts of the slide. In a TEM, an electron beam replaces the light beam and the slide is replaced by a specimen. Thicker or denser parts of the specimen allow fewer electrons through so that part of the image is darker.

Exam question

Using $\lambda = 10^{-11}$ m gives $V = 1.5 \times 10^4$ V. An STM does not damage the surface of the specimen and it does not need a vacuum to work.

Turning points in physics 2

Checkpoints

1 If a cyclist, travelling at 15 km h^{-1}, is overtaken by a car travelling at 30 km h^{-1}, she sees the car travelling past at 15 km h^{-1}. If the car turns around and drives towards her at 30 km h^{-1} it now appears to be moving at 45 km h^{-1}. Michelson and Morley thought that the speed of light would vary in a similar way.

2 7.99×10^{-30} kg

Exam question

At the speed of light c muons travel a distance ct_0 in their lifetime t_0 (where $t_0 = 2.2 \times 10^{-6}$ s). This distance is approximately 660 m. Because muons travel at speeds very close to c, from our point of view they seem to live for much longer than 2.2×10^{-6} s. The time they appear to survive for $= t_0/\sqrt{(1 - v^2/c^2)} = 5.047 \times 10^{-6}$ s, long enough for them to reach ground level.

Resources section

This section is intended to help you develop your study skills for examination success. You will benefit if you try to develop skills from the beginning of your course. Modern AS- and A2-level exams are not just tests of your recall of text books and your notes. Examiners who set and mark the papers are guided by assessment objectives that include skills as well as knowledge. You will be given advice on revising and answering questions. Remember to practise the skills.

Exam board specifications

In order to organize your notes and revision you will need a copy of your exam board's syllabus specification. You can obtain a copy by writing to the board or by downloading the syllabus from the board's website.

AQA (Assessment and Qualifications Alliance)
Publications Department, Stag Hill House, Guildford, Surrey GU2 5XJ (www.aqa.org.uk)

CCEA (Northern Ireland Council for Curriculum, Examinations and Assessment)
Clarendon Dock, 29 Clarendon Road, Belfast, BT1 3BG (www.ccea.org.uk)

EDEXCEL
Stewart House, 32 Russell Square, London WC1B 5DN (www.edexcel.org.uk)

OCR (Oxford, Cambridge and Royal Society of Arts)
1 Hills Road, Cambridge CB2 1GG (www.ocr.org.uk)

WJEC (Welsh Joint Education Committee)
245 Western Avenue, Cardiff CF5 2YX (www.wjec.co.uk)

Topic checklist

○ AS ● A2	AQA/A	AQA/B	CCEA	EDEXCEL	OCR/A	OCR/B	WJEC
Numbers and maths	○●	○●	○●	○●	○●	○●	○●
Errors and uncertainties	○●	○●	○●	○●	○●	○●	○●
Studying, revising, passing exams	○●	○●	○●	○●	○●	○●	○●

Numbers and maths

Basic numeracy is essential for a physicist, but it is no[t] enough. In physics, every measurement is real an[d] imperfect. We need to understand the implications o[f] our imperfect knowledge!

Examiner's secrets

You will be penalized if you give your answer to more significant figures than were used in the question.

Numbers

Every number you write down in any assessed work should be a significant number. No meaningless strings of decimal places!

Significant figures

In physics exams you should think in terms of significant figures rather than decimal places. You should always give your answers to the same number of significant figures as the data provided in the question. For example, the answer to $6.1 \div 3.1$ should be given as 1.0, rounded up to 2 significant figures.

- Significant figures are those numbers we are confident in.
- Every figure you write will be assumed to be significant – so don't give figures that cannot be justified.

$R = 3\ \Omega$ really means that R is $3\ \Omega$ to the nearest ohm; i.e. $R = 3 \pm 0.5\ \Omega$
$R = 3.0\ \Omega$ means R is $3.0\ \Omega$ to the nearest 0.1 ohms; i.e. $R = 3.0 \pm 0.05\ \Omega$ (greater precision) etc.

Test yourself

1 (a) If the PD across a component is 6.0 V and the current through it is 1.7 A, what is its resistance?
(b) Repeat the calculation given the following (more precise) measurements: $V = 6.00$ V and $I = 1.700\,0$ A.

2 Write the number 389.455 1 to
(a) 4 sig. fig.
(b) 2 sig. fig.

(*Answers* 1. (a) 3.5 Ω (b) 3.53 Ω.
2. (a) 389.5, or (better) 3.895×10^2 (b) 390, or (better) 3.9×10^2.)

Decimals

If a number has a decimal point, you should be able to infer the precision and the number of significant figures. For example, 13.45 clearly has 4 sig. fig.; 0.004 63 has 3 sig. fig. since leading zeros ignored; 3.80 has 3 sig. fig. since trailing zeros are given only if they are significant.

Whole numbers

Whole numbers are more tricky. For example 2 708 clearly has 4 sig. fig[.] but 4 600 could have 4, 3 or just 2 sig. fig. There is no way of knowing whether the zeros are significant figures or not. The whole mess is solved by the use of standard form.

Examiner's secrets

Calculator errors are still common. Never buy or borrow a calculator just before an exam. Get to know your calculator in good time for the exams!

Standard form

In standard form, a number is written as a decimal greater than or equa[l] to 1 and less than 10, multiplied by a power of 10. For example:

- 157 804 is written $1.578\,04 \times 10^5$
- 0.000 78 is written 7.8×10^{-4}

Standard form is very useful for:

- Dealing with very big and very small numbers.
- Rounding answers to an appropriate number of significant figures *without ambiguity*.

In standard form, 4 600 would be written 4.6×10^3, 4.60×10^3 or 4.600×10^3 according to the number of significant figures (2, 3 or 4).

- When you use standard form, the number of figures given in the decimal bit is always exactly the number of significant figures.

Algebra

Models and theories are often most concisely expressed as mathematical equations. We need to be able to interpret and manipulate them. Remember:

- whatever you do to one side of an equation, you must also do to the other side
- units must equate – you can work out the units of an unknown term, provided you know all other units
- you should also understand the meanings of inequality symbols: $>, \geqslant, <, \leqslant, \geq$ and $\leq$.

Action point

Find out what the inequalities mean.

Trigonometry and geometry

You must be competent at calculating trigonometric functions using degrees or radians as required. You must know the definitions of the trigonometric functions sine, cosine and tangent. Remember:

- 2π radians = 360°
- for small angles, θ (in radians) $\approx \sin\theta \approx \tan\theta$

Examiner's secrets

Choosing a suitable scale is not always easy. Neither axis necessarily *has to* start at the origin. You will get a better estimate of a graph's gradient if you stretch out the scale!

Graphs

- Graphs are a useful way of *showing* relationships.
- Graphs can be used to *test* relationships.

Straight-line graphs are best! You will often have to manipulate an equation to get it to predict a testable, straight-line graph (of the form $y = mx + c$). For example, the inverse-square law for radiation (of any kind) from a point source predicts that the intensity will be proportional to the inverse square of the distance from the source ($I = k/r^2$). So don't plot I against r (you will get a curve – bad idea); plot I against $1/r^2$). You should get a straight line through the origin.

Test yourself

Boyle's law states that the product of pressure and volume of a gas is a constant (provided temperature is constant). If you measured a range of values of P and V, what would you plot to get a straight line?

(*Answer* P = a constant/V, so plot P against $1/V$ and the gradient will be the constant of proportionality.)

Plotting graphs

Marks are awarded for:

- choosing a suitable scale, labelling axes, giving units, plotting points clearly and accurately, drawing a **line of best fit**

A line of best fit follows the underlying trend and ignores the scatter of experimental measurements. It may be a straight line or a curve. Any predictions should be based on the best fit. It smooths rough data.

If you are asked to plot A against B, then A should be plotted on the y-axis (vertical axis). If you have to decide how to plot your graph:

- the independent variable should be plotted on the x-axis and the dependent variable should be plotted on the y-axis. (y depends on x)

For example, distance travelled may depend on time (time never depends on distance travelled), and so distance is plotted up the y-axis.

- Gradient = increase in y-variable/increase in x-variable. Think big: draw big tangents or use big sections of a straight-line graph to find its gradient.

Examiner's secrets

When measuring gradients, credit is given for use of big triangles – large rises and runs. You should also be able to generate graphs using IT.

Errors and uncertainties

Every measurement has an associated uncertainty or error which you are expected to estimate and quote in practical work. Uncertainties are kept to a minimum by good experimental methods, but they never disappear.

Accuracy and precision

→ **Accuracy** is closeness to the truth.

→ **Precision** is the smallest change in value that can be resolved or measured reliably.

You may have to use your judgement to decide what this smallest resolvable change is – for a particular instrument in a particular experiment. The best most instruments can do is measure to the nearest graduation, but there are exceptions to this rule (thermometers and measuring cylinders can sometimes be read to the nearest half division).

Watch out!

It is quite possible to have an instrument or measurement which is highly precise, but not very accurate! Examples:

→ A *micrometer*, which has a precision of 0.01 mm, but has a zero error of 1 mm.

→ A *stopwatch*, which gives timings to the nearest 0.01 s, controlled by a person with a reaction time of 0.2 s.

Measurements, tolerances and uncertainties

A measurement should consist of two parts: a *number* (how many) and a *unit* (of what). In addition, all experimental measurements should include an estimate of uncertainty, called **tolerance** (e.g. length ±1 mm).

→ Experimental results should be tabulated, with units and tolerances given in column headings, wherever possible.

The minimum acceptable value for a measurement's tolerance is the instrument's precision. As a rule, you should quote the tolerance as ± the smallest graduation on the instrument. Better still, repeat measurements and use the spread in your results to estimate experimental error.

→ A good estimate of uncertainty is simply half the range.

The jargon

The terms *experimental error*, *uncertainty* and *tolerance* are used interchangeably here. Tolerance is normally used to describe the accepted give or take in any particular measurement. Uncertainty means exactly the same thing as error in this context, but it is a fairer term, since it does not imply a mistake has been made (uncertainties are present in *all* data).

Absolute and relative errors

→ Instrument tolerances are absolute errors – they have units

→ Relative error = absolute error/measurement (a fraction)

→ Percentage error = relative error × 100

Random and systematic errors

Systematic errors arise from poor experimental technique and poorly calibrated instruments (i.e. from a poor *system* of measurement). They can be hard to spot, since they do not increase the scatter in the experimental data. Zero errors and parallax (alignment) errors are systematic errors.

→ Systematic errors cause *inaccuracy* (consistent over- or under-estimation of a measurement).

Random errors are the errors that cause scatter in a set of data. Random errors can be reduced by repeating measurements and taking averages. (There is less uncertainty in an averaged value than in the raw data.) You can never get rid of random errors completely.

→ Random errors cause *imprecision*.

Test yourself

Find the period of a pendulum given the following timings for consecutive single swings: 6.2, 5.4, 6.0, 5.5, 5.1 s. Estimate the tolerance in your answer.
(*Answer* 5.6 ± 0.6 s.)

Errors in calculated quantities

When you *multiply* or *divide* factors to find a quantity, you should *add up* the relative (or percentage) error in each factor used. For example:

- the percentage error in electrical resistance is equal to the percentage error in current *plus* the percentage error in voltage
- to measure the volume of a cylinder, you use the formula $V = \pi r^2 l$, you measure r and l. The percentage error in V is just $2 \times$ percentage error in r (since r^2 is $r \times r$) plus percentage error in l

Never give a calculated answer more significant figures than the least precise data used to calculate it!

Test yourself

You measure a glass bead across three perpendicular diameters, using a micrometer. Your results are: 3.24, 3.15 and 3.10 mm. Calculate the volume of the bead and give the error in your answer.
(*Answer* $1.66 \pm 0.11 \times 10^{-8}$ m^3.)

Practical exams and assessed coursework

Practical skills can be assessed either through coursework or through exams. Both options should assess the same skills.

Coursework will be assessed on the following criteria:

Planning 8 marks
Implementing 7 or 8 marks
Analysing 8 marks
Evaluating 6 or 7 marks

Practical exams will aim to assess the same skills. It is worth reading the coursework mark criteria (given in the subject specifications) – even if you are taking the examined option.

Examiner's secrets

Practical exams require preparation. Be sure you know the form well in advance. Make sure you know instrument tolerances; get some practice with multimeters and CROs etc. (just in case!). You have to follow detailed instructions so don't let adrenaline take over. Read the instructions carefully before you do anything.

What is tested in practical exams and coursework?

- Can you plan an experiment with a purpose?
- Can you take accurate readings and measurements with a range of instruments (micrometers, vernier callipers, multimeters, CROs, etc.)?
- Can you tabulate your results, giving appropriate units and tolerances?
- Are you aware of sources of error – and can you see which sources are most important?
- Do you know how to estimate the magnitudes of errors?
- Can you minimize errors by good experimental technique?
- Can you draw logical conclusions from experimental data?

Watch out!

The list opposite is meant to help, but it is not a definitive guide to everything that should appear in the assessed practical component of your course! Check your subject specifications for greater detail of how marks are awarded.

Graphs

- Do you understand what graphs are for?
- Do you use a line of best fit when appropriate (and do you understand the need for best-fit lines)?
- Can you recognize an equation that should give a straight line?
- Can you manipulate equations and data to give the equation of a straight line?
- Can you choose appropriate scales for your graphs?
- Can you measure (accurately) the gradient of a curve?
- Can you infer correct units?

Test yourself

Why is *extrapolation* (extending a best-fit line to make predictions) more uncertain than *interpolation* (making predictions within the data range)?
(*Answer* Because the underlying relationship may have limits, a small error in gradient can result in a big error in values predicted far beyond the range tested.)

Studying, revising and passing exams

It is time to take control. You cannot learn passively so get organized, get educated and get qualified!

Studying

Learning is an active process so get *involved*.

Action point

Check your filing system right now. Try to make it as easy to use as possible. *Do not* have a section marked *miscellaneous*; you might as well throw anything deemed miscellaneous into the bin straightaway, because you'll never find it again!

Organize your notes

If you have switched from exercise books to file paper, beware: file paper gets dropped, gets lost, and gets muddled! You *must* get into good habits as soon as possible As a minimum:

- ➜ date every sheet of paper
- ➜ have a file for each examined unit of study
- ➜ subdivide your files into topics

If you are superefficient, you could even have your files ready before you start a topic (very few people do this, but it does help).

Action point

Make or revise your personal timetable. Let friends know about it (then they can organize their free time to suit you!). *Remember* free time is important too.

Organize your time

If you plan your use of time, you will get more done. Here are some suggestions.

- ➜ Make a timetable of everything you do on a regular basis.
- ➜ Allocate any free periods in the day to a subject or activity.
- ➜ Allocate time slots of a suitable length to A-level study (e.g. 4.30–6.0 or 6.30–9.00 p.m. etc.). If you have to see *EastEnders* or life is not worth living, treat yourself – block it into your timetable, but once you have a timetable, stick to it!

Watch out!

If you don't understand a topic first time around, you will have to learn it from scratch later – which is far more difficult.

Take responsibility for your learning

- ➜ Make use of all resources at your disposal.
- ➜ Every lesson has a point. Ask yourself at the end – did you get it? If not, do something about it!
- ➜ Ask and answer questions. Risk getting it wrong. Mistakes are good, we learn from them!
- ➜ Never accept that a concept is beyond you. You must assume that you are capable of learning everything. The brain's capacity for information may be infinite so don't impose limits!

Action point

Make a list of sources of further help and information. You may decide to incorporate time into your schedule for surfing the net or browsing books and magazines at the library.

Do all work set as soon as possible

Why?

- ➜ It's easier; it takes less time if the subject is fresh in your mind.
- ➜ If you get stuck, you'll have time to get help.
- ➜ A backlog of work is stressful. It's harder to relax when you're not o top of your work.
- ➜ If someone has to drag the work out of you, you never get to feel good about your achievement!
- ➜ You may find others come to you for help. Explaining your answers clarifies your understanding.

Revising

- Revision works only if you have studied actively and understood the subject matter when it was covered!

Revision takes time, so don't put it off until the last minute. Ignore any of your 'friends' who claim to do no work; they don't have your best interests at heart (they have their own).

- Plan your revision so that no topic gets missed out. (Don't leave out your least favourite subjects.)
- Don't change your work routine too much when revising.
- Your work space should be well lit, comfortable, uncluttered.
- All necessary resources (books, calculators, pens) must be to hand.
- No distractions (phone off, TV off; 'do not disturb' sign up).

Revise *actively*. Learn, test yourself, take notes and summarize them (use spider diagrams, mind maps, bullet lists, etc. – whatever works for you). Practise exam questions, check your answers for clarity, use of units, etc. Use a highlighter pen to make key facts stand out. This book is meant to be an interactive guide; make use of it!

- Short bursts of intensive revision are most effective. Divide a topic into chunks of suitable length (aim for 20–30 minute spells).
- Take (short) breaks between each burst. Leave the room, run about, speak to someone, have a drink, jump up and down, etc.
- Stop on time. Congratulate yourself and relax.

Don't overdo the coffee – especially late at night. The later you stay up, the more sleepy you will be the next day.

Action point

Buy a highlighter pen (or even a set of them). Start using them – on your notes and on this guide (assuming you own it!).

Action point

Time yourself while revising. You will soon find that 30 minutes is not very long. Don't be tempted to extend revision sessions to fit the content. Work hard to fit the content into the time slot!

Action point

Make twice as much use of your efforts by analysing where you lose marks in exam questions. It is often worth repeating a question you did poorly on. Make sure you learn from your mistakes.

Passing exams

- *Preparation* Make sure you know the format of the exam. Get a copy of the data and formulae you will be given and learn any essential formulae not provided. Learn definitions and units. Do as many practice questions as you can. Find out where you are dropping marks.
- *Answering the question* When you enter an exam, the adrenaline is flowing; you want to get as much information out of your head and onto your answer paper as you can, as soon as possible! You must fight the urge to start writing. Breathe deeply. Calm yourself. Now *read the question*. What is it asking for? (If it totally throws you, take another deep breath and do question 2 first.)
  - Write the question in short form (e.g. 'Find F').
  - Write down the key information (e.g. $m = 24$ kg).
  - Work out the answer one step at a time.
- *Layout of work* There are two essential steps to getting credit in exam answers: the first is to find the answer, the second is to communicate the answer to the examiner! Step two causes most problems. At the end of a difficult calculation you may be so relieved that you forget to check that your brilliant answer is actually legible!
  - *Remember* Strings of calculations are meaningless until you write down what they represent.

Examiner's secrets

- Write clearly, don't scrawl! Examiners are pressed for time. It is in your best interests to make it easy for them. Credit can only be given for a wrong answer if the workings are clear.
- Avoid misusing the '=' sign (don't string calculations together; check that the numbers on each side of an equation are actually equal!
- State the final answer clearly. Underline it if necessary. Give appropriate units.
- Do not cross out workings unless you are replacing them with something better.
- Never hedge your bets by writing two answers in the hopes that you will get credit for whichever one of them is right. Ambiguous answers are wrong answers.

Recommended reading

Breithaupt, J, *Up-grade A-level Physics*, Stanley Thornes (1996).
Brodie, D, *Introduction to Advanced Physics*, John Murray (2000).
Close, F, *The Cosmic Onion*, Heinemann Educational Books (1983).
Eastway, R, and Wyndham, J, *Why Do Buses Come in Threes?*, Robson (1999).
Ehrlich, R, *Turning the World Inside Out*, Princetown University Press (1990).
Fisher, I, *How to Dunk a Doughnut*, Weidenfeld and Nicholson (2002).
Holden, A, *The Nature of Solids*, Dover Publications (1992).
Hollins, M, *Medical Physics*, Macmillan (1990).
Homer, D, *Synoptic Skills in Advanced Physics*, Hodder and Stoughton (2002).
Honeywell, C, *Maths for Physics*, Hodder and Stoughton (1999).
Hunt, A, and Millar, R, *AS Science for Public Understanding*, Heinemann (2000).
Hutchings, R, *Physics*, Nelson (2000).
Johnson, K, et al., *Advanced Physics for You*, Stanley Thornes (2000).
Lockett, K, *Physics in the Real World*, Cambridge University Press (1990).
Muncaster, R, *Astrophysics and Cosmology*, Stanley Thornes (1997).
Powell, S, *Statistics for Science Projects*, Hodder and Stoughton (1996).

Index